Molecules
That Amaze Us

Molecules
That Amaze Us

Paul May
University of Bristol, U.K.

Simon Cotton
University of Birmingham, U.K.

CRC Press
Taylor & Francis Group
Boca Raton London New York

CRC Press is an imprint of the
Taylor & Francis Group, an **informa** business

CRC Press
Taylor & Francis Group
6000 Broken Sound Parkway NW, Suite 300
Boca Raton, FL 33487-2742

© 2015 by Taylor & Francis Group, LLC
CRC Press is an imprint of Taylor & Francis Group, an Informa business

No claim to original U.S. Government works

International Standard Book Number-13: 9781466589605 (Pbk)

Library of Congress Cataloging-in-Publication Data

May, Paul W., author.
   Molecules that amaze us / Paul May, Simon Anthony Cotton.
      pages cm
   Includes bibliographical references and index.
   ISBN 978-1-4665-8960-5 (paperback)
   1. Chemistry--Popular works. 2. Chemistry--Humor. I. Cotton, S. A. (Simon A.),
1946- author. II. Title.

QD37.M348 2015
540--dc23   2014034795

Visit the Taylor & Francis Web site at
http://www.taylorandfrancis.com

and the CRC Press Web site at
http://www.crcpress.com

To Natalie
and
to Michelle

CONTENTS

FOREWORD

After many years of almost complete neglect, the science of chemistry is finally starting to catch up with physics and biology in terms of popularity. When I go to bookstores I can typically find yards of books on physics for a general audience but almost nothing on chemistry, and what there is tends to be mis-shelved. But things are indeed starting to change. There has been a gratifying new interest in the periodic table and the elements among the general public, and even among artists and advertising people, both in the form of printed books and in the electronic media.

Perhaps the road is now open for more ambitious forays into the central science of chemistry? This new book is by two knowledgeable and expert popularizers of chemistry and deals exclusively with molecules and compounds rather than with the simpler atoms and elements. It is based on the very successful 'Molecule of the Month' website that was begun by Paul May fifteen years ago and to which his co-author Simon Cotton has been a frequent contributor. This also provides a curious new trend in that we frequently see print books later moving to the electronic medium but not so frequently a movement in the opposite direction from electronic to the print medium. The authors, who have been the biggest contributors to 'Molecule of the Month' over the years, strike an excellent balance between introducing the novice to the world of molecules while also keeping the expert chemist interested.

Here you will find as many as 67 different chapters presented in the form of a question and answer style that serves to make the material easier to understand. While avoiding jargon, Cotton and May still include the correct formula, molecular structure and complete references to all the molecules they report on. As somebody who has taught, researched and written on chemistry for over thirty years, I was surprised by how much new information and how many completely new molecules I discovered in the book.

Here are a few of the ones that were new to me: artemisinin, a new anti-malarial drug; kisspeptin, the molecule responsible for kick-staring puberty; epibatidine, a new painkiller derived from Amazonian poison-dart frogs; and psilocybin, the hallucinogen present in magic mushrooms. Not to mention skatole, tetrodotoxin, cisplatin, thujone and vancomycin. One wonders whether some enterprising individual might try to devise a periodic table of these molecules and maybe even a song set to the tune from Gilbert and Sullivan in the same way as the much imitated element song by Tom Lehrer.*

I highly recommend this book to all readers. It will vastly expand your knowledge and horizons of chemistry and the human ingenuity that surrounds it.

Dr. Eric Scerri
UCLA, Los Angeles
Website: www.ericscerri.com
Author of 'The Periodic Table, Its Story and Its Significance' and several other books on the elements and the periodic table.

---

* The original version of Tom Lehrer's song can be found at http://www.youtube.com/watch?v=DYW50F42ss8

PREFACE

In the mid 1990s, soon after the World Wide Web became a reality, several chemists decided that Molecule of the Month websites would be a good way of disseminating information about interesting molecules in an accessible way. One of these websites, based at Bristol University in the UK, was set up in 1996 by one of the authors (PWM) and administered by him ever since – making it one of the longest running chemistry websites on the web! Over the years, the other author (SAC) contributed over 50 articles for the site, and it has now become a valuable chemistry resource for schools, colleges, or the interested layperson alike.

At a time when more teaching materials are produced in an electronic format, the reader may wonder why the authors have bucked the trend, essentially proceeding in the reverse direction and turning electronic articles into a hardcopy book. Well, words in print still have life in them, and some readers do prefer to have a book in their hands. It also cannot be denied that there is a feel-good factor in seeing the articles in a more permanent form.

Some of these articles have appeared on the Bristol MOTM site before, but all have been updated for these pages, and many are brand-new. We hope that they appeal to a wide readership; the chemistry is meant to be intelligible to a high-school student, but the stories that lie behind the molecules can appeal to any interested person. Chemicals are morally neutral; they can be put to either good or bad uses. We have tried to be light-hearted in our approach to the subject matter, but rest assured, every fact stated here is believed by the authors to be utterly true.

Some of the molecules we describe are synthetic, but many others are products of nature. The layperson often is not aware of how much of the modern world around us is the product of clever chemists, who through hard work, inspiration, or blind luck, made the world safer by discovering cures for illnesses, or made the world more colorful by inventing synthetic dyes, or cleaner by inventing detergents, or warmer by creating synthetic clothing fibres, or more fragrant by inventing perfumes and so on. But on the flip side, chemistry is also responsible for recreational drugs, explosives, nerve gases, poisons and toxic waste. In either case, the stories behind these molecules really do amaze us, and we hope they'll amaze you too.

Although we are the authors of this book, it reflects the chemists who synthesise the molecules and also those who taught us and put us on the path to writing it. To them we express our thanks, which are also due to Eric Scerri, for his kindness in writing a Foreword, and particularly to the team at Taylor and Francis/CRC Press, notably Hilary Rowe and Jill Jurgensen. We also wish to thank the numerous people who gave us permission to use their photos or images, and the various companies and organisations who gave us permission to use photos of their trademarked products as illustrations.

Simon Cotton and Paul May

October 2014

AUTHORS

Simon Cotton obtained his BSc and PhD at Imperial College London, followed by research and teaching appointments at Queen Mary College, London, and the University of East Anglia. He subsequently taught chemistry in both state and independent schools for over 30 years; has lectured widely in the UK; and carries out research on the chemistry of iron, cobalt, scandium and the lanthanide elements. He has published widely, principally on the chemistry of the transition elements, the lanthanides and actinides, including six books, of which '*Every Molecule Tells A Story*' (CRC Press 2012) is the latest. For many years, he wrote 'Soundbite Molecules' as a regular column in the magazine *Education in Chemistry*, while he has written and narrated many 'Chemistry in Its Element' podcasts for the Royal Society of Chemistry's *Chemistry World* website. In 2005, he shared the Royal Society of Chemistry Schools Education Award. He is at present an Honorary Senior Lecturer in the School of Chemistry at the University of Birmingham, UK. In his spare time, he studies late-mediaeval churches and the original documents that date their construction.

Paul May grew up in Redditch, Worcestershire, and then went on to study Chemistry at Bristol University, graduating in 1985. He then joined GEC Hirst Research Centre in Wembley, where he worked on semiconductor processing for three years, before returning to Bristol to study for a PhD. His PhD was awarded in 1991, and he then remained at Bristol to co-found the CVD diamond research group. In 1992, he was awarded a Ramsay Memorial Fellowship and after that a Royal Society University Fellowship, and became a full-time member of staff in October 1999. He still runs the Bristol CVD Diamond Research Group, and is now a professor of physical chemistry. He has written nearly 200 scientific publications, and two books: one a collection of SciFi short stories called *Motorway Madness – and other short stories* (Amazon Kindle Press, 2011); the other a science humour book called *Molecules with Silly or Unusual Names* (Imperial College Press, 2008), based on the infamous website of the same name. He has maintained the *Molecule of the Month* website since 1996, upon which this current book is based. Recreational interests include table-tennis, science fiction and heavy metal music.

Chapter

1

ADENOSINE TRIPHOSPHATE (ATP)

I Don't Get the Joke?

The acronym for adenosine triphosphate is ATP, which sounds like 80p (short for 80 pence).

What Is ATP?

All living things, plants and animals, require a continual supply of energy in order to function. The energy is used for all the processes which keep the organism alive. Some of these processes occur continually, such as the metabolism of foods, the synthesis of large, biologically important molecules, *e.g.* proteins and DNA, and the transport of molecules and ions throughout the organism. Other processes occur only at certain times, such as muscle contraction and other cellular movements. Animals obtain their energy by oxidation of the foods they've eaten, plants do so by trapping the sunlight using chlorophyll (see p81). However, before the energy can be used, it is first transformed into a form which the organism can handle easily. This special carrier of energy is the molecule ATP.

How Does It Work?

The key to how it works is in its structure. The ATP molecule is composed of three components. At the center is a sugar molecule, ribose (the same sugar that forms the basis of DNA and RNA). Attached to one side of this is a base (a group consisting of linked rings of carbon and nitrogen atoms); in this case, the base is adenine. When joined together, the sugar and base are known as adenosine. The other side of the sugar is attached to a string of three phosphate groups. These phosphates are crucial to the activity of ATP.

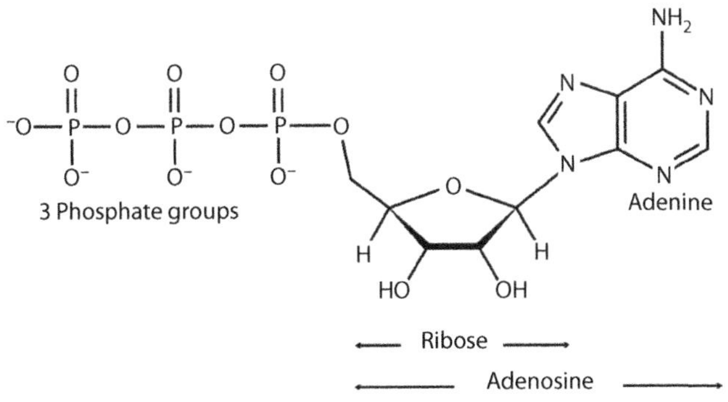

The structure of ATP showing the three components

How So?

ATP works by losing the endmost phosphate group when instructed to do so by an enzyme. This reaction releases a lot of energy, which the organism can then use to build proteins, contract muscles, generate heat, *etc.* The reaction product has one less phosphate group and so is called adenosine *di*phosphate (ADP), and the liberated phosphate group either ends up in solution as orthophosphate (HPO_4) or attached to another molecule such as an alcohol. Even more energy can be extracted by removing a second phosphate group to produce adenosine *mono*phosphate (AMP).

$$ATP + H_2O \rightarrow ADP + HPO_4 + \text{lots of energy}$$

$$ADP + H_2O \rightarrow AMP + HPO_4 + \text{lots of energy}$$

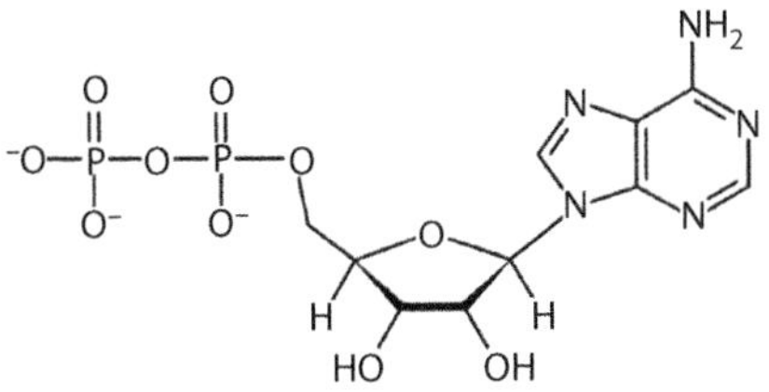

ADP 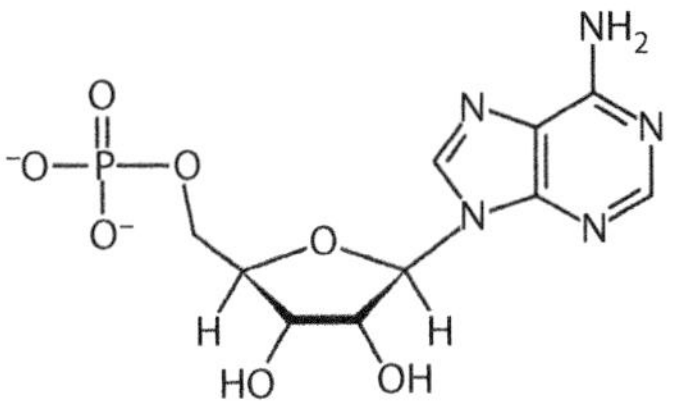   AMP

When the organism is resting and energy is not immediately needed, the reverse reaction takes place and the phosphate groups are reattached to the molecule, one at a time, using energy obtained from food or sunlight. Thus, the ATP molecule acts as a sort of rechargeable 'chemical battery', storing energy when it is not needed, but able to release it instantly when the organism requires it. It has been calculated that the human body contains only 250 g of ATP at any one time, which is roughly the equivalent energy of an AA battery. But it turns over more than its own weight in ATP in a day.

But I Thought Energy Was Stored as Fat?

For *long-term* storage, *i.e.* days or years, surplus energy from food is used to synthesize long-chained fatty acids (see p271) and stored as fat distributed around the body, or as glycogen (a form of polymerized glucose, see p193) in the liver. When energy is required by the body for a particular process, say to make a muscle contract, the stored fat or glycogen is removed from storage by enzymes and transported in the blood to the cells in question. There, it undergoes oxidation, reacting with the oxygen delivered by hemoglobin in the bloodstream (see p227), to produce the waste products water and CO_2, and releasing lots of energy. The energy converts ADP and AMP back to ATP (recharging the local 'battery'), which is then used as the power source for the cellular process. So ATP is a very temporary energy store localized within each cell.

In plants, the long-term energy store is another polymerized form of sugar called starch (see p193). In photosynthesis, the chlorophyll molecule traps energy from the sun and uses this to make ATP from ADP (see p81). The ATP is then transported to other parts of the cell which break it back down into ADP, and use the released energy to turn carbon dioxide and water into glucose, releasing oxygen. Enzymes then polymerize the glucose into starch and it is stored for later use. Hydrolysis of the stored starch using enzymes (called amylases) allows the plant to extract the glucose and use the freed energy to make ATP, which can then be used as a local energy source in cells for use in various biological processes such as growth. Amylase enzymes are also present in human saliva and allow us to digest starch. Foods that contain a lot of starch but little sugar, such as rice and potatoes, often taste slightly sweet because the amylase in saliva turns some of the starch into sugar as they are chewed.

So We Get ATP from Our Food?

We get the components from food, but these are metabolized into ATP in our body. The phosphate groups are the key ingredient, and these are part of what biologists call the Phosphorus Cycle. The fact that ATP is nature's universal energy store explains why phosphates are a vital ingredient in the diets of all living things. Modern fertilizers often contain phosphorus compounds that have been extracted from animal bones. These compounds are used by plants to make ATP. Animals then eat the plants, metabolize the

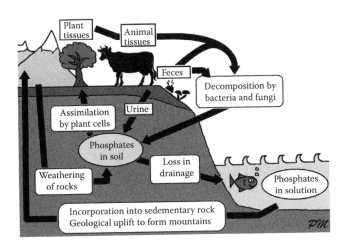

The Phosphorus Cycle in Nature

phosphates, and produce their own ATP. We also eat the plants and other animals, and convert their phosphorus into our own ATP. And when we die, our phosphorus goes back into the ecosystem to begin the cycle again.

WHERE IS THE ATP MADE?

In animals, ATP is recycled in the mitochondria, which are organelles found within animals cells, and which can make up to 25% of the total volume of the cell. Mitochondria resemble smaller cells trapped within larger animal cells. They also contain their own DNA, which is different from the DNA found in the nucleus of the larger cells. This observation has led scientists to speculate that mitochondria were once separate cells (a sort of primitive bacteria) which had evolved the neat trick of synthesizing ATP and thus storing energy. Other, larger, cells lacked this vital ability, but rather than waste a few million years evolving their own system to do so, they simply engulfed the mitochondrial cells and used them as internal power supplies. The mitochondria also benefit from this symbiosis, being protected from the dangers of the outside world, and nourished by a larger organism, for the small cost of some of their ATP.

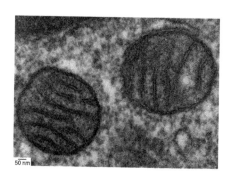

Transmission electron micrograph of mitochondria cells in lung tissue

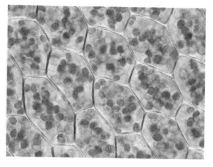

Rounded chloroplasts within the cells of the thyme moss

In plants, the ATP is recycled in the cell membranes of chloroplasts (cells specializing in photosynthesis found in the leaves and surface of plants). Light energy from the sun pumps H^+ ions across the cell membrane, and the potential difference this creates gives enzymes the power required to reattached phosphate groups to AMP or ADP. The energy stored in ATP is used to convert CO_2 into glucose and larger organic molecules. The ancient ancestors of chloroplasts were single-celled algae called cyanobacteria, which are still around today. It is believed that chloroplasts evolved the ability to photosynthesize early on, and – just as with mitochondria – were then engulfed by larger cells. A symbiotic relationship then developed between the two cells until the point came where they both could no longer survive without each other.

This 'endosymbiosis' theory (*endo* meaning one partner is inside the other) for both mitochondria and chloroplasts is believed to be one of the key steps in evolution. The abundance of energy in the form of ATP given to multi-celled organisms by their new 'internal power supplies' allowed some of the organism's other cells the freedom to specialize, and become light-sensitive (eye) cells, hairs for sensing or movement, antennae, *etc.*, vastly increasing the complexity of life on Earth.

SO HOW MUCH WOULD A PINT OF ATP ACTUALLY COST?

If you bought it as an aqueous solution of its disodium salt, the cost of a pint would be about $150,000 depending upon what concentration you wanted and what supplier you chose – significantly more than 80p, anyway!

Chapter

2

ADRENALINE/EPINEPHRINE
(NORADRENALINE/ NOREPINEPHRINE)

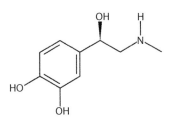

Epinephrine (adrenaline)

WHY THE TWO NAMES?

In 1901, the Japanese chemist Jokichi Takamine successfully
isolated and purified a hormone that he'd extracted
from the adrenal glands of sheep and oxen. He named it
'adrenaline', after the Latin for the adrenal glands (*ad* = above,
renes = kidneys), and it is still known as this in most countries of
the world – except the United States. This is because four years
before Takamine's discovery (and patent), an American chemist
called John Abel had also prepared extracts from animal adrenal
glands, but called it 'epinephrine', which is actually derived
from the Greek name for the adrenal glands (*epi* = above,
nephros = kidney). It is still unclear whether Abel's extract and
Takamine's pure hormone were the same thing, but to avoid
breaching the patent in the United States, Abel's formula was
sold and used there under the name epinephrine, which became
its accepted name – even when later purified versions of it were
shown to be identical to Takamine's adrenaline. Even today, the
same molecule is known by two different names (epinephrine
in the United States, and adrenaline almost everywhere else)
and is one of the few molecules where this double-naming
is allowed, along with the related molecule noradrenaline
(norepinephrine).

NOR?

In this context, 'nor' means the next lower example in the
homologous series, *i.e.* that the molecule is the same as
epinephrine except having one carbon-group less – the methyl
group attached to the nitrogen has been removed and replaced
by a hydrogen. 'Nor' may possibly be derived from the German:
'*N ohne Radikal*' meaning 'nitrogen without radical'.

Epinephrine (adrenaline)

Norepinephrine
(noradrenaline)

So What's the Difference between Epinephrine and Norepinephrine?

Well, apart from the structural difference, they play different roles in the body. They are both hormones and neurotransmitters, which means that (together with two other important neurotransmitters, dopamine and serotonin) they help transport impulses from one nerve to the next across the synaptic gap (see p169 and p451). Epinephrine is mainly produced in the adrenal glands, whereas norepinephrine is also made in the brain. Both molecules are involved in the so-called 'fight-or-flight' response, but act in different ways.

Fight or Flight?

The term 'fight or flight' is often used to characterize high-stress, physically exhilarating or life-threatening situations. Under these circumstances, the animal needs to prepare itself rapidly to either fight for its life or run away as fast as possible. To do this, fear signals from the brain cause the adrenal gland to pump epinephrine into the bloodstream, while norepinephrine is released by the brain cells. The main effect of epinephrine is to increase the force and rate of heart contractions, allowing more blood to flow round the body faster. The air passages also dilate, allowing the animal to take in more air, which they may require for the anticipated sudden increase in physical performance. It also narrows blood vessels in the skin, stomach and intestine, increasing the blood pressure, so that an increased flow of blood reaches the muscles and allows them to cope with the demands of exercise or stress. The feelings this redistribution of blood causes are sometimes poetically described as your 'skin crawling with fear', and having 'butterflies in the stomach'.

This gazelle chose flight rather than fight

At the same time, norepinephrine causes some of the other non-essential blood vessels to constrict, so increasing the blood pressure further, and also stimulates brain activity, increasing alertness and arousal. The muscles in the irises contract, sharpening the focus of the eyes onto potential dangers. Both molecules also trigger the conversion of stored glucose or glycogen into energy in the skeletal muscles. Other molecules are also involved in the fight-or-flight response, notably endorphins, which are the brain's natural painkillers. These are released allowing the animal to temporarily feel less or even no pain in moments of crisis. Injured victims have often reported running away from an accident site without feeling any pain, and only realized they had broken a leg several minutes later after the immediate danger was over and they'd calmed down (*i.e.* once the endorphins had dissipated) and the pain appeared. These effects all take place within a few milliseconds of the fear stimulus occurring, giving the animal the chemical boost needed to spring into immediate action.

WHAT OTHER DIFFERENCES ARE THERE?

It's a bit of a generalization, but you could say that the effects of epinephrine mostly concern the body, whereas those of norepinephrine mostly affect the brain. For example, norepinephrine is psychoactive, which means it can affect the way the brain responds and perceives the world, whereas epinephrine is not. Norepinephrine is involved in decision making and the learning process, as well as in regulating mood, and attention and focus levels. Those with low levels of noradrenaline generally have difficulty staying awake, concentrating and paying attention to tasks, and this can lead to conditions such as attention deficit hyperactivity disorder (ADHD) and depression. Simply giving these patients additional norepinephrine doesn't work, however, because the molecule cannot cross the blood–brain barrier. Instead, they are treated with drugs

Methylphenidate (*Ritalin*)   Venlafaxine (*Effexor*)

which mimic the action of norepinephrine, but which can cross into the brain. These include psychostimulant drugs such as *Ritalin* or amphetamines (such as *Adderall*) for ADHD, and antidepressants, specifically serotonin–norepinephrine reuptake inhibitors (SNRIs, see p451), such as *Effexor* and *Cymbalta*, for mood disorders.

You Mentioned Amphetamines ...

Yes, one of the reasons amphetamines (see p309), and other psychoactive recreational drugs, such as LSD (see p293) and magic mushrooms (see p427), have become so widely used is that they mimic the effect of neurotransmitters, such as norepinephrine, serotonin and dopamine, creating abnormal responses in the brain. These can be feelings of pleasure or euphoria, or audible or visible hallucinations. Moreover, because neurotransmitters are linked to the learning response, people become addicted to the perceived 'rewards' given by the drug, despite the usually dire health consequences.

Is That Why We Get 'Adrenaline Junkies'?

Not really. Epinephrine itself is not psychoactive and there's no evidence that it's chemically addictive. However, there are people who seem to thrive on dangerous activities, but this may be less to do with the so-called 'adrenaline rush', and more to do with the release of endorphins (natural painkillers that act in similar way to morphine (see p347) and may also be addictive), and of other neurotransmitters (noradrenaline, serotonin and dopamine) which are linked to learning. The idea is that if the activity doesn't hurt or kill the creature, the levels of these molecules in the brain increase and the creature is rewarded with feelings of pleasure or satisfaction, and this positive feedback makes the creature more likely to repeat the risky activity. Suck risk-taking may be dangerous, but on the other hand it could arguably be what makes higher creatures, especially humans, curious and exploratory. No risk, no reward, as it were.

A bungee jumping 'adrenaline junkie'?

SO, APART FROM 'FIGHT OR FLIGHT', IS EPINEPHRINE USEFUL?

As a hormone, epinephrine has an effect on nearly all body tissues. The effects vary depending on the type of receptor those tissues have. For example, in the airways it causes the muscles surrounding the bronchioles to relax, allowing more air through. Epinephrine itself, or variations of the molecule, have therefore been used to treat asthma – the variations are to retain the bronchodilation

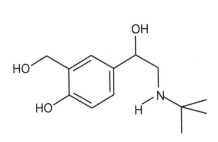

Salbutamol (*Ventolin*)

An asthma inhaler

Salmeterol (*Serevent*)

effect without the unwanted increase in heart rate. Such drugs include salbutamol (*Ventolin*) and salmeterol (*Serevent*), which are usually administered via an inhaler.

Epinephrine is often used to treat anaphylactic shock resulting from an extreme allergic reaction. Anaphylaxis has two immediate dangers, loss of blood pressure and constriction of the airways. Epinephrine relaxes the breathing passages so that the patient doesn't suffocate and also increases the blood pressure by peripheral vasoconstriction and accelerating the heart rate. People who know they have a risk of anaphylaxis, such as those with a severe allergy to common substances like peanuts, often carry around an 'EpiPen'. This is a single shot of epinephrine which is usually injected into the thigh muscle as soon as the symptoms appear.

An old-style EpiPen

Epinephrine can also be used in extreme circumstances to restart the heart after a cardiac arrest. You may remember the infamous scene in the movie *Pulp Fiction* where Vincent (John Travolta) gives Mia (Uma Thurman) an epinephrine injection directly into her heart to revive her after a drug overdose. Needless to say, this is a Hollywood exaggeration, as the would-be doctor would need to get the needle through the chest, past the ribs, and into the left ventricle, a task which has been described as 'trying to hit a kiwi fruit inside a turkey with a wild stabbing motion at your first attempt'. The standard procedure from a paramedic would have been to give an anti-opiate like *Narcan* to counteract the effects of the heroin, followed by CPR to restart and then keep the heart beating until the *Narcan* took effect. If epinephrine was given, it would be by normal injection into a vein and then pumped around the body by the actions of chest compressions.

BUT I'VE SEEN THE 'SHOT TO THE HEART' SCENE MANY TIMES?

Yes, it's now become a popular dramatic plot-device in TV shows and movies when the action needs spicing up. But since 1990, no doctor would *ever* treat a patient by stabbing them in the heart with a giant needle – this would be a good way to kill them. Before 1990, an intra-cardiac injection *was* used (rarely) ... but only by trained medical personnel ... and only if the heart was *completely* stopped ... and only if every other option was exhausted.

Nikki Sixx and Mick Mars of *Mötley Crüe* on stage in 2005, eighteen years after Sixx's heart was kickstarted with an intra-cardiac epinephrine injection

That has not stopped Hollywood writers, though, who have used the shot-to-the-heart in films such as *The Rock*, *Robocop* and *Get Him to the Greek*, and TV programmes such as *Firefly* (in which it is self-administered!), *Alias*, *Smallville* and even *CSI*.

And on the musical side of Hollywood, a medic allegedly gave Nikki Sixx, the bass player from *Mötley Crüe*, two epinephrine injections directly into his heart after a heroin overdose in 1987. This incident led to one of Crüe's most famous songs 'Kickstart My Heart':

> *'When I'm enraged or hittin' the stage,*
> *Adrenaline rushing through my veins, and I'd say we're still kickin' ass.*
> *Ooo, ahh, kickstart my heart, hope it never stops ...'*

CAN SOMEONE BE SCARED TO DEATH?

Yes, in theory. The fight-or-flight response to a stressful or terrifying situation is to flood the body with epinephrine, making the heart speed up. But if there is too much epinephrine in a sudden rush, the system of nerves and muscles which regulates the heart rhythm can become overloaded. If this happens, the heart can go into spasms, and the person will drop dead. And it is not just fear that can bring this on – any strong emotional response, such as happiness or sadness, will do. There are reports of people who have died in religious passion, or while having sex (irreligious

"Good news, Mr Jones. We've successfully treated your heart condition. But the bill is rather large, I'm afraid.... Mr Jones...?

passion, you might say!). A study in Germany showed that there was an increase in sudden heart attacks on days when the German soccer team were playing in the World Cup. And for about a week after the 9/11 terrorist attacks in 2001 there was an increase of sudden cardiac death among New Yorkers. These sudden cardiac failures can happen to anyone, of all ages, but they are usually linked to people with existing heart problems or a predisposition to heart disease.

Chapter

3

AMMONIUM NITRATE

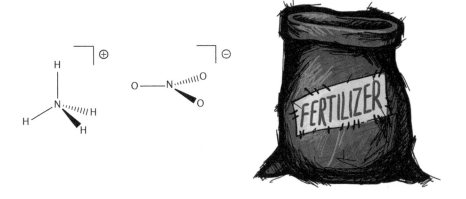

If Ammonium Nitrate Is So Explosive, Why Is It Used as a Fertilizer?

It contains 35% nitrogen, much of it as nitrate, so it is readily taken up by plants.

What Makes It Explosive?

Ammonium nitrate is a bit of a contradiction. It is made up of two ions, tetrahedral ammonium ions and flat, triangular, nitrate ions.

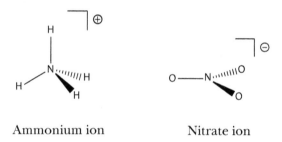

Ammonium ion   Nitrate ion

In the ammonium ion, nitrogen is in the −3 oxidation state, and is bound to hydrogen, while in the nitrate group, nitrogen is in the +5 oxidation state, and is bound to oxygen. You might think that these reactive components would undergo a redox reaction, destroying both ions, but at room temperature, they do not.

And at Higher Temperatures?

On fairly gentle heating, to around 200°C, ammonium nitrate decomposes, forming nitrous oxide (laughing gas, see p371).

$$NH_4NO_3(s) \rightarrow N_2O(g) + 2H_2O(g)$$

Above around 240°C, it can explode, especially when molten and under pressure. The presence of impurities, such as organic molecules, makes this happen more readily, as in mixtures of ammonium nitrate and fuel oil (ANFO), which is widely used as an explosive in mining and quarrying.

How Does That Happen?

Ammonium nitrate can also release oxygen on heating to higher temperatures:

$$NH_4NO_3(s) \rightarrow N_2(g) + 2H_2O(g) + \tfrac{1}{2}O_2(g)$$

Hydrocarbon fuels with long carbon chains need oxygen to burn, supplied by the ammonium nitrate. Thus, with octane:

$$C_8H_{18}(l) + 25NH_4NO_3(s) \rightarrow 25N_2(g) + 59H_2O(g) + 8CO_2(g)$$

When detonated, this releases around 4288 joules per gram of the ANFO mixture, much more energy than ammonium nitrate by itself, and producing an enormous quantity of rapidly expanding hot gas, generating the explosive force.

It can be used to make other explosive mixtures; thus a mixture with TNT (see p543) was known as *Amatol*, widely used in the two World Wars.

And Terrorists Have Used It?

Terrorists used it for the Bali bombing on October 12th 2001 and in bombing the British Consulate in Istanbul on November 26th 2003. It killed 12 people in a bombing outside the High Court in Delhi on September 7th 2011 and 17 people in two bombings in Hyderabad on February 21st 2013. Anders Breivik used it in a car bomb in Oslo on July 22nd 2011, killing eight people, some two hours before he murdered 69 people on the island of Utøya. The IRA used ammonium nitrate in the Omagh bombing on August 15th 1998 and at London's Baltic Exchange on April 10th 1992.

The remains of the Federal Building in Oklahoma City after it was destroyed by an ammonium nitrate bomb in 1995

Perhaps most infamously, Timothy McVeigh used it at 9 am on April 19th 1995, when the peace of Oklahoma City in the sleepy American Midwest was shattered by a bomb that killed 168 people when it went off outside a Government building.

JUST TERRORISTS?

No, there have been a lot of accidents involving ammonium nitrate. The first major one occurred in the United Kingdom, on April 2nd 1916 at an explosives factory at Faversham, in Kent, killing 120 people. A disaster at an explosives factory in Oppau, Germany, killed 561 on September 21st 1921, and on April 16th 1947 an ammonium nitrate explosion on board the USS Grandcamp in the harbor of Texas City killed 581 people and destroyed much of the town. The ship's anchor, which weighed over a tonne, was thrown two miles by the force of the explosion. This is still regarded as one of the worst industrial accidents in US history, and one of the largest ever non-nuclear explosions in the world.

More recently, on September 21st 2001, over 300 tonnes of ammonium nitrate exploded and flattened the factory of the AZF fertilizer company in Toulouse, killing 28 people as well as

Rescue workers search through the debris after the 1947 Texas City explosion (Photo used with permission of Moore Memorial Library, and the City of Texas City, USA)

destroying many other buildings. On April 22nd 2004, a train carrying fertilizer exploded, obliterating the railway station at Ryongchon in North Korea and killing 154 people. The cause was believed to be an accident, though it was rumored to be the result of an assassination attempt on Kim Jong-Il, the North Korean leader. Most recently, a fire followed by an explosion at the West Fertilizer Company in West, Texas, killed 15 people and injured some 200 others on April 17th 2013.

WHY DID THESE ACCIDENTS HAPPEN?

Part of the problem is poor regulation, and storing ammonium nitrate near other materials or close to sources of fire. The Texas City disaster in 1947 was linked to a fire caused by a seaman smoking nearby. Once ammonium nitrate starts decomposing, the (very) exothermic process is impossible to stop, not least because it produces oxygen, and the reaction gets out of control.

So Making Ammonium Nitrate Is Dangerous?

It involves a fairly safe neutralization reaction. Ammonia is neutralized with nitric acid (itself manufactured from ammonia), creating a solution of ammonium nitrate. The solid can be obtained by crystallizing hot solutions, or in industry by spraying molten ammonium nitrate into a tower where drops solidify as they fall into lumps known as prills. These can be used as a fertilizer, or in making ANFO explosive.

$$NH_3(aq) + HNO_3(aq) \rightarrow NH_4NO_3(aq)$$

Why Do Terrorists Use Ammonium Nitrate?

Terrorists often use ANFO or similar mixtures, mainly because ammonium nitrate and the organic ingredient (gasoline or diesel) are usually fairly widely available commercially and legally. It's tricky for authorities to do much about this, though, as banning ammonium nitrate sales would seriously hurt the farming industries in many developing areas of the world where terrorism is often prevalent, such as Afghanistan, Pakistan and the Middle East – countries that rely on fertilizers to feed their population.

Can Nothing Be Done?

Chemistry may have found an answer. Sandia National Labs in the United States have developed a cheap mixture of ammonium nitrate and iron sulfate which is a cheap and effective fertilizer, but which, crucially, when mixed with fuel oil does not become an ANFO explosive. As soon as the mixture is mixed with a solvent, such as fuel oil, a metathesis reaction occurs and the two metal ions swap partners, leading to the inert compounds iron nitrate and ammonium sulfate.

$$2NH_4NO_3 + FeSO_4 \rightarrow (NH_4)_2SO_4 + Fe(NO_3)_2$$

There is hope that this mixture can be distributed to world hot-spots to enable fertilizers to be used for growing crops, and not as 'fertilizer bombs'.

Chapter

4

ARTEMISININ

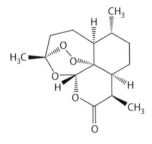

Artemisinin

A frustrated
mosquito!

THAT'S THE CHINESE ANTIMALARIAL DRUG, ISN'T IT?

Yes, that's right.

WHY IS IT SO IMPORTANT?

Something like 200 million people are infected with malaria each year, and over a million will die, mainly children aged under five, in sub-Saharan Africa. New treatments are badly needed.

AREN'T THERE OTHER DRUGS FOR TREATING MALARIA?

Until the 1930s, quinine was the antimalarial drug of choice. After war broke out in the Pacific at the end of 1941, the Dutch quinine plantations in the East Indies became inaccessible, and synthetic alternatives like chloroquine (see p433) came into use. These drugs were successful for some years, but after a while the *Plasmodium falciparum* and *Plasmodium vivax* parasites developed resistance to chloroquine and other antimalarial drugs, meaning that new treatments were needed.

HOW DOES DRUG RESISTANCE COME ABOUT?

Not all the parasites are the same. Some will possess genetic differences that make them resistant to a particular drug molecule, which can be passed on to their offspring.

WHERE DOES ARTEMISININ FIT IN?

For the past century, all the alternative treatments for malaria have been synthetic drugs. Artemisinin was discovered due to the Vietnam war. Because there were many casualties in the North Vietnamese army, Ho Chi Minh, the North Vietnamese leader, got in touch with Zhou en Lai, the Premier of the Chinese Republic, as North Vietnam had China's backing. Possibly because Zhou en Lai himself had experienced malaria some 40 years earlier, the Chinese leadership swung into action, setting up a secret military project, known as 'Project 523' from the date on which the programme was launched, May 23rd 1967.

SO THEY INVENTED A NEW DRUG?

Not quite; they reinvestigated an old one, by looking at a 2000-year-old Chinese pharmacopoeia, Ge Hong's *The Handbook of Prescriptions for Emergency Treatments*. This identified the herb known as *qinghaosu*, obtained from sweet wormwood (*Artemesia annua*) as a remedy for fevers. They tried extracting the herb with hot solvents, but success only followed

Sweet wormwood (*Artemesia annua*)

when they realized that the ancient text ran '*Soak a handful of qinghao in water, wring out the juice and drink it all*', so they tried extraction with cold ether, which worked. Animal tests were a success; then the team tried it on themselves before it went into clinical trials, where it was found to be better than chloroquine. Artemisinin itself, the active molecule, was isolated in 1972. When its structure was determined in 1975, it was found that its activity stemmed from a novel endoperoxide bridge. For some years, artemisinin was not known in the West, until it was revealed at a conference in 1981.

WHO WAS RESPONSIBLE FOR THE DISCOVERY?

A number of scientists were involved. The two main protagonists are reported to be Zhenwing Wei and Youyou Tu. In 2011, the latter received the prestigious Lasker award in clinical sciences for her work.

Dr. Youyou Tu (Courtesy of the Albert and Mary Lasker Foundation. With permission)

SO ARTEMISININ CAME ALONG AT THE RIGHT TIME

At just the moment that the malaria parasites were becoming resistant to all the existing drugs.

HOW DOES ARTEMISININ WORK?

At first it was thought that the peroxide group combined with iron in infected red blood cells and generated very reactive free radicals that destroyed key molecules and led to the death of the parasite – a kind of 'dirty bomb'. However, in 2003, a research team headed by Dr. Sanjeer Krishna at St George's Hospital Medical School in London discovered that artemisinin inhibited an enzyme called *Plasmodium falciparum* ATP6 (PfATP6) responsible for a 'pump' transporting calcium ions across cell walls in the parasite, though it seems that both iron and the peroxide group are important to its action.

SO NOW EVERYTHING IN THE GARDEN IS LOVELY?

Unfortunately, no. Increasingly, people are worrying about the possibility of the malaria parasite becoming resistant to artemisinin. In 2005, researchers found that a mutation of one amino acid meant that the parasite became resistant. The World Health Organization is now pressing for artemisinin to be used only in combination with another drug. They are urging the adoption of ACT (Artemisinin Combination Therapy) in which artemisinin is mixed with another drug such as mefloquine, the thinking being that while a particular parasite might be resistant to one of the drugs, it is exceedingly unlikely to be resistant to two or three of them, so drug-resistant parasites will be eliminated. At the moment, approved combination therapies involve artemether or artesunate in particular. These and some other derivatives have better medicinal properties than artemisinin.

Artesunate

Artemether

Although artemisinin itself has drawbacks as a medication (*e.g.* weak oral activity, poor solubility either in oil or water), it is readily reduced by $NaBH_4$ to dihydroartemisinin, which in turn is readily converted into useful derivatives such as artemether and artesunate.

AND?

Another problem with artemisinin is simply due to its success. It is in very short supply and is also relatively expensive. Artemisinin is

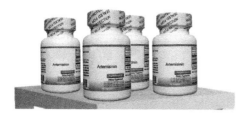

Artemisinin is now available over-the-counter, and even as health food supplements

largely obtained from sweet wormwood, which contains less than 1% artemisinin. Making artemisinin from scratch ('total synthesis') requires many steps, and is also expensive. Two promising routes have been described; one uses 'engineered' yeast (*Saccharomyces cerevisiae*) to produce artemisinic acid, via farnesyl pyrophosphate, and amorpha-4,11-diene. Others have reported a single continuous-flow process that makes artemisinin from dihydroartemisinic acid, itself easily made from artemisinic acid. It is possible that these processes could be scaled up to produce artemisinin industrially in large quantities, thereby reducing the cost of ACT to a level where it could be afforded by Third World economies.

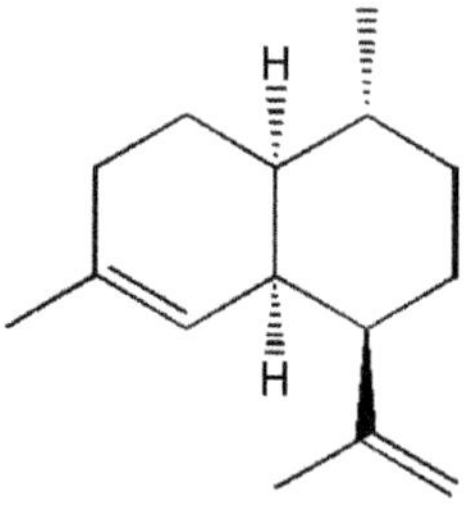

Amorpha-4,11-diene

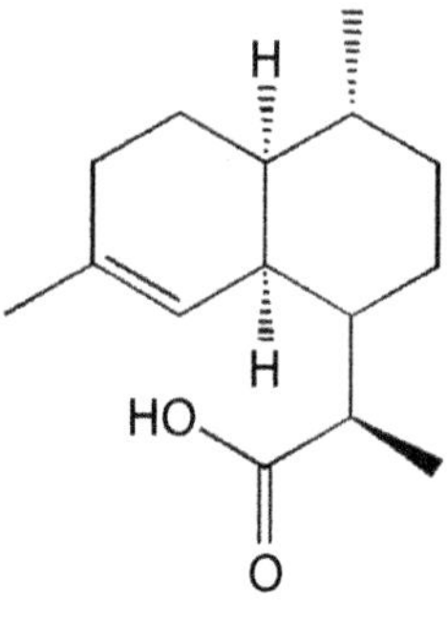

Dihydroartemisinic acid

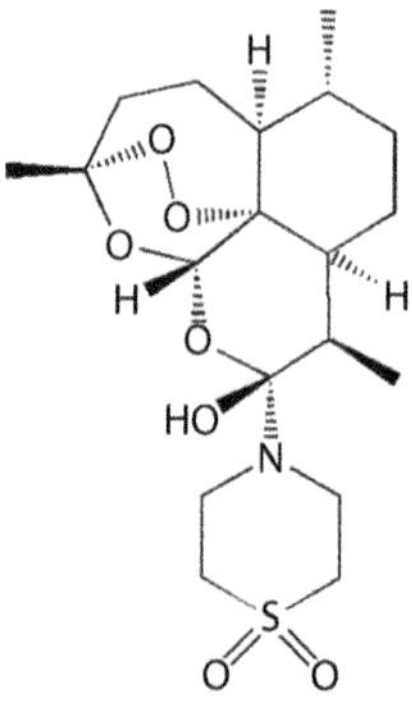

Artemisone

Another artemisinin-type molecule known as artemisone was reported in 2006. It combines high activity against the malarial parasite with other desirable features, including low lipophilicity (solubility in fat) and negligible neuro- and cytotoxicity (*i.e.* it's not poisonous to either cells or the nervous system).

So It Is Back to 'Natural Drugs'?

Yes, but maybe not for good. Scientists have synthesized molecules containing one or two peroxide bridges. Some of these are so new they haven't been given proper names yet, and just have code names; these include OZ439 and RKA 182, and they are currently being tested to see how effective they are.

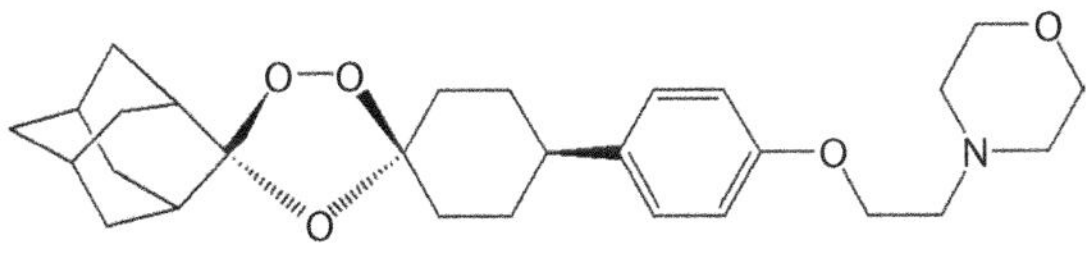

OZ439

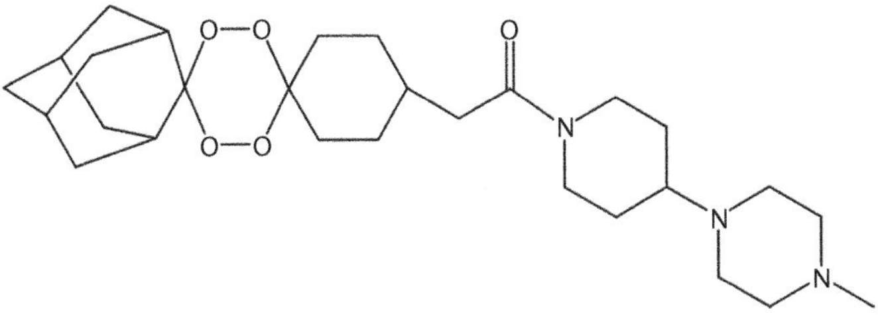

RKA182

In 2013, a new class of antimalarial drugs based on a quinolone-3-diarylether structure was developed at the Oregon Health & Science University, which attacks the malaria parasite at different stages of its life cycle. This approach may make it much harder for the parasite to develop resistance to these new molecules.

The most promising of these so far only has a code number (ELQ-300) rather than a name, but results of clinical trials are eagerly awaited.

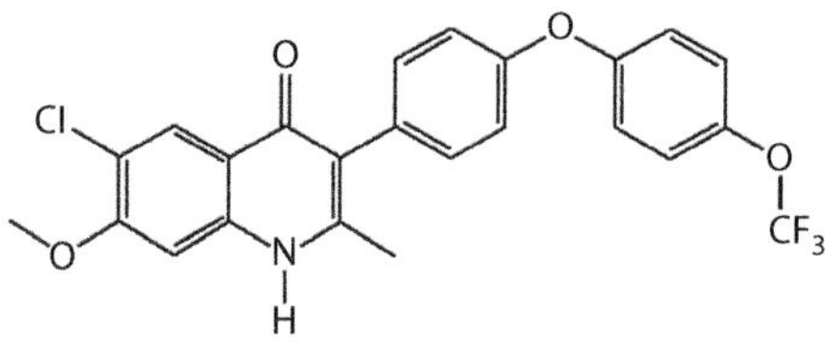

ELQ-300

Chapter

5

ASPIRIN

Acetylsalicylic acid (aspirin)

IS IT TRUE THAT ASPIRIN IS CONSIDERED TO BE THE FIRST MODERN DRUG?

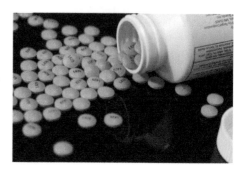

Yes, but it goes back a long way. The association of willow bark with fever-reducing properties was mentioned by the ancient Greek writer Hippocrates (440–377 BC). In the 1750s, an Oxfordshire clergyman named Edward Stone tested powdered willow bark and found that it relieved feverish symptoms associated with malaria, and

Generic aspirin tablets. These are coated with an orange enteric coating which prevents the tablets from dissolving in the acidic stomach causing irritation, but breaks down rapidly in the alkaline (pH 7–9) environment present in the small intestine

he published his discovery in 1763. Other scientists built on his observations. The molecule present in the willow, named *salicin* from the Latin *salix*, meaning willow tree, was isolated in 1828. Ten years later, it was found that salicin could be converted to salicylic acid, which had even more potent pain-relieving properties.

I'VE READ THE NAME 'ACETYLSALICYLIC ACID' ON ASPIRIN PACKETS. WHAT IS IT AND HOW DOES IT DIFFER FROM SALICYLIC ACID?

The common chemical name of aspirin is acetylsalicylic acid, which is the acetyl ester of salicylic acid.

SO WHO MADE ASPIRIN FIRST?

What we now call aspirin, acetylsalicylic acid, was first synthesized in 1853 by Charles Frederic Gerhardt, but it was not until

Salicin

Salicylic acid

Acetylsalicylic acid
(aspirin)

1897 that chemists at the German *Bayer AG* company found a
reliable way of making it from salicin, derived this time from the
meadowsweet plant. A side effect of salicylic acid as a medication
to treat fevers and pain was that it caused gastro-intestinal
complications. One version of the story says that the father of the
chemist Felix Hoffmann (1868–1946) suffered from rheumatism,
and that the sodium salicylate he took irritated his stomach, so
the son had the bright idea of making an acetylated version of
salicylic acid to try on his father. But after Hoffmann's death,
another chemist named Arthur
Eichengrün claimed that he was
the man responsible, and more
recent analysis has supported
this story. At any rate, *Bayer AG*
named their new drug 'aspirin'
(after the old botanical name for
meadowsweet, *Spiraea ulmaria*),
and put it on the market in 1899
as a painkiller that reduced both
temperature and inflammation.
Nowadays, around a billion
aspirin tablets are taken
worldwide each year.

Felix Hoffman

IS THAT ALL THAT ASPIRIN DOES?

It has several roles. Apart from being an analgesic (painkiller), an antipyretic (fever reducer) and an anti-inflammatory agent, there is evidence that long-term use of aspirin contributes to less risk of certain cancers, including lung, colon, prostate, bowel and breast cancers. It has also been shown that middle-aged men taking a tablet of aspirin a day can reduce the risk of heart attack by maybe 50%, as aspirin can reduce the likelihood of blood clotting.

Copyright 2002 by Randy Glasbergen.   www.glasbergen.com

"An aspirin a day will help prevent a heart attack if you have it for lunch instead of a cheeseburger."

IS ASPIRIN SAFE?

Relatively safe. The toxic dose for adults is about 10–30 g, which is many times greater than the dosage in a standard tablet of ~300 mg. Both salicylic acid and aspirin can breach the stomach's protective lining, causing bleeding, but aspirin causes much less stomach irritation. Back at the time of the 1918–1919 influenza ('Spanish flu') pandemic, aspirin was widely used as a medication, and it has been suggested that high doses of aspirin may have contributed to the fatalities. Its use with children has since been discouraged, since aspirin taken during viral illness can produce fatalities in children suffering from Reye's syndrome.

How Does Aspirin Work?

In 1971, the British pharmacologist Sir John Vane led the research team that showed that aspirin inhibits the production of prostaglandins (see p419). These are hormones produced locally in the body when tissue is damaged or infected, affecting blood flow as they cause the disease-fighting white blood cells to race to the damage site (which is what causes the swelling and inflammation). One of these hormones, called prostaglandin E_2 (PGE_2), affects temperature regulation, making the body's temperature rise and causing fever. Aspirin works by blocking prostaglandin synthesis, thereby preventing swelling, inflammation and also reducing fever.

How Exactly Does Aspirin Do This?

Prostaglandins are made by an enzyme called cyclooxygenase, or COX for short. The active site on COX contains a serine group,

Aspirin blocks the active site on the COX enzyme by transferring its acetyl group which binds a molecule called arachidonic acid and converts it via some clever biochemistry into the required prostaglandin. However, if aspirin is present in the bloodstream, it can rapidly transfer its acetyl (CH_3CO) group to the serine residue in the enzyme, thus blocking the active site, and rendering the enzyme inactive.

In fact, there are three cyclooxygenase enzymes, COX-1, COX-2 and COX-3 (a variant of COX-1). Aspirin inhibits both COX-1 and COX-2. COX-1 occurs widely in mammalian cells; COX-2 is generated at the sites of inflammation and cell damage. It is the blocking of COX-1 that is responsible for the increased risk of stomach irritation when aspirin is taken, as it stops the formation of a protective prostaglandin (prostacyclin), in cells in the stomach lining, which reduces gastric acid production.

DOES BLOCKING COX-1 DO ANY GOOD AT ALL?

Yes, certainly. Another effect of blocking COX-1 is to cut down the formation of thromboxane A2, which induces platelet aggregation and blood clotting, and this is where the cardioprotective effect of aspirin comes from.

WHAT DOES ASPIRIN DO TO COX-2?

COX-2 has a role in making the prostaglandin PGH_2, a precursor to other prostaglandins, which mediate pain, fever and inflammation. Therefore, inhibiting COX-2 reduces pain, fever and inflammation.

SO CAN YOU MAKE ASPIRIN BLOCK COX-2 BUT NOT COX-1, SO YOU DON'T GET SIDE EFFECTS?

This was actually tried with a class of 'super-aspirin' drugs. The best known of these was Vioxx (*Rofecoxib*), which had great promise

Vioxx (*Rofecoxib*)

in treating long-term arthritic complaints. It rapidly became a best-selling drug, but in 2004 it became apparent that people taking Vioxx for long periods were more likely to suffer from heart attacks and strokes, so the manufacturers withdrew it from the market.

SO TAKING ASPIRIN IS ALL GOOD?

It sometimes seems like it, but there are risks. Recent research from the University of Wisconsin has indicated that long-term use of aspirin may carry a small but significantly increased risk of age-related macular degeneration, causing loss of eyesight. It is a question of what meets the needs of a particular patient.

Chapter

6

CAFFEINE

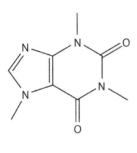

Caffeine

Café-au-lait

Is Caffeine a Drug?

It is the most widely ingested psychoactive drug in the world, but it's legal. And it is safe, at least to drink coffee or tea, within reason.

Why the Qualification?

If this giant can on the back of a mini was really full of *Red Bull* it would contain ~270 g of caffeine (Courtesy of © Red Bull Media House. With permission)

Health Canada says that the recommended daily limit for a healthy adult is around 400 mg a day, while the UK *Food Standards Agency* recommends a limit of half that for a pregnant woman. Caffeine can readily cross the placental barrier so it is sensible for a pregnant woman to restrict caffeine intake to about two cups a day (or less). An 8-oz cup of 'instant' coffee contains around 90 mg of caffeine, but if you drink a '*Grande*' cup from *Starbucks*, with over 300 mg, this is coming close to the recommended daily amount. 8-oz cans of drinks like *Red Bull* and *Monster Energy* contain about the same amount of caffeine as a cup of 'instant' coffee, so all of these are safe – it just depends on how much you are drinking.

There have been cases where consuming a lot has been linked with health problems, like the 28-year-old Australian motor-cross rider who drank seven to eight cans of an energy drink over a seven-hour period, and whose heart stopped (happily, he recovered). Part of the problem is that the safe amount can vary a lot from person to person depending upon how quickly they metabolize it, and also upon factors like gender (on average, men metabolize it faster than women) and whether you are a smoker or not, together with genetic factors.

BUT THE LETHAL DOSE?

Again this would vary from person to person, but it would normally be in the range of 5–10 g.

THAT MEANS A LOT OF CUPS OF COFFEE!

Agreed, you would have to drink well over 50 cups of coffee in rapid succession to get near that, but the real problems seem to be linked with caffeine 'energy tablets', which are sold under a number of trade names such as *Proplus*, *Vivarin*, *No-Doz* and *Stay Awake*.

Caffeine pills are very popular with students staying up all night to cram for exams

WHY IS THAT?

They represent a very concentrated form of caffeine, and while there is a physical limit to how much tea or coffee can be drunk, it is possible to consume a lethal dose of caffeine by eating these tablets. A coroner reporting on the death of a 23-year-old man in England in 2010 said that the man's blood caffeine level of 251 mg per liter was 70 times more than that usually obtained from a high-energy caffeine drink. Apparently the man had consumed two spoonfuls of caffeine dissolved in an energy drink!

How Does Caffeine Work on the Body?

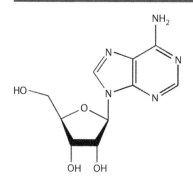

Adenosine

Caffeine acts as a stimulant because it is an antagonist of adenosine, an inhibitor of the central nervous system. Structurally, caffeine resembles adenosine, and therefore binds to the adenosine receptor, blocking its action. With nothing to slow them down, there's greater dopamine and glutamate activity leading to increased neuron activity in the brain and more adrenaline production. Caffeine is slightly addictive because it is a cardiac stimulant.

And Then?

Caffeine (a 'trimethylxanthine') is metabolized in the human liver by cytochrome P450 oxidase enzymes, which rip methyl groups off the caffeine molecule, producing three 'dimethylxanthines' called paraxanthine (84%), theobromine (12%) and theophylline (4%) – followed by further demethylation and oxidation, leading to urates and uracil derivatives.

Paraxanthine   Theobromine   Theophylline

WHAT DO THEY DO?

There is particular interest in paraxanthine, the main metabolite.
Studies on rats – which metabolize caffeine much faster than we
do – show that paraxanthine has a greater stimulating effect on
motion than caffeine itself or the other two metabolites. Scientists
have evidence that paraxanthine in particular may have a protective
role against Parkinson's disease, though, on the downside, extremely
high serum paraxanthine concentrations can be associated
with spontaneous abortion in pregnant women. Interestingly,
theobromine is one of the main active ingredients in chocolate,
which may be why coffee and chocolate go so well together.
Theobromine is also toxic to dogs, which is why you should never
feed chocolate to your dog.

DOES CAFFEINE JUST COME FROM COFFEE?

No, in addition to coffee plants (*Coffea arabica* or *Coffea canephora*),
the tea bush (*Camellia sinensis*) is another major source. There are

others, like the kola nut (*Cola acuminata* and *Cola nitida*), while there are small amounts in cocoa beans.

Red Catucaí coffee, a variety of *Coffea arabica*, showing the beans in different stages of maturity

Roasted coffee beans

Why Do Plants Make Caffeine?

Plants make caffeine to protect themselves against insects because methylxanthines inhibit insect feeding. They are also pesticides, owing to their ability to inhibit the activity of some insect enzymes.

How Do These Plants Make It?

The biosynthesis of caffeine begins with xanthosine, which is converted into 7-methylxanthosine, 7-methylxanthine, theobromine and caffeine, in that order. The genes encoding three distinct *N*-methyltransferase enzymes used to make caffeine from xanthosine have been isolated and then expressed in tobacco plants. The resulting caffeine-containing tobacco plants are unpalatable to tobacco cutworms (*Spodoptera litura*).

So You Can Get Decaf-Coffee Plants?

Cameroon produces a caffeine-free Coffea species, *Coffea charrieriana*, but in the West solvents such as dichloromethane, ethyl acetate or even supercritical CO_2 are used to dissolve the caffeine from the coffee beans to make decaffeinated coffee. Scientists can now engineer low-caffeine coffee plants, too.

Is Coffee Good for You?

Coffee and tea contain polyphenolic antioxidants believed to be beneficial to health. The smell of your tea and coffee has nothing to do with the caffeine, which is quite an involatile substance.

The Smell?

More than 1000 volatile organic compounds are found in roasted coffee, but only around 50 of them are important contributors to the overall aroma. When coffee beans are roasted, complex reactions occur, including the Maillard reaction, which uses sugars,

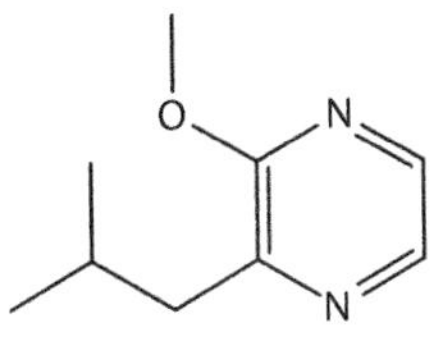

2-Isobutyl-3-methoxpyrazine 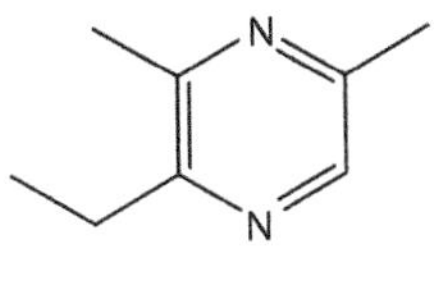   2-Ethyl-3,5-dimethylpyrazine

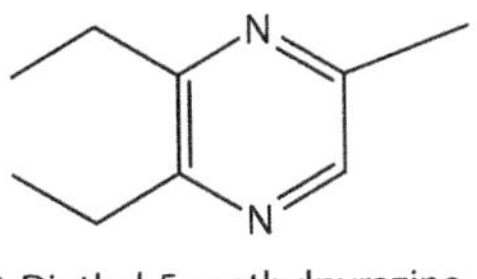

2,3-Diethyl-5-methylpyrazine 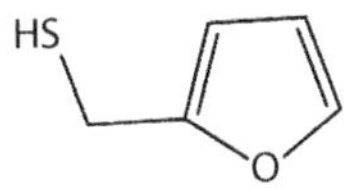   Furfuryl mercaptan

amino acids and peptides to form a wide range of volatile and smelly molecules that cause the fragrance of the beans. Important odorants in coffee include substituted pyrazines, such as 3-isobutyl-2-methoxypyrazine, 2-ethyl-3,5-dimethylpyrazine and 2,3-diethyl-5-methylpyrazine, as well as furfuryl mercaptan, which contribute the 'burnt, roasted' note.

IS IT JUST HUMANS WHO GET A BUZZ FROM CAFFEINE?

Caffeine can be toxic to other animals, including birds and dogs, probably because they have less ability to metabolize it. Famously, NASA scientists tested a variety of drugs on ordinary house spiders, including LSD, mescaline, amphetamine and marijuana, as well as caffeine. Strikingly, caffeine was the drug with most effect on the spiders' ability to spin webs, the results having no patterns and just consisting of a few random threads.

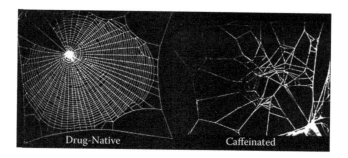

Enjoy your coffee.

Chapter

7

CAPSAICIN

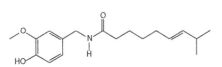

Capsaicin

That's the Molecule in Chilli Peppers and Curries, Isn't It?

Chilli peppers contain capsaicin and a few other molecules with similar structures, known as capsaicinoids. Capsaicin and the very similar dihydrocapsaicin together make up some 90% of the capsaicinoid content of a chilli, with capsaicin accounting for some 70%; it makes up around 0.14% of the mass of a chilli.

And Indians Have Eaten Curries for Thousands of Years, Right?

Chilli plant *Capsicum* – green chilli

Archeologists have found that cooks in the ancient Indus River civilization over 4000 years ago used ginger and turmeric, but not chillis. Starch grains from human teeth, like residues inside a cooking pot, contained these spices. So spicy food yes, but curries no.

Why Not?

Chilli peppers originated in pepper plants of the genus *Capsicum* in Central and Southern America and have been used there for perhaps 6000 years, but it was not until the time of Christopher Columbus 500 years ago that they were brought to Europe and found their way to India.

Why Does Capsaicin Make Chilli Peppers So Hot?

In 1997, David Julius and colleagues at the University of California, San Francisco, reported that capsaicin binds to a receptor protein on the surface of pain- and heat-sensing cells. When capsaicin binds, the channel is opened and Ca^{2+} ions enter the cell, triggering a nerve signal. This receptor is also activated by 'noxious heat' (temperatures above 42°C), and is sensitized by acid. The brain gets the same response whether the channel is opened by heat or by capsaicin, so it interprets it in the same way, as 'hot'.

WHY DOES THE PLANT MAKE CAPSAICIN?

Capsaicinoids are made by pepper plants of the genus *Capsicum*; they are secondary metabolites produced by the peppers as a defense against some pathogens, like fungi, and also against herbivores. Experiments have shown that capsaicinoids protect chilli seeds from the *Fusarium* fungus. The plants use an enzyme called *Capsaicin synthase* to make capsaicin from vanillylamine and 8-methyl-6-nonenoic acid, but the biosynthetic pathway is not yet worked out.

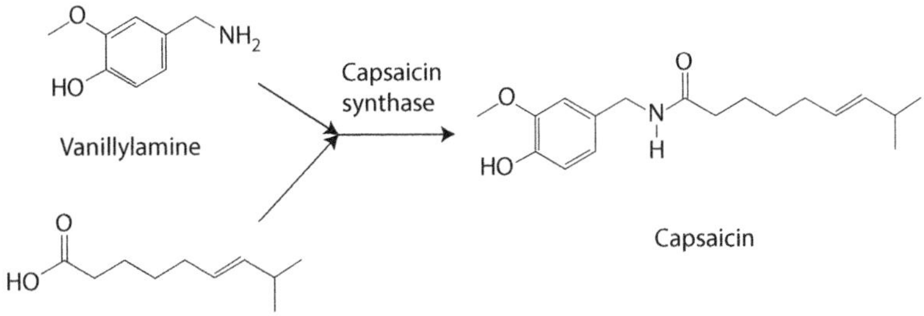

Vanillylamine

Capsaicin synthase

Capsaicin

8-Methyl-6-nonenoic acid

How Do Other Capsacinoids Compare with Capsaicin Itself?

Some of the variants are shown below, where the asterisk indicates the attachment point of the R-group.

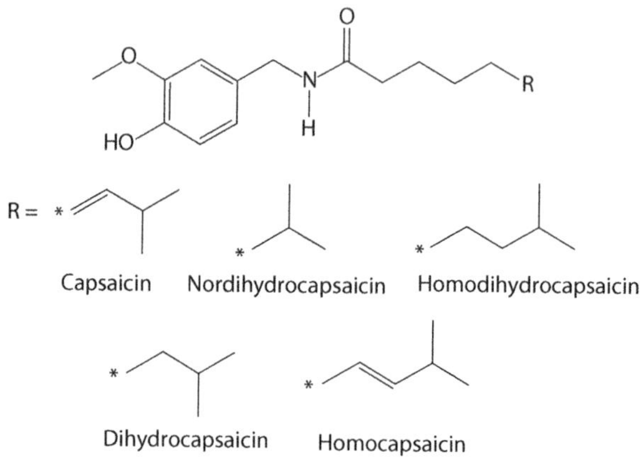

They Seem to Have Similar Structures

Scientists have found that for a molecule to bind to the 'capsaicin receptor' and produce a response, it has to have certain features. Capsaicinoids need to have an aromatic ring with substituents in positions 3 and 4, with a phenolic group at carbon 4 being essential; a long hydrophobic chain is also important, as well as the amide bond (peptide-type linkage).

And They Are All Pretty 'Hot'?

Capsaicin is the hottest; other capsaicinoids have similar structures but are not quite as 'hot'. The hotness scale is measured in Scoville units (named after the inventor, Wilbur Scoville). Pure capsaicin has a Scoville value of 16,000,000, dihydrocapsaicin slightly less at 15,000,000, with the other capsaicinoids clocking in between 8.6 and 9.1 million.

Wilbur Scoville   Dorset Naga

How Can You Measure Hotness?

The Scoville scale means how many times the pepper source has to be diluted before the pepper can no longer be detected. Instrumental methods are now used to measure capsaicinoid content; most recently, scientists have found that they can measure capsaicinoid content by adsorbing capsaicin onto multi-walled carbon nanotubes and measuring the current when the capsaicin undergoes electrochemical oxidation.

How about Peppers?

The Dorset Naga is one of the hottest yet bred, at 1.6 million, and the hottest habanero around 600,000; mild peppers like Jalapeno average around 5000 and the capasaicinoid-free green bell peppers are zero. Tabasco sauce scores about 5000.

Is Capsaicin Hot for Everyone?

Mammals find capsaicin hot, but not birds. If you don't want the squirrels to get the seeds or nuts that you put out for the birds, coat them with capsaicin; the squirrels will leave them well alone, but the birds will still tuck in.

Why Isn't Drinking Water a Good Idea If a Curry Is Too Hot?

Much of the capsaicin molecule is non-polar, so it is not soluble in a polar solvent like water. You need something non-polar to dissolve it, like fat or the casein protein in milk, after the 'like dissolves like' rule. So drink milk, rather than beer or lager when you have that Bangalore Phall down at the local curry house.

Are the Receptors Just in the Tongue?

No, they are also found in many other areas, like your finger tips and face, as well as some more 'sensitive' areas of the body. If you've just been chopping chillis, it is a very good idea to wash your hands before you go to the toilet.

So Apart from Curries, Is Capsaicin Too Hot to be Useful?

In high concentrations, capsaicin is unpleasant; inhaling the vapors from the ultrahot Dorset Naga is said to make your nose tingle, and chefs are reported to wear gloves as touching it is painful. It is used in pepper sprays to subdue criminals or fight off muggers (or grizzly bears, for that matter). Police pepper sprays are the hottest application of capsaicin at around 2 million Scoville units.

Police 'pepper' sprays contain capsaicin

Perhaps surprisingly, capsaicin has been used for pain relief for hundreds of years, with the Mayan people using them to treat asthma, coughs and sore throats, while the Aztecs employed them for toothache. This application depends on the fact that application of capsaicin to the skin initially causes pain, but the pain neurons desensitize and thus the skin becomes desensitized after a while. Nowadays, capsaicin is an ingredient in pain-relieving creams, which contain up to 0.0075% capsaicin; these can alleviate the pain associated with arthritis and shingles as well as post-operative pain. A mixture of capsaicin and a lidocaine derivative (QX-314) has been suggested as a local anesthetic; capsaicin binds to its receptor and holds pores open for the QX-314 to enter, so that it can block Na^+ channels from the inside.

Chapter

8

CARBON DIOXIDE

$$O = C = O$$

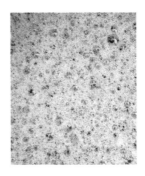

Carbon dioxide bubbles released
from brewing beer

THAT'S THE GREENHOUSE EFFECT GAS, ISN'T IT? THERE MUST BE A LOT OF IT IN THE ATMOSPHERE!

Yes and no, and what's more, there is more than one type of Greenhouse Effect.

EXPLAIN, PLEASE

Carbon dioxide (CO_2) absorbs energy strongly in two regions of the infrared part of the electromagnetic spectrum (as do water, methane and ozone). If there was no carbon dioxide in the Earth's atmosphere at all, the Earth's temperature would be around −20°C. Life would be possible, but it would not be exactly *hospitable*. There is a 'natural greenhouse effect', reducing the loss of long-wavelength infrared heat by radiation, due to 'natural' amounts of CO_2 in the atmosphere, just under 300 parts per million (ppm).

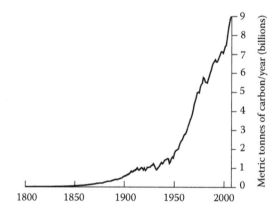

Amount of CO_2 emitted into the air in the last 200 years. For most countries, large-scale use of fossil fuels began when they industrialized, around 1850. Most of the CO_2 release is due to burning of petroleum, natural gas and coal

Sun

Carbon dioxide and
other 'greenhouse' gases

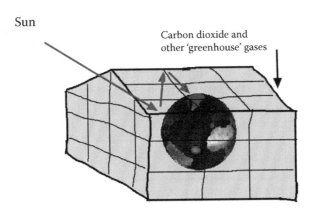

The Greenhouse Effect is caused when heat from the Sun is trapped in
the Earth's atmosphere by molecules such as CO_2, rather than being
re-radiated back out into space. This process keeps the Earth warm,
but too much CO_2 in the atmosphere can lead to the Earth overheating
(global warming), with consequences such as extreme, erratic weather
and climate change

When people talk about *the* 'Greenhouse Effect', what they
are referring to is something extra; the effect on the Earth's
temperature caused by carbon dioxide concentrations rising above
the 'historic' level. Before the Industrial Revolution, the CO_2 level
was around 280 ppm, 0.028% of the atmosphere. Daily average
CO_2 levels (measured at Mauna Loa Observatory in Hawaii) have
increased from around 310 ppm (0.031%) in 1958 to 400 ppm
(0.040%), first reached on May 9th 2013.

WHAT IS CAUSING THIS?

CO_2 gets into the atmosphere in various ways, and at present
there is more being added to the atmosphere than is being
removed. There are 'natural' processes that add CO_2 to the

atmosphere – wild fires and volcanoes are two examples. However, the recent rapid increase in CO_2 levels is due to man-made processes, such as burning fossil fuels, whether for transportation, providing heat in cold weather, or generating power (*e.g.* coal-fired power stations). In some parts of the world, clearing areas like forests to free up land is an important cause.

AND THERE IS RESPIRATION OF COURSE

Indeed, this happens in all living cells and is used to obtain energy from glucose to keep life processes going (see p193). We should really refer to it as aerobic respiration, because it requires oxygen (see p387).

$$C_6H_{12}O_6(aq) + 6O_2(g) \rightarrow 6CO_2(g) + 6H_2O(l)$$

When this happens in the human body, the oxygen required is transported in the blood from the lungs, and the CO_2 produced is removed by the blood, to the lungs, with both molecules using hemoglobin as the 'taxi' molecule (see p227).

Of course, carbon dioxide is also removed from the atmosphere – photosynthesis in green plants, algae and cyanobacteria is a major contributor to this (see p81), trapping CO_2 as sugars such as glucose or glucose polymers (see p193).

$$6CO_2(g) + 6H_2O(l) \rightarrow C_6H_{12}O_6(aq) + 6O_2(g)$$

Growth of plants is an important means of removing CO_2 from the atmosphere (it roughly balances what is added by plant decay and 'natural' forest fires). But over 40% of the CO_2 emissions is thought to dissolve in the Earth's oceans.

Surely If the Oceans Are Soaking Up CO$_2$, That Removes the Problem?

Unfortunately, this leads to a slight decrease in the pH of seawater. It becomes more acidic because carbon dioxide dissolves in water (reversibly) to form a solution containing the weak acid, carbonic acid (H$_2$CO$_3$).

$$CO_2(g) + H_2O(l) \rightleftharpoons H_2CO_3(aq)$$

Actually, most dissolved CO$_2$ remains as solvated CO$_2$ molecules, rather than carbonic acid molecules, so it is more helpful to write the equilibrium as

$$CO_2(aq) + H_2O(l) \rightleftharpoons HCO_3^-(aq) + H^+(aq)$$

(This process helps control the pH of blood.)

It is because the oceans are gradually becoming slightly more acidic that some scientists are concerned about the effects of this ocean acidification upon coral reefs and shellfish.

Does the Acidity of CO$_2$ Have Any Other Consequences?

Because of CO$_2$ dissolving in it, rainwater is not neutral; it is slightly acidic, with a pH of 5.6. This means when it gets into the ground, it can slowly dissolve certain rocks, such as limestone. So rainwater is naturally slightly acidic, and 'acid rain' is defined as rainwater with a pH below 5.6, which happens when the more acidic oxides of sulfur and nitrogen – from sources like coal-fueled power stations and automobiles – enter the atmosphere.

Acidic rain can dissolve carbonate rocks such as limestone (calcium carbonate) to produce the soluble calcium hydrogen carbonate, causing slow erosion, and creating 'hard' water.

$$H_2O(l) + CO_2(g) + CaCO_3(s) \rightarrow Ca(HCO_3)_2(aq)$$

Over a period of time, this creates caves underground. If the water evaporates, the process is reversed, leading to stalagmites and stalactites.

$$Ca(HCO_3)_2(aq) \rightarrow H_2O(g) + CO_2(g) + CaCO_3(s)$$

The same process occurs in a kettle when 'hard' water is boiled, causing a 'fur' in the kettle.

In a similar way, acidified rain also leads to erosion and decay of buildings and statues. Many ancient stone statues on

Acid rain erodes stone statues

The Statue of Liberty is green due to oxidation of the copper sheets from which it is made

medieval cathedrals are now unrecognizable due to erosion by acid rain, and even more recent metal statues, which are usually made from bronze, often look green due to the reaction of the copper content with acidified CO_2 to form copper carbonate (*verdigris*).

ARE THERE OTHER WAYS OF REMOVING CARBON DIOXIDE FROM THE AIR?

One way is to use carbon capture and sequestration (a.k.a. carbon capture and storage, CCS). This can involve removing CO_2 from the chimney gases of a power station or factory, sometimes by reaction with a basic amine. Ultimately, these schemes depend upon storing the CO_2 in underground cavities of porous rock (often worked-out oil fields) under a 'cap' of impermeable rock.

IS THERE A LINK BETWEEN CO_2 LEVELS AND TEMPERATURE?

The Earth's temperature undergoes periodic variations – around AD 1000, there was what is known as a 'Medieval Warm Period', the warmest epoch in the past 2000 years, until today, while more recently there was a 'Little Ice Age' centered around 1700. Average temperatures today are warmer than at any time in the past two millennia (around 0.8°C warmer in the past century and a half), and the trend is upwards. Many scientists think that the rise in CO_2 levels does indeed contribute to that, not least because every transition from a glacial to a warm period in the past 420,000 years was accompanied by a substantial rise in the CO_2 content of the atmosphere.

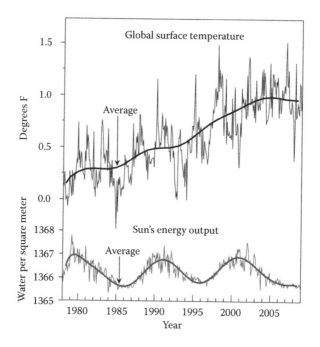

The graph shows the global surface temperature (top) and the Sun's
energy received at the top of the Earth's atmosphere (bottom).
Although the amount of solar energy received by the Earth has followed
its natural 11-year cycle, with small variations up and down, over the
same period, average global temperature has risen significantly. Thus,
the increasing temperature cannot be put down to solar effects, but
instead the Greenhouse Effect caused by increasing CO_2 levels is far
more likely (From NOAA National Climatic Data Center)

WHAT ABOUT THE IMPORTANT PROPERTIES OF CO_2 ITSELF?

CO_2 is a colorless gas that is denser than air and which does not
support combustion. This is why it is used in fire extinguishers.
Some fire extinguishers create their carbon dioxide when activated
by pressing the button, using a reaction between an acid (like
tartaric acid) and sodium hydrogen carbonate, though this type

has fallen out of use. Dry extinguishers use sodium bicarbonate powder – when spread on the fire, it decomposes, releasing carbon dioxide:

$$2NaHCO_3(s) \rightarrow Na_2CO_3(s) + CO_2(g) + H_2O(g)$$

Other extinguishers are just cylinders containing CO_2 at about 60 times atmospheric pressure. At this pressure, the CO_2 is liquid, and can be heard sloshing around when the extinguisher is moved. When the nozzle is opened and the pressure suddenly drops back to 1 atmosphere, the liquid CO_2 instantly turns into a gas and rushes out to quell the fire. Being heavier than air, CO_2 tends to hug the ground and form an invisible blanket between the fire and the source of oxygen that the fire needs.

A CO_2 fire extinguisher being used in a training exercise by the US Army

There are parts of the world where CO_2 has been released suddenly with tragic consequences, most famously on August 21st 1986 at Lake Nyos in Cameroon, when the lake suddenly emitted a huge cloud of CO_2 which suffocated 1700 people and 3500 livestock.

A similar event two years earlier at Lake Monoun had killed 37 people. There are caves like the *Grotta del Cane* (Cave of the dogs), near Naples, where a carpet of CO_2 at floor level will suffocate a dog, but not its (taller) owner. Levels of CO_2 in air much more than 5% are dangerous to health.

At normal pressures, carbon dioxide has no liquid phase, so when it's cooled down below −78.5°C, carbon dioxide solidifies. When warmed, the solid carbon dioxide sublimes back into a gas. Solid CO_2 is used as a refrigerant, especially to keep frozen food cold, and is known as dry ice – it gets that name as there's one big advantage when solid CO_2 is used to refrigerate frozen foods – the packaging doesn't become wet. Dry ice used to be used to create smoke effects during rock concerts or theatrical performances, by simply adding chunks of it to a bucket full of warm water. Nowadays, most smoke effects are produced by 'smoke machines', which vaporize a glycol/water mixture and spray it out to form a fog or haze. Dry ice is, however, sometimes used in combination with these to produce low-lying 'fog' which stays a few feet above the ground.

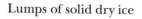

Lumps of solid dry ice   Looks like they've overdone the dry ice
a bit …

What's CO_2 Used for – Apart from Fire Extinguishers?

Back in 1772, Joseph Priestley noted that CO_2 solutions had a pleasant sharp taste, and the ability of CO_2 to dissolve well under pressure proved to be the basis of the soda-syphon and of the carbonated (fizzy) drink industry. Natural carbonated waters like *Perrier* and *San Pellegrino* are also familiar.

Carbon dioxide is of course also formed when yeast uses the enzyme zymase to catalyse the fermentation of sugar into alcohol and carbon dioxide.

$$C_6H_{12}O_6(aq) \rightarrow 2C_2H_5OH(aq) + 2CO_2(g)$$

This process is also used when baker's yeast produces fermentation in leavened bread – the gas produced makes the bread 'lighter'. Baking soda is used to release CO_2 in a dough or batter and cause it to rise. This employs a mixture of sodium bicarbonate and a weak acid like tartaric acid to produce the gas:

$$H^+ + HCO_3^- \rightarrow CO_2 + H_2O$$

A similar mixture is used to make the traditional sweet material, sherbet powder, sometimes using citric acid as an alternative to tartaric acid. Sugar and sometimes colors or flavorings are also added.

Chapter

9

β-CAROTENE

Is It True That Eating Carrots Helps You See at Night?

There may be some slight foundation in that, but that's not the reason the story got around. Early in World War II, Adolf Hitler was on a roll. He'd occupied Austria and Czechoslovakia without even a shot being fired. The German Army had invaded Poland, Denmark, Norway, Holland, Belgium, and France by the end of June 1940. But the roll ended when the Royal Air Force (RAF) denied the Luftwaffe control of the air in the Battle of Britain, preventing Hitler from unleashing the invasion barges. So in autumn 1940, the Luftwaffe's campaign shifted to night bombing, into what became known as the Blitz on London and other cities. The only aircraft available to the RAF for night fighting was the Bristol Blenheim, a converted bomber, slower than some of the bombers it was trying to intercept, and armed only with four machine guns, which lacked 'stopping power' against the German armored bombers.

So?

Bristol Beaufighter night-fighter

Then in October 1940, Number 29 and 604 Squadrons were equipped with the new Bristol Beaufighter. These twin-engine aircrafts could do well over 300 mph, carried four 20 mm Hispano cannons

underneath, and, importantly, had
the first AI (airborne interception)
radar sets. The AI operator
monitored the echoes from the
enemy aircraft and vectored the
pilot onto it. Once in range, those
four cannons made the Beaufighter
a mean bomber-killer. By the end of
the Blitz in May 1941, one pilot, John
Cunningham, with his observer, C.F.
('Jimmy') Rawnsley, had shot down
14 German bombers using AI.

John 'Cat's Eyes' Cunningham
photographed in 1995 (From
© Press Association. With
permission)

WHAT'S THIS GOT TO DO WITH CARROTS?

Pilots like Cunningham and Bob Braham became famous, but
Britain didn't want to let the Germans know that their success was
due to the secret radar system. So the story was promoted that the
success of these night-fighter crews was due to their having eaten lots
of carrots for several years, giving them enhanced night vision. John
Cunningham got the nickname 'Cat's Eyes' Cunningham. This story
had some plausibility and also helped Lord Woolton, the British food
minister, to promote the value of vegetables in the wartime diet.

SO THERE IS A LINK BETWEEN CARROTS AND EYESIGHT?

Yes, it's all due to the molecule *beta*-carotene. Carrots are only
one of a number of fruits and vegetables that are good sources of
β-carotene; others include broccoli, spinach, apricots and nectarines.

WHY ARE CARROTS ORANGE?

That is due to their high β-carotene content. β-Carotene absorbs
light in the green-blue region of the visible spectrum (400–500 nm),

reflecting red and yellow, so that it appears orange in color. As a result, β-carotene is widely used as a safe natural food colorant (E160). However, if you eat too many carrots, you can turn yellow or even orange, because the excess of fat-soluble carotene gets deposited in the skin; the complaint is called *carotenemia*. A few years ago, a four-year-old child's hands and face went bright orange and yellow after the child had been drinking lots of a soft drink fortified with large amounts of β-carotene colorant. When a student spent a month on a wholly carrot-based diet in 2007, the palms and soles of her feet turned orange!

IF THIS IS A β-CAROTENE, IS THERE AN α-CAROTENE, TOO?

Indeed there is, though it is less abundant than the β-isomer. It only differs from β-carotene in the position of one double bond. The structure of β-carotene was worked out in 1930–1931 by Paul Karrer, who was awarded the 1937 Nobel Prize in Chemistry for his achievements.

α-Carotene. Can you spot which double bond has switched position from that in β-carotene shown on the first page of this chapter?

HOW COULD β-CAROTENE POSSIBLY HELP YOUR NIGHT VISION?

It is a precursor to Vitamin A. Enzymes in the body, notably in the small intestine, can cleave β-carotene molecules, producing two molecules of retinol (also known as Vitamin A) which is then stored in the liver. Oxidation of retinol generates retinal, and when this is linked to a protein called *opsin*, it forms rhodopsin, which is the light-sensitive pigment in the retina.

Retinol (Vitamin A) – two of
these molecules are formed
when β-carotene is cut in half
by enzymes

Retinal – part of the light-
sensitive pigment in the
retina of the eye

WOULDN'T IT JUST BE EASIER TO EAT VITAMIN A?

Too much vitamin A is bad for you, as high levels of vitamin A cause a complaint known as hypervitaminosis A. This can be caused by eating the liver of some animals, notably the polar bear, husky, walrus and seal. Although this isn't usually a problem for most people in the modern world, it can be an issue for Eskimos or polar explorers. In 1913, the Arctic explorer Xavier Mertz died after eating the liver of his huskies. Instead, the body produces Vitamin A as it is needed, from stored carotene.

ARE THERE OTHER WAYS OF GETTING β-CAROTENE?

Because rice is part of the staple diet in much of the world, researchers added two β-carotene biosynthesis genes in making a variety of *Oryza sativa* rice, producing rice with a greatly enhanced β-carotene content that would help combat Vitamin A deficiency, a worldwide problem. The enhanced β-carotene content gives the rice a golden-yellow color, hence its name 'golden rice'. Tests have shown that golden rice is as good a source of

Golden rice

Vitamin A as food supplements, as its carotene content can be converted into Vitamin A in the human body. But because it is considered to be a genetically modified food, as usual, it is not regarded favorably by environmentalists.

How Do Plants Make It?

Well, the full biosynthetic pathway is very complicated, but in effect it involves stitching together a whole bunch of smaller units, called isoprenes.

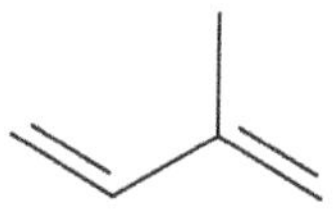

An isoprene unit. Bonding many of these together in the correct way yields many biological molecules, such as terpenes and steroids. If they are bonded in the way shown below, the result is β-carotene

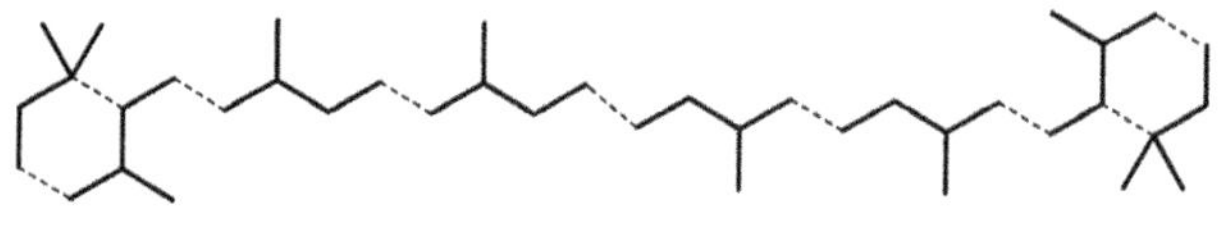

Why Do Plants Make It?

There is still an ongoing debate about this. One possible explanation is that β-carotene works in tandem with chlorophyll (see p81), helping to absorb wavelengths of sunlight that chlorophyll cannot. Chlorophyll absorbs most strongly in the blue and red and reflects green (which is why most leaves look green), and the energy in the green part of the spectrum would be wasted if it were not for molecules such as β-carotene also present in the leaves, which can absorb green – although this process is nowhere near as efficient as for chlorophyll.

So Why Don't All Plants Look Orange?

In those green plants that contain β-carotene, the orange color is masked by the stronger pigment from green chlorophyll. However, in Autumn, when the plants withdraw nutrients from the leaves to conserve energy over the winter, the chlorophyll molecules break down first, while the more robust β-carotene molecules survive longer. Thus, the leaves

In Autumn, leaves turn from green (the color of chlorophyll) to the oranges and reds of carotene

gradually lose their green pigments and the red-orange color of the carotenes start to appear, and we get the glorious reds and oranges of Autumn (Fall).

Chapter

10

CHLOROPHYLL

ISN'T THAT WHAT MAKES LEAVES GREEN?

Aerial view of the Amazon rainforest

Yes, but it's much more than just a pigment. Chlorophyll is the molecule that absorbs sunlight and uses its energy to synthesize carbohydrates from CO_2 and water.

This process is known as photosynthesis and is the basis for sustaining the life processes of all plants. Since animals and humans obtain their food supply by eating plants, photosynthesis can be said to be the source of our life also.

SO WE ARE SOLAR POWERED?

Effectively, yes.

WHO DISCOVERED THIS?

A whole bunch of bewigged European scientists solved the puzzle piece by piece over a period of about 100 years. In 1780, the English chemist Joseph Priestley found that plants could '*restore air which has been injured by the burning of candles*'. He used a mint plant, and placed it into an upturned glass jar in a vessel of water for several days. He then found that '*the air would neither extinguish a candle, nor was it all inconvenient to a mouse which I put into it*'. In other words, he discovered that plants produce oxygen (luckily for the mouse!).

A few years later in 1794 the French chemist Antoine Lavoisier discovered the concept of oxidation, but before he could make any further advances he was executed during the French Revolution for

Joseph Priestley   Antoine Lavoisier   Jan Ingenhousz

being a Monarchist sympathizer. The judge who pronounced the sentence said '*The Republic has no need for scientists*'.

So it fell to a Dutchman, Jan Ingenhousz, who was a court physician to the Austrian empress, to make the next major contribution to the mechanism of photosynthesis. He had heard of Priestley's experiments, and a few years later spent a summer near London doing over 500 experiments, in which he discovered that light plays a major role in photosynthesis.

> '*I observed that plants not only have the faculty to correct bad air in six to ten days, by growing in it … but that they perform this important office in a complete manner in a few hours; that this wonderful operation is by no means owing to the vegetation of the plant, but to the influence of light of the sun upon the plant*'.

Very soon after, more pieces of the puzzle were found by two chemists working in Geneva. Jean Senebier, a Swiss pastor, found that 'fixed air' (CO_2) was taken up during photosynthesis, and Theodore de Saussure discovered that the other reactant necessary was water. The final contribution to the story came from a German

Jean Senebier   Theodore de Saussure   Julius Robert Mayer

surgeon, Julius Robert Mayer, who recognized that plants convert solar energy into chemical energy. He said:

> *'Nature has put itself the problem of how to catch in flight light streaming to the Earth, and to store the most elusive of all powers in rigid form. The plants take in one form of power, light; and produce another power, chemical difference'.*

SO WHAT ACTUALLY HAPPENS IN PHOTOSYNTHESIS?

The actual chemical process which takes place is the reaction between carbon dioxide and water, catalysed by sunlight, to produce glucose and a waste product, oxygen. The glucose sugar is either directly used as an energy source by the plant for metabolism or growth, or is polymerized to form starch, so it can be stored until needed (see p193). The waste oxygen is excreted into the atmosphere, where it is utilized by plants and animals for respiration. The chemical equation for photosynthesis is

$$6CO_2 + 6H_2O \rightarrow C_6H_{12}O_6 + 6O_2$$

where $C_6H_{12}O_6$ is glucose, with sunlight being the energy source and chlorophyll the catalyst.

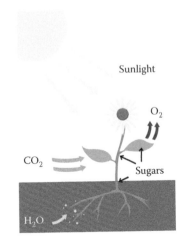

How Does Chlorophyll Trap Sunlight?

Chlorophyll is the photoreceptor molecule that traps this 'most elusive of all powers'. It is found in the chloroplasts of green plants, and is what makes green plants, green. Its name actually comes from the Greek *chloros* meaning 'green', and *phyllon* meaning 'leaf'. The basic structure of a chlorophyll molecule is a porphyrin ring (a conjugated flat ring-shaped molecule based on porphene), co-ordinated to a central metal atom.

This is very similar in structure to Vitamin B12 and the heme group found in hemoglobin, the main difference being that in Vitamin B12 the central atom is cobalt, in heme it's iron (see p227), whereas in chlorophyll it's magnesium.

Plant cells are green because they contain chlorophyll

There are actually several types of chlorophyll, which differ only slightly in the composition of a side chain. The two most common types are chlorophyll *a* where the side group is $-CH_3$, and chlorophyll *b* where it is $-CHO$. Both of these two chlorophylls are very effective photoreceptors because they contain a network of alternating single and double bonds, and the orbitals can delocalize, stabilizing

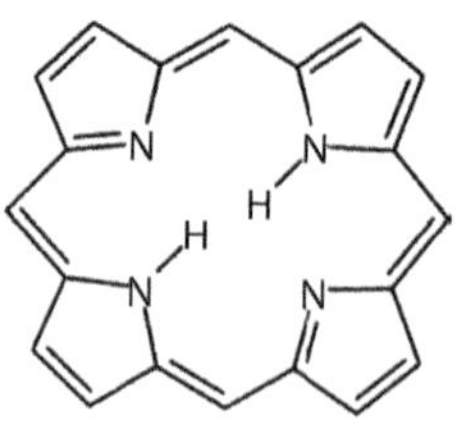

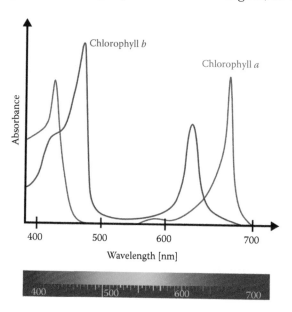

Porphene

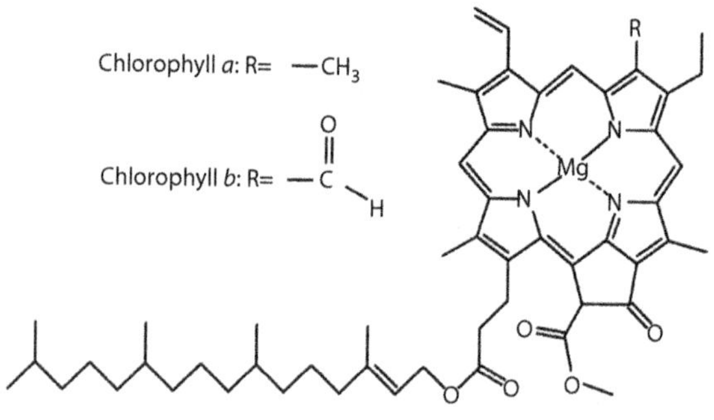

Chlorophyll a: R= —CH₃

Chlorophyll b: R= —C

Chlorophyll

the structure. Such delocalized polyenes have very strong absorption bands in the visible regions of the spectrum, allowing the plant to absorb the energy from sunlight.

The different side groups in the two chlorophylls 'tune' the absorption spectrum to slightly different wavelengths, so that light

Absorption spectra of chlorophyll *a* and *b*

that is not significantly absorbed by chlorophyll *a*, at, say, 460 nm, will instead be captured by chlorophyll *b*, which absorbs strongly at that wavelength. Thus these two kinds of chlorophyll complement each other in absorbing sunlight. Plants can obtain all their energy requirements from the blue and red parts of the spectrum; however, there is still a large spectral region, between 500 and 600 nm, where very little light is absorbed. This light is in the green region of the spectrum, and since it is reflected, this is the reason plants appear green. In fact, chlorophyll *a* is used as a green dye (called 'Natural Green 3') that's often found in soaps and cosmetics.

SO WHY AREN'T LEAVES BLACK?

Good question – a black pigment would be much more efficient because *all* the visible light energy from the sun would be absorbed, rather than some being reflected back and wasted, as the green light is at present. Scientists are still unclear about the reasons for this. One idea is that because all modern plants are believed to have evolved from a common ancestor – a sort of green algae – they all inherited the same chlorophyll. If this molecule did its job of absorbing light well enough, there would be no need to evolve a more efficient version. Other reasons state that evolution is not a directed process, and is not capable of thinking like an engineer. Given the same problem, an engineer might design a photoreceptive molecule that absorbs as much of the spectrum as possible, including ultraviolet and infrared (*i.e.* a black pigment). But evolution can only work with what it has, and slowly change the current system into a better version. If there's no driving need for a better version, then the molecule won't evolve. Even if there is a need, there may not be an evolutionary mechanism to achieve it. This is a good argument to use on Creationists – a truly 'Intelligent Designer' would not have chosen an inefficient molecule such as chlorophyll to harvest solar energy; a far better choice would have been a molecule such as retinal (see p73) which absorbs much more of the spectrum.

BUT I'VE SEEN LEAVES WITH COLORS OTHER THAN GREEN AS WELL?

Chlorophyll is not the only pigment in leaves. But chlorophyll

absorbs so strongly that it can mask other less intense colors. Some of these more delicate colors (from molecules such as carotene and quercetin) are revealed when the chlorophyll molecule decays in the Fall, and the woodlands turn red, orange, and golden brown. Chlorophyll can also be damaged when vegetation is cooked, since the central magnesium atom is replaced by

In Fall (Autumn), the fragile chlorophyll molecules break down first and the reds and yellows of other more robust pigment molecules, such as carotenoids, show through

hydrogen ions. This affects the energy levels within the molecule, causing its absorbance spectrum to alter. Thus, cooked leaves change color – often becoming a paler, insipid yellowy green.

I STILL DON'T SEE HOW ABSORBING LIGHT POWERS PHOTOSYNTHESIS

The chlorophyll molecule is the active part that absorbs the sunlight, but just as with heme (see p227), in order to do its job (synthesizing carbohydrates), it needs to be attached to the backbone of a very complicated protein. This protein may look haphazard in design, but it has exactly the correct structure to orient the chlorophyll molecules in the optimal position to enable them to function efficiently manner.

In fact the energy absorbed by one chlorophyll molecule is not sufficient to catalyse the reaction to make carbohydrates. So, many chlorophyll molecules work together, like a big radar dish, to collect light energy and funnel it to a central 'master' chlorophyll molecule by way of a series of electron transfer processes. With time, the master chlorophyll molecule becomes more and more positively charged. Once the master chlorophyll has stored up sufficient charge, it can lose its charge by grabbing an electron from a nearby water molecule: the water is thus oxidized to O_2 gas and H^+ ions. The H^+ ions then diffuse through the cell membrane and in doing so set up a potential difference. An enzyme called ATP synthase then collects the H^+ ions and shuttles them back across the membrane to where they came from, and the potential energy that's released is used by the enzyme to make ATP molecules from ADP (see p1). ATP is the molecule that stores energy for short-term use in plants and animals. The ATP is transported around the plant to where the energy is needed, such as places where the plant is growing, and other enzymes then use the energy stored in the ATP to fix CO_2 into sugars in a complicated process called the Calvin Cycle. The sugars are either oxidized immediately to release their energy for various metabolic processes or polymerized into starch (a long-term energy store) or cellulose which makes up the structure of the plant cells.

SO YOU COULD SAY THAT PLANTS ARE MADE UP OF SOLID CARBON DIOXIDE?

Yes, in a way. It's amazing to think that when you look at plants, all the leaves, stems, tree bark, and even the fruit and nuts were once just carbon dioxide and water. This is becoming more important when we consider the effect that atmospheric CO_2 has upon global warming. One of the best ways to 'mop up' CO_2 gas is to fix it into plants. The amount of CO_2 removed from the atmosphere each year by photosynthetic organisms is enormous – estimates suggest something like one billion tonnes of carbon per year. Green plants and algae help to mitigate the excess CO_2 in the atmosphere put there by mankind burning fossil fuels. So deforestation, which is occurring all over the Earth, is one of the worst things we can do if we want to prevent CO_2 build up and a possible runaway Greenhouse Effect. The second worst thing we can do is to burn those forests, releasing the stored CO_2 back into the air – but guess what, we're doing that too!

Chapter

11

CHOLESTEROL

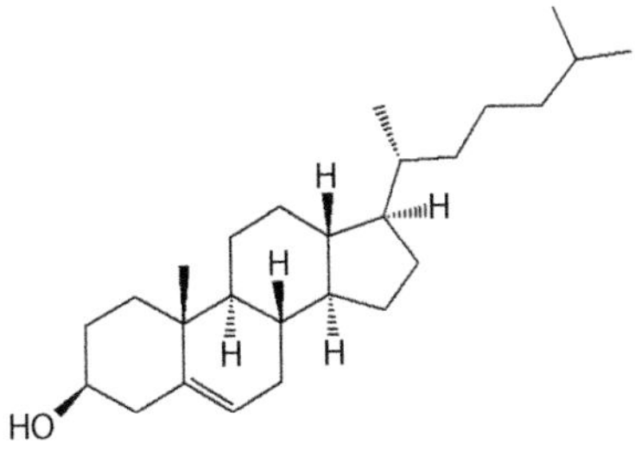

Cholesterol

THAT'S THE STUFF THAT CAUSES HEART ATTACKS, ISN'T IT?

Yes, if there's too much of it in your bloodstream, but it's also an essential ingredient to all animal life.

HOW SO?

It performs a number of vital functions, such as forming and maintaining cell membranes, and helps the cell to resist changes in temperature, and protects and insulates nerve fibers. Within the liver, cholesterol is the starting material for the production of bile salts which are necessary to aid digestion. In the skin it converts into Vitamin D when exposed to sunlight. And elsewhere it is the vital precursor to the sex hormones progesterone, testosterone (see p489), estradiol (see p185) and cortisol.

WHAT DOES IT LOOK LIKE?

Michel Eugéne Chevreul

Cholesterol is a soft, waxy compound belonging to the steroid family of molecules and is found among the lipids (fats) in the bloodstream as well as in all cells of the body. The French chemist François Poulletier de la Salle is credited with first identifying it in solid form from gallstones around 1769. However, it wasn't until 47 years later at a meeting of the French Academy of Sciences that another chemist, Michel Chevreul, suggested that this substance, which had fat-like properties, should be named *cholesterine*, from the Greek *chole*, meaning bile, and *stereos*, meaning solid. It later became known as cholesterol.

Cholesterol is formed in the liver via a series of complicated (37 steps!) biochemical reactions at a rate of approximately 1 g a day. To a lesser extent, it is created by cells lining the small intestine and by individual cells in the body. All the cholesterol required for biological functions is produced by the body and this makes up approximately 85% of the total blood cholesterol level.

WHAT HAPPENS TO THE EXCESS?

It's recycled. The liver converts it into bile, which then goes into the digestive tract to help digestion. About 50% of this is then reabsorbed by the bowel back into the bloodstream. In fact the total amount of cholesterol in the body remains roughly constant, no matter how much extra you eat. The body simply slows down or speeds up its own cholesterol production depending on what you eat.

GLASBERGEN   Copyright 2008 by Randy Glasbergen.

"With this new drug, cholesterol forms *outside* of the body, where it can't clog the arteries."

SO WHAT'S THE PROBLEM WITH EATING CHOLESTEROL, THEN?

In the Western diet, the typical daily intake of cholesterol for a healthy adult is 200–300 mg, which is about one quarter to a third of the amount made naturally by the body. These usually come from eating animal fats, in foodstuffs such as cheese, eggs, beef, pork, poultry, shrimp and fish, and these fats are complex mixtures of triglycerides (see p207) plus cholesterol. The problem is that nearly all of this extra ingested cholesterol is in the form of a cholesterol-ester, which is poorly absorbed by the body and so ends up circulating around in the bloodstream where it can cause problems. But it's actually the 'type' of cholesterol that's crucial.

Is There More than One Type?

Not really. But cholesterol can be transported in the blood in two different ways, one good, one bad.

How So?

Cholesterol is not soluble in water (or blood), and so to be transported around the body in the bloodstream it has to hitch a ride in a 'taxi'. This 'taxi' molecule is called a lipoprotein, which is a mixture of proteins and fats. Lipoproteins often form

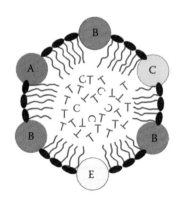

Schematic diagram of a lipoprotein. A, B, C and E are types of proteins connected together by the oval-shaped phospholipids forming a spherical, water-soluble droplet. The long hydrocarbon 'tails' from the lipids point into the middle of the droplet, forming an oily central region which can solvate cholesterol (C) and other hydrophobic molecules such as triglycerides (T)

spherical structures, with the outer surface being hydrophilic (water-liking) allowing them to dissolve easily in water (or blood), and an inner region containing all the long-chained hydrocarbon tail groups, which is hydrophobic (water-hating) and can solvate non-polar molecules, such as cholesterol and fats (triglycerides). These lipoprotein 'taxis' allow the normally insoluble cholesterol molecules to be transported in the bloodstream, and help them to move into and out of cells.

However, there are two types of lipoprotein, low density and high density. Low-density lipoprotein (LDL) is the major cholesterol carrier in the blood, and is often called 'bad cholesterol' because it has been linked with heart disease. If too much LDL circulates, it can slowly build up on the walls of arteries that feed the heart

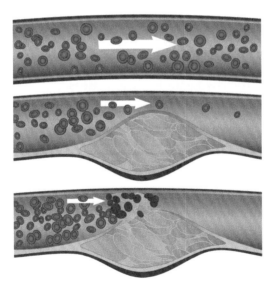

An illustration of cholesterol build-up in an artery. In the top picture the artery is healthy, but in the middle one cholesterol has begun to build up to form a plaque, partially blocking the artery. In the lower one the plaque has ruptured, causing blood to clot and block the artery completely

and brain. Together with other substances, it can form 'plaque' – a hard, thick deposit that can clog those arteries (a condition called *atherosclerosis*). If the deposit becomes too thick, or the turbulent blood flow in its vicinity causes a blood clot to form there, the artery can be partially or completely blocked, starving the heart or brain of its blood supply, leading to a heart attack or stroke.

WHAT ABOUT THE OTHER TYPE?

High-density lipoprotein (HDL), or 'good cholesterol', carries about 30% of cholesterol in the blood. They are called high density because they contain a higher proportion of protein compared to cholesterol than LDL. It is now believed that HDL carries cholesterol away from the arteries and back to the liver, where it is metabolized and removed. It may even help remove excess cholesterol from the plaques, slowing their growth. Blood tests typically report the amount of cholesterol in LDL and HDL, and it's the LDL:HDL ratio that's crucial in the risk factor for heart disease.

WHAT DOES THIS HAVE TO DO WITH EATING FATTY FOODS?

The cholesterol in fatty foods ends up being transported in LDLs, and the body then reduces its rate of production of HDL to keep the overall concentration of cholesterol constant. Thus, the LDL:HDL ratio increases, and with it the cardiac risks.

WHAT CAN BE DONE ABOUT IT?

Changes in diet, lifestyle, and exercise can help lower and maintain low levels of LDL blood cholesterol. However, in some cases, other factors such as genetics make the use of drugs necessary to treat the problem.

WHAT TYPE OF DRUGS?

The most commonly prescribed drugs are statins, which help to reduce the levels of LDL cholesterol, although they don't have much effect on the HDL levels. Statins work by inhibiting the enzyme in the liver which produces cholesterol, and are therefore a preventative medicine which needs to be taken regularly. They were first developed by the drug company Merck, which came up with the terms 'good' and 'bad' cholesterol. When statins first became available, Merck needed to market them effectively, and to do so they needed to convince the public that high cholesterol levels were dangerous, and that statins could prevent this. A high-profile marketing campaign by Merck and other drug companies in the early 1990s succeeded to such an extent that by 2008 one statin, atorvastatin (marketed as *Lipitor* or *Torvast*), became the best-selling pharmaceutical in history, with its manufacturer Pfizer reporting sales of $12.4 billion.

Atorvastatin

Chapter

12

CISPLATIN
THE ANTI-CANCER DRUG

Cisplatin

ISN'T IT STRANGE TO HAVE A METAL IN A MEDICINE?

People have taken all sorts of metal compounds in the past, like iron compounds to treat anemia or gold compounds in the treatment of arthritis. It's just that a metal anti-cancer drug was unheard of 30 years ago.

SO IT HAS GOT PLATINUM IN IT, AND IT'S A *CIS*-ISOMER?

Yes, *cis*-, meaning on the same side, just like cisalpine means on the same side of the Alps. So the two chlorines are on the same side of the molecule and the two ammonias are also next to each other.

HOW WAS IT DISCOVERED?

It depends upon what you mean by 'discovered'. cis-$[Pt(NH_3)_2Cl_2]$ was first made by Michele Peyrone in 1845, but the discovery that led to it becoming an anti-cancer drug was not made until 1964.

WHY DID THEY CHOOSE TO STUDY THIS COMPOUND?

Barnett Rosenberg

They didn't; it is a perfect example of a serendipitous scientific discovery. Barnett Rosenberg was a physicist-turned-biophysicist working at Michigan State University who wanted to see if electric currents had any effect upon cell division. So he and his research technician, Loretta Van Camp, took a solution of growing *Escherichia coli* cells in an ammonium chloride buffer, inserted what he took to be inert platinum electrodes, and switched on the current. To their surprise, the cells did not divide, but grew very long, up to 300 times their normal length. They established that platinum was dissolving, and Van Camp found that some platinum salts (and not the current) affected the cells. In 1966 Andrew Thomson made *cis*- and *trans*-$[Pt(NH_3)_2Cl_2]$ and *cis*- and *trans*-$[Pt(NH_3)_2Cl_4]$ for testing, the team finding that *cis*-$[Pt(NH_3)_2Cl_2]$ was by a long way the most effective compound at producing this filamentous growth of the cells.

SO WHAT DID THEY DO NEXT?

Rosenberg had the idea that, since the *cis*-isomer (now abbreviated simply to cisplatin) inhibited cell division in bacteria like *E. coli*, then it might kill higher cells. So they injected platinum compounds into mice carrying the *Sarcoma 180* tumor. Cisplatin caused remission, the *trans*-isomer did not, as was reported in 1969. By 1971, the first tests on humans were taking place in the United Kingdom, as the (U.S.) National Institutes of Health were not keen on metal-based drugs. It was found to be especially effective against patients with testicular cancer, and despite problems with kidney toxicity, as well as nausea and vomiting, this was controlled with the use of other drugs, as well as hydration therapy to counter the kidney problem. Cisplatin received FDA approval in 1978.

AND IT IS A SUCCESSFUL DRUG?

Hundreds of thousands of people have been successfully treated with cisplatin or with its successors carboplatin (which has fewer side effects) and oxaliplatin, used in the treatment of colorectal cancer.

HAS IT HELPED ANYBODY FAMOUS?

Soon after cisplatin was approved, the British jockey Bob Champion was diagnosed with stage-3 testicular cancer in July 1979. He had a course of cisplatin-based therapy in late 1979 and early 1980. He returned to the saddle and won the 1981 Aintree Grand National on the horse *Aldaniti*. This was made into a movie '*Champions*' starring John Hurt in 1983.

Bob Champion winning the 1981 Grand National on *Aldaniti* (© Press Association Images. With permission)

Lance Armstrong at the team presentation of the 2010 *Tour de France* in Rotterdam

The most famous person to be treated with cisplatin is the American cyclist Lance Armstrong, who in October 1996 was also diagnosed with stage-3 testicular cancer. In his book '*It's Not about the Bike*', Armstrong wrote: '*I thought, Bring it on, give me the platinum'*. He received a cocktail of various anti-cancer drugs, including vinblastine, etoposide, ifosfamide, and cisplatin, with the chemotherapy finishing successfully on December 13th, 1996. Within three years he was finishing first in the *Tour de France*. Ironically, perhaps it was his familiarity with taking such a drug cocktail when he was ill that made him later take more serious performance-enhancing drugs (such as cortisone, steroids, and testosterone, see p489) that led to his fall from grace in 2013.

HOW DOES CISPLATIN WORK?

Cisplatin is usually given intravenously in saline solution; under these conditions, the high chloride concentration ensures that it remains as *cis*-[Pt(NH$_3$)$_2$Cl$_2$] molecules. But when it enters cells where the chloride concentration is much lower, water molecules can substitute chloride in hydrolysis reactions:

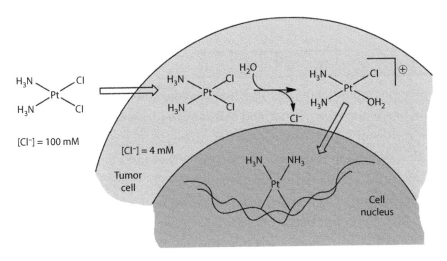

Diagram showing how cisplatin enters a tumor cell, where one of its Cl's is substituted by a water molecule. This new molecule then enters the nucleus where it cross-links to DNA strands, distorting their structure (Based on R. A. Alderden, M. D. Hall and T. W. Hambley, *J. Chem. Educ.*, **83**, (2006) 728.)

$$[Pt(NH_3)_2Cl_2] + H_2O \rightleftharpoons [Pt(NH_3)_2(H_2O)Cl]^+ + Cl^-$$

$$[Pt(NH_3)_2(H_2O)Cl]^+ + H_2O \rightleftharpoons [Pt(NH_3)_2(H_2O)_2]^{2+} + Cl^-$$

The DNA in the cancer cell is the main target of the cisplatin. $[Pt(NH_3)_2H_2O(Cl)]^+$ binds to a purine base in one strand of the DNA, losing a water molecule, whereupon a base in the second strand can replace the chloride, cross-linking the two strands. This distorts the DNA, so that it is not recognized by the cell's repair mechanism, and the cell dies.

WHAT ABOUT THE *TRANS*-ISOMER?

It is not active, as its geometry prevents it cross-linking with DNA. Incidentally, the existence of both *cis*- and *trans*-$[Pt(NH_3)_2Cl_2]$ proves that these complexes have square-planar geometries. If $[Pt(NH_3)_2Cl_2]$ were tetrahedral, only one isomer would exist.

AND THE MESSAGE OF CISPLATIN?

Two: one to politicians and decision makers and one to researchers. The first is that you cannot make discoveries to order – after all, if the results of experiments could always be predicted, you wouldn't bother doing the research. The second is that just because your results are unexpected and unpredictable, one should not assume that something has gone wrong and abandon them, instead of following them up. In the quotation famously attributed to Pasteur, '*In the field of observation, chance only favours the prepared mind*' (*le hasard ne favorise que les esprits préparés*).

Chapter

13

COCAINE

Who Said That?

It's usually attributed to Robin Williams, the actor, making a joke about cocaine.

It's God's way of telling you you've got too much money!

Why?

It has acquired a jet-set image. Popular in the 1920s Hollywood film circles, it was superseded by amphetamines, and then came back into fashion with films like *Easy Rider* (1969) and glam-rock. Conspicuous consumption spread through yuppies in the Western world. The supply was fueled by South American drug cartels, most notoriously originating in Medellín and Cali in Colombia. Then 'crack' cocaine came into use, as a cheaper form that introduced cocaine among less affluent consumers.

But Surely It's Been around for Much Longer than That?

Indeed. The coca bush *Erythroxylum coca* grows wild in the Andes;

it makes cocaine ultimately from the amino acid L-glutamine to stop insect predators from eating it – with its leaves containing around 0.3–0.7% cocaine. In fact, the coca plant is a spectacular organic chemist as it makes just the one isomer out of the 16 possible from the four chiral centers.

Coca bush (*Erythroxylum coca*)

People in Bolivia, Peru, Ecuador and Chile have chewed coca leaves for 8000 years or more. The Incas made pellets of coca leaves with alkali (lime) or calcite and chewed them, giving a slow release of cocaine that they absorbed from their saliva. Coca chewing suppressed their hunger and they believed that it increased their strength and endurance. Teams of coca-chewing runners could cover up to 300 km a day with messages to and from the Inca. From the mid-sixteenth century, the Spanish *conquistadors* gave the Inca slave-workers coca leaves to make them work harder for longer in the silver mines, while during his Polar expedition in 1909, the Arctic explorer Sir Ernest Shackleton used 'Forced March' coca tablets to do what it said on the bottle.

SURELY COCAINE WAS BANNED BY THEN?

Actually, it was regarded as a useful stimulant in the nineteenth century and at one stage it was even available in over-the-counter medications. A Corsican chemist named Angelo Mariani made the discovery that the alcohol in wine extracted the cocaine from coca leaves and put the result, *Vin Mariani*, on sale. He understood the value of publicity, sending bottles to famous people and getting them to endorse his product. Jules Verne, H.G. Wells and the great inventor,

A poster advertising *Vin Mariani* in 1894

Thomas Alva Edison, willingly testified, as did Pope Leo XIII. Not only did the Pope endorse it, he awarded *Vin Mariani* a gold medal! Imitation being the sincerest form of flattery, a Georgian pharmacist named John Styth Pemberton created his own version of Mariani's concoction, called 'French Wine Cola'. Prohibition came to Atlanta in 1886, so Pemberton created a new non-alcoholic version of the drink which he called *Coca-Cola*. In 1903, the cocaine extract was removed from the recipe, being replaced by sugar and caffeine to supply the stimulant, although the name *Coca-Cola* remained unchanged.

So They Thought That Cocaine Was Good for Health?

Photo of Sherlock Holmes tiles on the wall of Baker Street tube station in London. Maybe that is not just tobacco in Holmes' pipe!

Well, an American surgeon William Stewart Halsted used cocaine injections to invent nerve-block anesthesia in 1884. Sadly, he was to become a cocaine addict. Sigmund Freud, inventor of psychoanalysis, pioneered the use of cocaine as a local anesthetic for eye surgery (now replaced by *lidocaine* or *benzocaine*), and cocaine became briefly available in over-the-counter medications for treatments of complaints like toothache or catarrh. But cocaine was generally banned in North America and Europe toward the end of the nineteenth century. Around this time, the fictional detective Sherlock

Holmes became the first celebrity with a cocaine problem in the books *A Scandal in Bohemia* and *The Sign of Four*.

WHAT'S THE DIFFERENCE BETWEEN COCAINE AND CRACK?

Traditional cocaine is a salt, while crack cocaine is molecular in structure. Both ionic and molecular forms are involved in cocaine processing.

Illicit cocaine is extracted from coca leaf in three stages. This is usually done in three different laboratories that can be hundreds of miles apart, by a solvent-extraction route. In the final step at a 'crystal lab', the cocaine molecules are dissolved in an organic solvent such as ether (ethoxyethane) and insoluble impurities are filtered off. To this filtrate is added a mixture of concentrated HCl and propanone, then cocaine hydrochloride crystallizes out. Other solvents like butanone are also used.

Cocaine hydrochloride powder

Cocaine hydrochloride salt is what is commonly sold as cocaine powder, which has traditionally been injected or snorted. If the cocaine hydrochloride salt is treated with alkali, it reverts into the neutral molecular base. This can be done using baking soda ($NaHCO_3$) to make crack, one form of free cocaine. If the hydrochloride salt is smoked, heating it up destroys most of the cocaine, whereas the molecular base 'crack' form is more volatile, so many more cocaine molecules survive to be absorbed. The name 'crack' comes from the cracking or popping noise that the baking soda makes when it's heated up both during manufacture and when smoked.

How Does Cocaine Work?

It is a central nervous system stimulant, preventing reuptake of neurotransmitters like dopamine (see p169), serotonin (p451), and noradrenaline (p9). This impacts on the pleasure center of the brain, explaining its stimulant effects. It is also a sodium-channel blocker, which causes its anesthetic effects, because it interferes with the transmission of nerve impulses.

Are the Effects of All Forms of Cocaine the Same?

How cocaine is taken makes a big difference; the faster the blood concentration changes and the faster it reaches the brain, the more effect the cocaine has but the shorter the effect lasts. If swallowed, most of the cocaine is destroyed in the liver before it gets to the brain, whereas if injected or smoked it reaches the brain in seconds (like nicotine, see p363). Since crack cocaine has a covalent molecular structure, its melting point is low (98°C) so that it has a high vapor pressure at quite low temperatures. Crack addicts heat the solid drug and inhale the volatile vapors, which are easily absorbed, giving the user a rapid and intense sensation. However, the effects are short-lived, which means that users quickly get addicted and they also have to keep increasing the dosage.

Studies, particularly in rodents, indicate that cocaine use affects neurons in the *nucleus accumbens* region of the brain. Brain scans of cocaine users show widespread loss of gray matter, directly related to the length of their cocaine abuse, which is also associated with greater compulsivity to take cocaine. And of course there is the real risk of cardiac arrest, even in people in excellent physical health.

DOES COCAINE MAKE OTHER ANIMALS 'HIGH'?

Many animals will self-administer cocaine. Studies on honey bees (*Apis mellifera L.*) show that cocaine intake increases both the frequency and vigor of their 'waggle dance' to indicate the direction of food sources to their nest mates. They also suffer from 'withdrawal symptoms'. Sound familiar?

HOW DOES THE BODY DEAL WITH COCAINE?

The human liver uses carboxylesterase enzymes hCE1 and hCE2 to metabolize cocaine. Only about 1% of the cocaine is excreted unchanged. Within four hours of ingesting it, the cocaine metabolites can be detected in urine; the presence of products like benzoylecgonine has implicated sportsmen such as Diego Maradona, Kieron Fallon and Martina Hingis in cocaine abuse. If cocaine is taken in combination with ethanol, hCE1 can turn cocaine into cocaethylene (the ethyl ester); the combination of cocaine and ethanol is believed to be more toxic than either on their own.

Benzoylecgonine can also be detected in a user's sweat, notably in their fingerprints, which shows that the person concerned has taken cocaine into their body.

Is Ethanol the Only Substance with Which Cocaine Reacts Badly?

There have been a number of sudden deaths in the United States of 'delirious and combative' males sprayed with pepper spray by law-enforcement officers. Cocaine was implicated in nearly half the cases investigated, and animal testing suggests that exposure to the capsaicin (see p53) in the pepper spray increases the drug's lethality.

Is It True That a Lot of Banknotes are Contaminated with Cocaine?

Yes. In 2007, a Dublin City University team found that virtually all the Euro notes they examined in Dublin had detectable levels of cocaine. Similar results have been found using £10 and £20 notes in the London area, and with dollar bills in the United States.

How Does It Get There?

Bank notes get rolled up into a tube to 'snort' cocaine, so those notes will be very contaminated. When they touch other notes, for example, in a counting machine at a bank, some of the cocaine is transferred to other notes. Just because drugs can be detected on notes in your wallet does not mean that you are a drug user.

So How Do We Know If It Is Accidental Contamination or Not?

Most of the Irish notes examined had about 2 ng of cocaine, but 5% of them had 200 ng or more. It is the amount of cocaine there that tells you if the drug got there accidentally.

And Not Just Banknotes, I Gather?

No, indeed. Scientists can detect cocaine or its metabolites in the environment. Its use is so widespread in the Western world that it has been detected in waste waters of numerous cities in countries, including Ireland, Belgium, Italy, Spain and Switzerland, as well as in city air (*e.g.* Milan, Rome, Oporto, São Paulo in Brazil and Santiago in Chile).

How Is the Cocaine Detected?

The quickest technique for checking banknotes, developed by a group in Bristol, involves tandem mass spectrometry. If the note is heated up to 285°C for a second, the cocaine molecules are released and carried on a stream of air into the spectrometer. Once inside the spectrometer, the cocaine molecules are ionized and also get protonated. If this ion bumps into other ions, the cocaine molecular ion can split into smaller fragment ions. If the spectrometer is looking for cocaine, it is programed to pick out the peaks with m/z values of 182 and 105, the two main fragments of the molecular ion, and also to measure their intensities. The technique is sensitive and very rapid; it can check

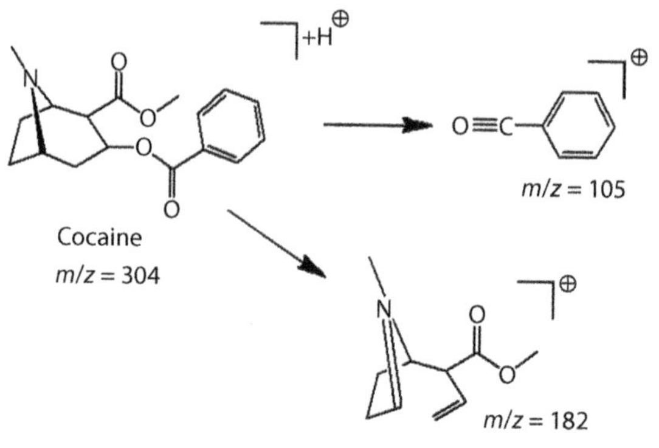

Cocaine
m/z = 304

m/z = 105

m/z = 182

50 banknotes in about 4 minutes. For more accurate results, solvent extraction is used recovering 90–100% of the drug from the banknote.

But that's not all that analytical chemists can do.

How So?

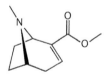

Methylecgonidine

They can distinguish between someone who has been smoking crack or merely snorting cocaine. When crack cocaine is smoked, some of the cocaine is converted into a pyrolysis product, methylecgonidine (anhydroecgonine methyl ester) – it is not found in 'normal' ingestion of cocaine.

Do We Know Where All This Cocaine Comes From?

Cocaine originating from different coca-growing regions has different isotopic ratios of $^{13}C:^{12}C$ and $^{15}N:^{14}N$ (these appear to depend upon factors like soil type and local weather conditions). So law-enforcement officers can use this information, together with the (variable) levels of the trace alkaloids truxilline and trimethoxycocaine, to say from which region of Bolivia, Colombia and Peru the drug originates. Traditionally, the drugs trade has been linked with South American criminal gangs. The most famous of these were probably the Colombian *Medellín* and *Cali* cartels. The most celebrated narcoterrorist was Pablo Emilio Escobar Gaviria (1949–1992); he was 43 when he died in a shoot-out with Colombian National Police. At one time, he was estimated to be the seventh richest man in the world; his fortune amounted to 25 billion American dollars.

Pablo Escobar (© Press Association Images. With permission)

Chapter

14

DEET

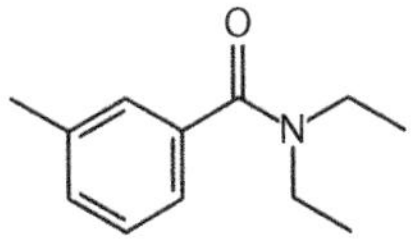

DEET (*N,N*-Diethyl-*meta*-toluamide),
the spray-on insect repellant

WHY MAKE AN INSECT REPELLANT?

Insect bites and stings, as well as being painful and annoying, can be lethal. Mosquito bites, in particular, transmit malaria to humans, and it is estimated to cause approximately 3 million deaths every year worldwide. Mosquito bites can also infect people with other nasty illnesses, such as West Nile virus, various types of encephalitis, and dengue fever. Insect bites and stings can also cause allergic reactions in people which can sometimes be fatal. So a chemical spray which can ward off insects is a potentially life-saving invention.

SUCH AS?

Over the years, various types of substances have been used to repel mosquitoes. These include things such as smoke, plant extracts,

"Here, try this, it's more effective!"

oils, tars and muds. However, the first really effective ingredient used in mosquito repellents was citronella oil – a herbal extract derived from the citronella plant. The insect-repelling properties of citronella oil were only discovered by accident in 1901 when it was used as a hairdressing fragrance. It is believed that the chemical citronellal and other terpenes from which it is composed are responsible for its repellent activity. Unfortunately, citronella oil is very volatile and evaporates too quickly from surfaces to which it is applied. Also, large amounts are needed to be effective.

So It Wasn't Much Use?

It has its limitations, but is still in use today in so-called 'natural' insect repellents. But its disadvantages inspired researchers to search for more effective alternative compounds. Many of the early attempts at creating synthetic insect repellents were initiated by the US military. Out of this research came the discovery of the repellent dimethyl phthalate in 1929. This material showed a reasonable level of effectiveness against certain insect species, but it was ineffective against others. Two other materials were developed as insect repellents: Indalone was found to repel insects in 1937, and Rutgers 612 (2-ethyl-1,3-hexane diol) was synthesized soon after. Like dimethyl phthalate, these materials had certain limitations which prevented their widespread use.

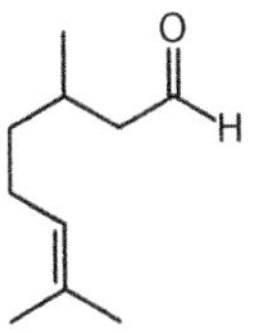

Citronellal

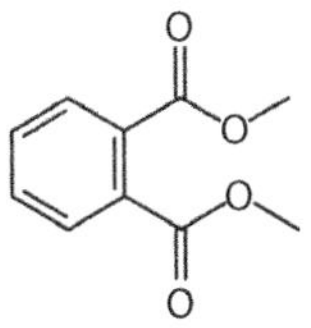

Dimethyl phthalate

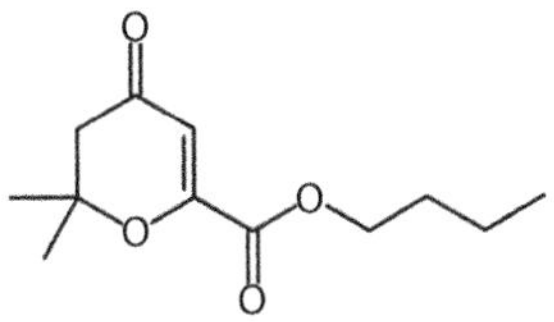

Indalone

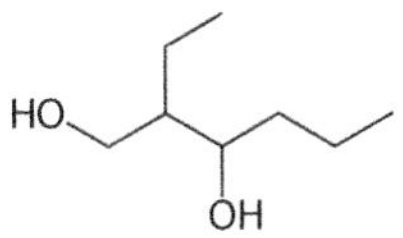

Rutgers 612

Following the experience with jungle warfare in World War II, scientists in the US Army developed a new repellent based on *N-N*-diethyl-*meta*-toluamide, and named it DEET (after d.t., short for diethyl toluamide). It was first used by soldiers in 1946 and then released for civilian use in 1957. DEET is currently the most widely used active ingredient for mosquito repellents, and is found in creams, lotions and aerosols. It is estimated that one-third of the population of the United States use DEET in any given year.

DEET

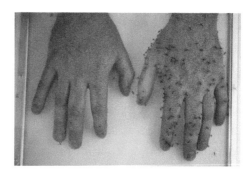

Guess which hand has been sprayed with DEET? (Photo courtesy of Greg Allen, US Dept of Agriculture, Agricultural Research Service)

How Does It Work?

The exact mechanism is still a matter for debate, but most repellent chemicals are thought to work by interfering with the mosquito's homing system. A mosquito's antennae contain a number of chemical receptors which are sensitive to different compounds, such as CO_2 and octenol exhaled by animals (see p381) and lactic acid, which naturally evaporates from the skin of warm-blooded animals. The receptors allow the mosquitoes to follow these emissions back to their source, and hence 'home in' on their prey. However, when a repellent is applied to the skin, it evaporates and masks these scents. Thus, the mosquito is effectively 'blinded'. More recent research has shown that mosquitoes, as well as being 'blinded' by the chemical, actively dislike the smell of DEET.

The DEET-based anti-mosquito lotion Care-Plus (Courtesy of Care-Plus. With permission)

In fact, Jonathan Day of the University of Florida, who has studied the effect of DEET upon mosquitoes in the lab, said in a recent article in National Geographic News: *'repellent is a bit of a misnomer. Mosquitoes will come right up and actually touch the skin that has the repellent on it. But then they become noticeably agitated and they are unable to land and initiate blood feeding'.* He says that the mosquitoes usually then go to the side of the cage and stay there for hours, probably waiting for their sense of smell to return before starting the hunt for another meal.

HOW MUCH IS NEEDED?

DEET is usually applied to the skin, but because the active ingredient must evaporate from the surface to work, the repellent activity only lasts for a limited time depending upon the concentration used. 100% DEET was found to offer up to 12 h of protection while several lower-concentration DEET formulations (20–34%) offered 3–6 h of protection. The US Center for Disease Control recommends 30–50% DEET to prevent the spread of pathogens carried by insects.

DOES IT ONLY WORK ON MOSQUITOES?

No, DEET works on many other biting insects as well, including flies, gnats, midges ('no-see-ums') and ticks (which can spread Lyme disease).

HEALTH ISSUES

Products that use DEET or citronella oil as the primary active ingredients have been reported to causes rashes in some people, and there have been increasing debates as to whether DEET is a safe substance to use on human skin. DEET has extensive independent research into its safety for decades and has proved safe to use as long as usage instructions are followed.

Like all chemicals, however, if it is misused, the results can be tragic. In 2012, two Canadian sisters died in Thailand after drinking a cocktail containing DEET. Apparently DEET is sometimes used as an ingredient to add an extra kick to a euphoria-producing cocktail called a '4 × 100' which is popular among young holidaymakers in Thailand. The 4 × 100 contains cough syrup, cola, ground-up *kratom* leaves and ice, plus a splash of DEET. It isn't known whether the DEET actually produces the alleged kick, but autopsies of the two girls showed that a contributory cause of death was an overdose of DEET in their systems.

CAN MOSQUITOES BECOME RESISTANT TO DEET?

It is certainly possible. Researchers from the London School of Hygiene and Tropical Medicine recently placed some mosquitoes in a room with a human arm covered in DEET. The first time, the mosquitoes were tempted with the arm but they were put off by the smell of the DEET. However, when they eventually came back for a second attempt the researchers found the DEET was less effective. This may be a cause for concern, because if mosquitoes worldwide develop a resistance to DEET we will be left with using other, less effective repellents.

But now that scientists are beginning to understand how molecules interact with the mosquito's receptors, it is becoming possible to design newer, more effective repellent molecules by modifying their structure. *N*-acylpiperidines are a group of molecules that show great promise in this area, and the goal is that by tailoring the chemical structure of these molecules, an insect repellent can be developed that is safe, non-toxic and 100% effective.

WHAT ABOUT MOSQUITO COILS?

These can be reasonably effective, and last up to 8 h. They are usually made from a dried paste of pyrethrum powder, obtained from the dried flower heads of Kenyan chrysanthemum plants, shaped into a spiral. The end of the coil is set alight, and the coil gradually burns away toward the center, releasing a smoke that dissuades mosquitoes from entering the vicinity. There

A mosquito coil

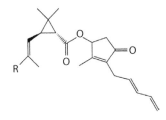

Pyrethrin 1,
R=CH$_3$; pyrethrin 2,
R=CO$_2$CH$_3$

can be many active ingredients in the coils, but the main ones are pyrethrins which are a pair of compounds containing a cyclopropane ring. In low doses, they repel insects, but at higher doses, they kill them. They are effectively a 'nerve gas' for insects – they affect the flow of sodium out of nerve cells, causing the insect's nerves to fire repeatedly. The insects 'twitch' themselves to death!

Chapter

15

DIFLUORODICHLOROETHANE, CF$_2$Cl$_2$

(FREON-12, CFC-12 OR R-12) AND RELATED COMPOUNDS

If It Contains Carbon, Fluorine and Chlorine, It Is a Chlorofluorocarbon, Right?

Right, and CFC is a shorthand for chlorofluorocarbon.

Why Is It Known as CFC-12?

There is a code for naming molecules like this. Think of them as CFC-xyz, where x is the number of carbon atoms in the molecule minus one; y is the number of hydrogen atoms plus one; and z is the number of fluorine atoms. So because there is one C atom, $x = 1 - 1 = 0$; there are no H atoms, so $y = 0 + 1 = 1$; there are two F atoms, so $z = 2$. Really, it should be CFC-012, but people don't usually bother showing the zero. *Freon* is a trade name used by the du Pont chemical company for many different types of CFC refrigerant, with the number after being used as an identity code.

Why Were They Banned?

It's a long story; we need to start with why they were invented. CFCs do not occur naturally; they were invented to do a job in refrigerators that no known gas would do. Refrigerators work by transferring heat from one area to another, based on the principle that when a liquid evaporates it cools its surroundings (which is why we sweat to cool down in summer).

So for a good refrigerant you want a chemical that is a gas at room temperature, but which liquefies under pressure around room temperature. Before CFCs, ammonia and sulfur dioxide were used as the refrigerant gas. Both are toxic and corrosive, so by no means ideal for use in the home.

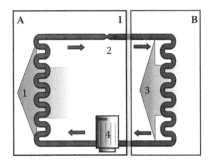

Refrigerator vapor compression cycle. **A**: Kitchen. **B**: Refrigerator box. **I**: Insulating layer. 1: Radiator/condenser, the hot CFC (or other refrigerant) cools down by transferring its heat to the air in the kitchen, and partially liquefies. 2: Expansion valve, the pressurized liquid expands rapidly into the volume behind the valve, and ~50% of it vaporizes, cooling down as it does so. 3: Evaporator unit, heat from the air inside the fridge causes more of the liquid to evaporate, cooling the fridge contents further. 4: Compressor, the now warm gas is compressed back to an even warmer liquid, and then pumped back under pressure to stage 1

CF$_2$Cl$_2$ met pretty well all the criteria for a refrigerant gas. The strong C–F bonds help make it very unreactive, and it has a boiling point around −29.8°C. It is non-flammable and non-toxic, as its inventor Thomas Midgley demonstrated in 1930 at a meeting of the American Chemical Society. He filled his lungs with Freon, exhaling over a burning candle and extinguishing it. CF$_2$Cl$_2$ is also cheap, is immiscible with lubricating oils and has a detectable ether-like smell, all desirable properties for a refrigerant.

HOW DID MIDGLEY MAKE IT?

There is more than one method. Midgley seems to have used the reaction of CCl$_4$ with SbF$_3$ (the Swarts reaction), producing CF$_2$Cl$_2$ with some CFCl$_3$. Using a suitable catalyst, HF can react

Thomas Midgley, Jr. As well as inventing CFCs for use as refrigerants, he is credited with creating tetra-ethyl-lead, the anti-knock additive used in gasoline

with CCl_4 forming a mixture of chlorofluoromethanes.

After that, lots of uses beckoned. Apart from refrigerators and air-conditioning systems, CFCs were good solvents for substances like perfumes and deodorants in aerosol cans, and good blowing agents to create polymer foams with cellular structures. CF_2Cl_2 was also the solvent used for 'Silly String' (aerosol string).

So Everything in the Garden Was Rosy?

Indeed, right up into the 1970s, CFCs like CF_2Cl_2 were ideal for their jobs, but then it was realized that they were major contributors to the breakdown of the ozone layer.

How Did This Come About?

Ozone molecules (see p387) are produced in the upper atmosphere through the action of UV-C radiation (light with a wavelength less than 242 nm) on oxygen gas.

$$O_2 + \{UV\text{-}C\} \rightarrow O + O$$

$$O_2 + O + M \rightarrow O_3 + M$$

(where M is a neutral molecule like O_2 or N_2)

CFCs at one time were used as
propellants in aerosol sprays …

… and as blowing agents to
create expanded polystyrene
used as building insulating, food
packaging, coffee cups, *etc.* …

… and in air-conditioning units (in
houses and in cars) …

Ozone molecules are also destroyed by UV-B radiation (with a
wavelength between 240 and 320 nm):

$$O_3 + \{UV\text{-}B\} \rightarrow O + O_2$$

$$O_3 + O \rightarrow 2O_2$$

so a steady-state concentration of ozone is reached, with the overall
reaction being

$$2O_3 \rightarrow 3O_2$$

... and, of course, in refrigerators

Unlike most chemicals which are attacked and broken apart by a myriad of reactive species in the environment (OH, H_2O, O atoms, sunlight), CFCs are so unreactive and stable that they are not destroyed and hang around for a long time – long enough to diffuse upward to reach the stratosphere, where UV light supplies the energy to break the weakest bond, between C and Cl.

$$CF_2Cl_2 \rightarrow CF_2Cl + Cl$$

This generates Cl atoms, which catalyze the breakdown of ozone, partially removing the protective ozone layer that blankets the Earth – especially in the polar regions creating the so-called 'ozone hole'. This allows more dangerous UV radiation to reach the Earth's surface, which hugely increases the risk of skin cancer, as well as being damaging to plants and animals.

$O_3 + \{UV\text{-}B\} \rightarrow O + O_2$ (This is a useful reaction, the ozone is removing harmful UV)

$O_3 + Cl \rightarrow ClO + O_2$ (But this is an unwanted reaction, protective ozone is destroyed)

$O + ClO \rightarrow Cl + O_2$ (This is another unwanted reaction; O atoms are needed to make ozone, if they're removed then the amount of ozone decreases)

Overall: $O + O_3 \rightarrow 2O_2$ (*i.e.* Ozone is turned back into oxygen, but the Cl catalyst is unaffected, and so can catalyze another reaction, then another ...)

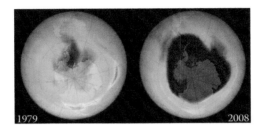

NASA satellite images of the ozone hole growing over the Antarctic over a 30-year period, as a direct consequence of CFC use

So What Happened Next?

CFCs were rapidly banned under the Montreal Protocol (1987) – in fact the manufacture of CF_2Cl_2 is now banned – and substitutes were introduced in the early 1990s, such as hydrocarbons like butane or less stable CFCs like chlorodifluoromethane (HCFC-22 or R-22).

Chlorodifluoromethane
(HCFC-22)

1,1,1,2-Tetrafluoroethane
(HFC-134a)

The next step was to introduce hydrofluorocarbons, HFCs, like 1,1,1,2-tetrafluoroethane (HFC-134a). Unlike CFCs, it does not contain any chlorine.

Explain, Please

Hydrofluorocarbons contain no chlorine and so do not destroy ozone.

Is There Any Reason Why HFC-134a Is Chosen?

Yes, it has become the most widely used HFC, not least because of its very low toxicity. It has a boiling point of $-26°C$, very similar to that of CF_2Cl_2, so it can be liquefied under pressure, making it well suited to refrigerant or air-conditioning systems. It has zero ozone-depletion potential, but it has a high global-warming potential, so it is not perfect either.

What Does That Mean?

Global warming potential (GWP) measures how much heat a 'greenhouse gas' traps in the atmosphere, compared to the same mass of CO_2 (which is given a GWP of 1, see p61). The GWP of tetrafluoroethane is 1430.

Wow, So What Are They Doing about That?

The European Union banned the use of HFC-134a in air-conditioners in new vehicle types from 1st January 2011 (with a view to a complete ban in all new cars from 1 January 2017), and in June 2013 China and the United States – the two biggest users – agreed to phase out HFCs.

What Alternatives Are There to Using 1,1,1,2-Tetrafluoroethane?

2,3,3,3-Tetrafluoropropene (HFO-1234-yf) is the preferred alternative at present. It has a much lower GWP (4) and much shorter lifetime in the atmosphere than HFC-134a, so is likely to be adopted by many auto manufacturers in the EU, though it is slightly flammable. Because of this problem, people are looking at using mixtures of HFO-1234-yf with HFC-134a.

2,3,3,3-Tetrafluoropropene
(HFO-1234-yf)

DOES 1,1,1,2-TETRAFLUOROETHANE HAVE ANY OTHER USES?

Being volatile and chemically inert, it is a useful solvent. It has been used to extract various organic molecules, like the antimalarial drug artemisinin (see p29) from *Artemisia annua*, in place of flammable hydrocarbons and supercritical CO_2. It is a propellant gas in aerosols, notably in drug delivery (though this can cause interference problems with infrared anesthetic gas monitors), as well as in the duster gas used to clean computers.

WHAT'S THE FUTURE FOR THE OZONE HOLE?

CFCs have lifetimes of several decades, so even though their use has been banned in most countries for over 25 years (and are due to be banned in *all* countries by 2020), they remain prevalent in the atmosphere and still doing damage to the ozone layer. However, scientists believe that the problem has just about 'peaked', and that within the next 25 years or so the ozone layer should begin to recover, although it may not reach its pre-CFC levels for 50–100 years. So this is one environmental story that has a happy ending (sort of), in that the dangers were recognized just in time to do something about the problem. And, for once, the politicians actually agreed on a solution. Now all we need to do is tackle global warming … (see p61).

Chapter

16

DDT

THE CONTROVERSIAL INSECTICIDE

DDT, Surely No One Uses That Now?

In some parts of the world, such as India, it is used as an
agricultural pesticide. Some countries like Ecuador have continued
to use it successfully to fight malaria.

I Thought It Was Banned?

Greenpeace has declared war on organic compounds containing
chlorine. This largely ignores the fact that there are over 5000
known natural organohalogen compounds (*e.g.* see p177), with
some being produced in nature in huge quantities. Thus some
99% of chloromethane in the atmosphere is formed in the seas
when organic biomass reacts with chlorine under the influence of
sunlight.

Why Did People Use DDT?

DDT was a runaway success when introduced after World War II. It
was especially important in combating typhus. This is a bacterial

Widespread but uncontrolled spraying of DDT in the United States in
1958 led to a huge drop in malaria cases, but also led to worries over the
environmental impact

infection associated with lice, which was responsible for deaths on a large scale in the nineteenth century, notably in Napoleon's retreat from Moscow and the Irish famines, as well as in World War I. It was a big killer in the Nazi concentration camps in World War II, Anne Frank being one famous victim.

Even soldiers would get personally sprayed with DDT to protect them from lice

As the Allied armies liberated Western Europe in 1943–1945, the scene was set for more typhus epidemics, with sick and malnourished refugees scurrying everywhere. Take Naples in the 1943–1944 winter as an example; a typhus epidemic was halted in its tracks by dusting everyone with DDT. DDT won, lice lost. Broadcasting on the radio on September 28th 1944, Winston Churchill famously referred to it as the *'excellent DDT powder which had been fully experimented with and found to yield astonishing results'.* DDT was also used at Iwo Jima.

So Why Do People Say It Is Such a Bad Substance?

Rachel Carson (US Fish & Wildlife Service National Digital Library)

The most influential factor was the publication of Rachel Carson's book *Silent Spring* in 1962. This claimed that indiscriminate use of DDT was causing damage and in particular threatening wildlife, notably certain birds. There were also suggestions that it was a carcinogen. This chimed with a burgeoning environmental movement and the use of DDT was banned in the United States in 1972 – something that Carson hadn't suggested. The Stockholm Convention of 2004 limited the use of DDT to only controlling malaria, but it is still recommended by the World Health Organisation for 'the indoor use of DDT in African countries where malaria remains a major health problem'. Evidence suggests that, if anything, DDT has *anti*-carcinogenic effects. Indeed, the *o,p′*-isomer of DDD, 'mitotane' – derived from DDT – is used medicinally to treat inoperable cancers of the adrenal gland.

o,p′-DDD (mitotane)

WHAT DOES DDT STAND FOR, ANYWAY?

Its full name is 1,1,1-trichloro-2,2-di(4-chlorophenyl)ethane, but that's a bit of a mouthful. A shorter form of that crops up in one of the best-known chemical limericks:

> A mosquito was heard to complain
> that a chemist had poisoned his brain.
> The cause of his sorrow
> was *para*-dichloro-
> diphenyltrichloroethane.

In fact, the DDT that is used commercially contains mainly the *p,p'*-isomer but also around 15% of the *o,p'* isomer, as well as some breakdown products like DDE (dichlorodiphenyldichloroethylene) and DDD (dichlorodiphenyldichloroethane).

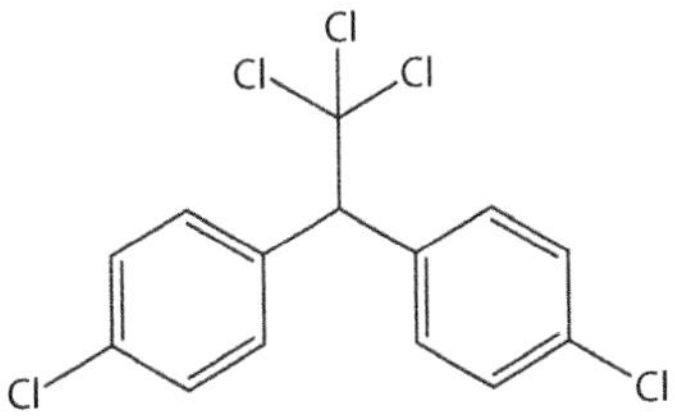

p,p'-DDT

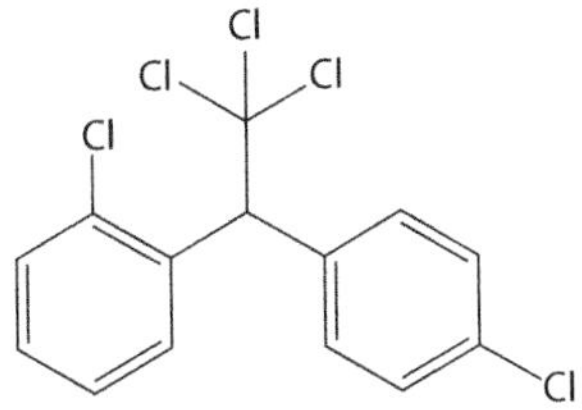

o,p'-DDT

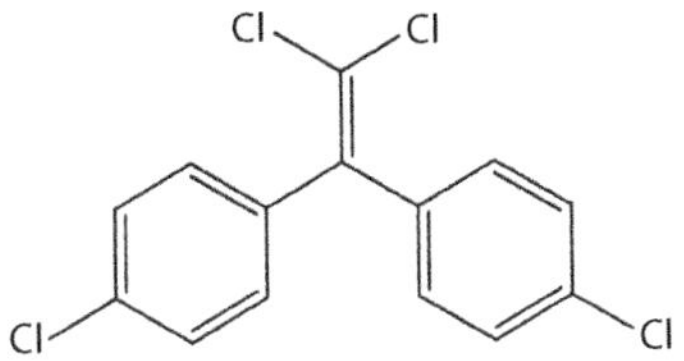

p,p'-DDE

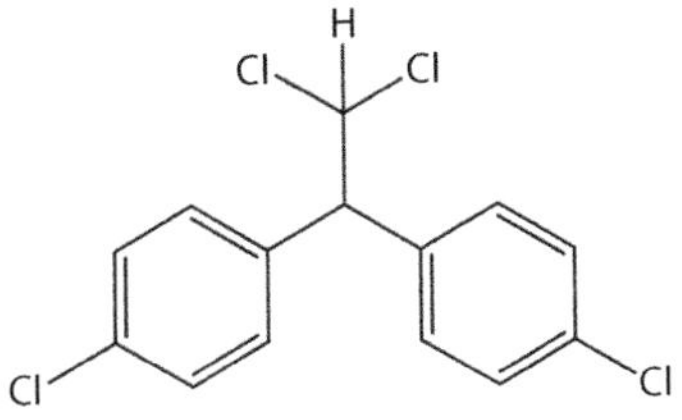

DDD

WHY HAS DDT GOT THE BLAME?

Being a covalent molecule, DDT tends to be soluble in non-polar solvents, such as body fat. Because of this – and its slowness in being metabolized – it bio-accumulates up food chains into the tissues of higher organisms, such as birds of prey. DDT and its metabolites have also been blamed for the thinning of birds' eggshells, though the evidence is not totally clear, and the effect may vary from one bird to another. It has been suggested that a metabolite, *p,p'*-DDE, can inhibit the enzyme calcium ATPase, which reduces transport of calcium from the blood to the eggshell gland, leading to thinner shells. Environmental scientists claim that the banning of DDT is one of the main reasons (along with the passing of the Endangered Species Act in 1973) that brought the symbol of the United States, the bald eagle, back from the brink of extinction in the United States.

SO WHAT WENT WRONG WITH DDT?

The problem arose because people used DDT quite indiscriminately. If you use it on mosquito breeding grounds and inside homes, and in conjunction with other sensible measures (*e.g.* mosquito nets) then it is extremely effective and can significantly curb malaria.

In 1948, Ceylon was faced with over 2 million cases of malaria a year, so a campaign began to eradicate malaria by spraying houses with DDT. By 1962, malaria had been virtually vanished. But then, because of ecological concerns, spraying was stopped in 1964, and by 1969, malaria was back to pre-1948 levels. This pattern has been found in other countries that abandoned the use of DDT.

DOES IT KILL ALL INSECTS?

There is a species of Brazilian bee, *Eufriesea purpurata*, whose males are attracted by DDT and actually collect DDT from the walls of houses where it has been sprayed for malaria control. These bees accumulate DDT up to levels of 2 mg per bee (up to 4% of its body weight) with no apparent ill effects. No one knows why these bees like DDT, one obvious suggestion is that they mistake it for a sex pheromone, or maybe an area marker. And, in fact, DDT is relatively non-toxic to ordinary honey bees (*Apis mellifera*).

WHAT IS THE FUTURE OF DDT?

DDT is reckoned to have saved more than 25 million lives, so it makes sense to use it responsibly. The answer would lie in properly discriminated use targeted at specific pests, notably in the buildings and swamps where the mosquitoes responsible for spreading malaria

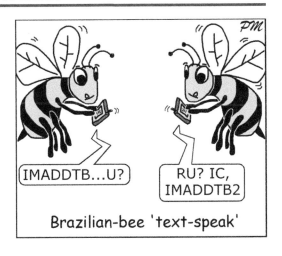

IMADDTB...U?   RU? IC, IMADDTB2

Brazilian-bee 'text-speak'

breed or threaten humans, along with companion strategies like mosquito nets. The obvious place for its use is sub-Saharan Africa, where mortality due to malaria is unchecked.

Chapter

17

DIGITALIS

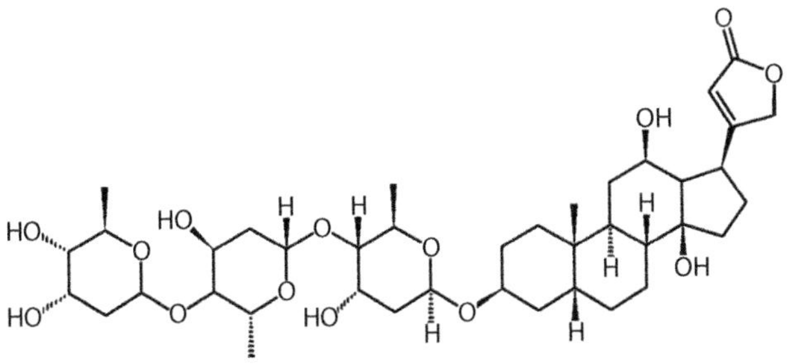

Digoxin – one of the components in digitalis extract

That's a Heart Drug, Isn't It?

Yes, it's an example of a cardio-active or cardiotonic drug, in other words, a steroid which has the ability to exert a specific and powerful action on the cardiac muscle in animals. The term digitalis is used for drug preparations that contain cardiac glycosides, in particular digoxin. This works by increasing the intensity of the heart muscle contractions but diminishing the rate, and has been used in the treatment of heart conditions ever since its discovery in 1775.

That's a Surprisingly Long Time Ago for a Modern Drug?

That's because the active ingredient is based upon an extract of the common foxglove (*Digitalis purpurea*), found commonly all over Europe.

William Withering
(1741–1799)

How Did They Find Out That Foxgloves Could Treat Heart Disease?

In the eighteenth century, an English doctor called William Withering was working as a physician in Staffordshire, England. As a hobby, he became interested in plants and botany in general, and he became such an expert in the local flora that he published a huge textbook whose title begins '*A botanical arrangement of all the vegetables growing in Great Britain ...*' and continues for a further 24 lines! By the age of 46, he'd become the richest doctor outside of London, and bought Edgbaston Hall in Birmingham, which is now Edgbaston Golf Club.

AND WHERE DO THE FOXGLOVES COME IN?

In 1775, one of his patients came to him with a very bad heart condition, but at that time there was no effective treatment for this so the odds didn't look good for the patient's survival. However, the patient – not discouraged by the inability of conventional medicine to treat his ailment – went instead to a local old woman (some reports say she was a gypsy), who gave him a secret herbal remedy. Astonishingly, the remedy seemed to work, and the patient got much better!

Purple foxglove (*Digitalis purpurea*)

When Withering heard about this, he became quite excited and searched for the old woman throughout Shropshire. When he eventually found her, and after much bargaining, the gypsy finally told her secret – the herbal remedy was made from a concoction of over 20 different ingredients, one of which was an extract of the purple foxglove. Withering was not surprised by this, as the potency of digitalis extract had been known since the dark ages, when it had been used as a poison for the medieval 'trial by ordeal', and also used as an external application to promote the healing of wounds. Digitalis has also been a remedy for dropsy, which was the old term for swelling of soft tissue in the legs and arms. Unknown to the physicians at the time, these swellings were often caused by poor blood circulation due to heart problems.

WHAT HAPPENED NEXT?

Well, over the next nine years, Withering tried out various formulations of digitalis plant extracts on 160 patients with heart ailments. He found that if he used the dried, powdered leaf, without boiling them in liquid (which destroyed the active ingredients) he got amazingly successful results, and the painful angina symptoms of the patients went away. He introduced its use officially in 1785.

Withering's memorial plaque in Edgbaston, Birmingham, UK, decorated both with a foxglove and the plant *Witheringia solanacea* which was named after him

WAS IT SAFE?

Digitalis purpurea plants contain a mixture of several cardiac-active molecules related to digitalis in amounts and proportions which vary with locality and with season. Before modern synthetic methods, therefore, digitalis preparations varied considerably in potency and quality. Because of this, and the fact that the therapeutic dose is so small (as low as 0.3 mg daily are all that is needed), it was very easy to exceed the safe dosage. Indeed, Withering recommended that the drug be diluted and administered repeatedly in small doses until a therapeutic effect became evident. This procedure was very effective in experienced hands, but was also very time-consuming.

WHAT ABOUT NOWADAYS?

Even today, drugs based on digitalis extract, such as *Digitoxin* and *Digoxin*, are some of the best-known treatments to control the heart rate. Nowadays, preparations from digitalis leaves are made using modern recrystallization methods and are carefully standardized by bio-assay.

WHAT IS THE ACTIVE MOLECULE IN THE EXTRACT?

Well, it's actually several molecules. Hydrolysis of the digitalis extract produces many related molecules, all of which contain an α,β-unsaturated lactone ring (or cyclic ester) in addition to other structural similarities. Both the $C_{14}(\beta)$ hydroxyl group and the unsaturated lactone are essential to its activity as a drug. As well as digoxin (shown above), three other common ones are shown below.

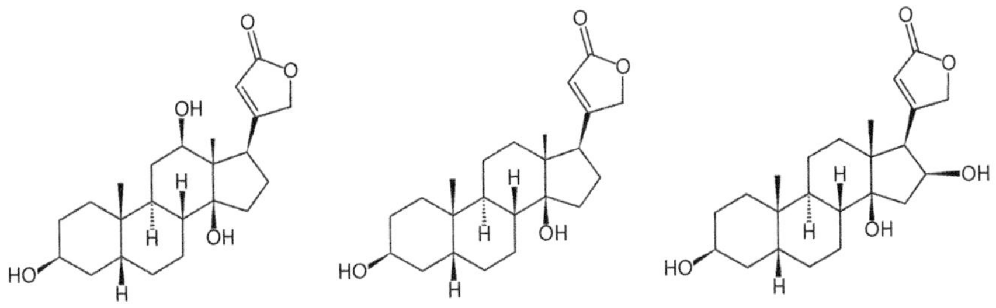

Digoxygenin   Digitoxygenin   Gitoxygenin

WHAT HAPPENED TO WITHERING?

In 1799, Withering became very ill and it looked as though he was going to die. One of his friends, who was noted for his black sense of humor, wrote in a letter to a friend *'The flower of Physic is withering …',* which is a very clever pun – if a bit tasteless. When Withering finally died, his friends carved a bunch of foxgloves on his memorial.

Chapter

18

DIMETHYLMERCURY
AND THE KAREN WETTERHAHN
STORY

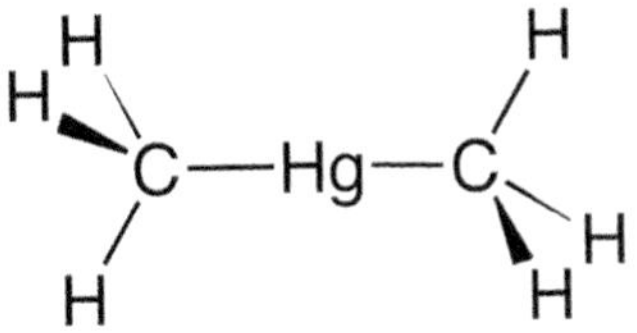

Professor Karen Wetterhahn

WHO'S SHE?

Karen Wetterhahn was an internationally respected Professor of Chemistry in New Hampshire in the 1990s (actually the 'Albert Bradley Third Century Professor in the Sciences' at Dartmouth College). She was an expert researcher in the field of the effects of heavy metals on living systems, especially in their role in causing cancer. By an exquisite irony, she herself became a victim of a heavy-metal poison.

HOW DID IT HAPPEN?

On August 14th 1996, Wetterhahn was doing something few professors do – working at the lab bench. She was studying the way mercury ions interacted with DNA repair proteins, and was using dimethylmercury, $Hg(CH_3)_2$, as a standard reference material for ^{199}Hg NMR measurements.

Wetterhahn knew about the high toxicity of dimethylmercury and took very reasonable precautions, donning safety glasses and latex gloves, doing manipulations in a fume cupboard and only working with very small quantities behind the fume cupboard sash. Dimethylmercury, supplied by a chemical manufacturer, came in a sealed glass vial. A colleague of hers cooled the vial in ice–water to reduce the volatility of the $Hg(CH_3)_2$ and then cut off the glass top to open it. Wetterhahn pipetted a small sample into an NMR tube and transferred the rest into a storage container, sealed and labeled the tubes and cleaned up, disposing of the latex gloves. Nevertheless, less than a year later, she was dead from the effects of mercury poisoning.

So Why Did She Die?

Wetterhahn later recalled spilling a drop, possibly more, of dimethylmercury on her gloved hand. Tests subsequently showed that this would have penetrated the glove and started entering her skin within 15 seconds. It is now accepted that the only safe precaution to take when handling this compound is to wear highly resistant laminated gloves underneath a pair of long-cuffed neoprene (or other heavy-duty) gloves.

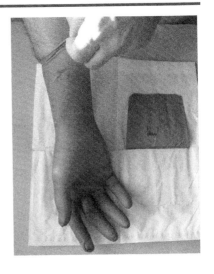

Neoprene gloves

In January 1997, she began to notice definite symptoms that worried her – tingly fingers and toes, slurred speech. She began to have problems with her balance and her field of vision started to shrink. Mercury poisoning was diagnosed on January 28th 1997. Tests revealed that she had a blood mercury level of 4000 micrograms per liter, 80 times the toxic threshold. Two weeks later, she slipped into a coma from which she never recovered, dying on June 8th 1997.

WAS SHE THE FIRST VICTIM OF DIMETHYLMERCURY POISONING?

Dimethylmercury was first made in 1858 by George Buckton of the Royal College of Chemistry (now Imperial College) in London, who somehow managed to avoid being poisoned by it. In contrast, two English chemists who made it in 1865 were dead within months. And, in 1971, a Czech chemist who had been making dimethylmercury died in less than a month. Buckton must have been a very careful laboratory worker, as he lived up to the age of 87, only dying in 1905. Mind you, in 1865, that same year that two people were fatally poisoned by $Hg(CH_3)_2$, Buckton gave up chemistry and spent the last 40 years of his life as an entomologist.

WHAT'S DIMETHYLMERCURY LIKE?

It looks like water, as it is a colorless liquid at room temperature, with apparently a faint sweet smell (it's not a good career move to try smelling this!). It has a boiling point of 92°C at atmospheric pressure and a density of 2.96 g/cm³. Like many HgX_2 systems, it has a linear structure, with Hg–C = 2.083 Å.

Dimethylmercury

WHY IS DIMETHYLMERCURY TOXIC?

It is one of the most potent neurotoxins known, now referred to as 'supertoxic'. Once in the body, it is metabolized to the methylmercury ion, $CH_3Hg^+(aq)$, which can bind to cellular proteins. It causes *ataxia* (lack of coordination), sensory disturbance and changes in the mental state. It inhibits several stages of neurotransmission in the brain. It probably crosses the blood–brain barrier as a methylmercury–cysteine complex, causing severe brain damage.

After the death of Karen Wetterhahn, tests showed that dimethylmercury could go through ordinary laboratory rubber gloves within seconds; anyone using dimethylmercury (which itself is severely discouraged) must now wear highly resistant laminated gloves underneath a pair of heavy-duty gloves. The scientists and doctors who treated Karen Wetterhahn concluded:

> 'Dimethylmercury appears to be so dangerous that scientists should use less toxic mercury compounds whenever possible. Since dimethylmercury is a "supertoxic" chemical that can quickly permeate common latex gloves and form a toxic vapour after a spill, its synthesis, transportation, and use by scientists should be kept to a minimum, and it should be handled only with extreme caution and with the use of rigorous protective measures'.

WHY IS THE FORMATION OF A CYSTEINE COMPLEX IMPORTANT?

Mercury is a 'soft acid' so that it binds to easily polarizable donor atoms in 'soft' bases. This gives the mercury ion a high affinity for sulfur and sulfur-containing ligands. It will thus attack the thiol groups of enzymes and inhibit them. Zeise made the first mercaptan ligands (R-SH) and coined their name, based on the Latin expression *mercurium captans*, capturing mercury.

SO MERCURY LIKES SULFUR?

The main ore of mercury is *cinnabar*, HgS. It has been mined for 2500 years in places like Almaden in Spain. Other important

Red cinnabar ore

mines are at Idria in Slovenia and Monte Amiata in Italy. In Roman times, criminals sentenced to work in quicksilver (the old name for mercury) mines had a short life expectancy because of the toxicity of the mercury in cinnabar. It was regarded as a death sentence, and the Roman scribe Pliny described the symptoms of mercury poisoning as far back as the first century AD. Cinnabar was widely used as a pigment in the ancient world and its red color is recognized even today (from 1982 to 1987, one of the authors of this book drove a Mini whose color was described as 'cinnabar' by the manufacturers).

Have There Been Other Cases of Mercury Poisoning?

Unfortunately, there have been lots. Once upon a time, people used mercury and its compounds to treat syphilis. Although not successful, it enjoyed a vogue for a considerable time until penicillin (see p409) was found to be an effective cure for this disease in its earlier stages. One important use of mercury compounds was in hat making. Two hundred years ago, the furs used to make beaver felt hats (such as the 'top hats' that were fashionable then) were dipped into mercury(II) nitrate solution as a preservative and to soften the animal hairs. Unfortunately, the workers in the felt-hat trade absorbed mercury through their skins. The resulting mercury poisoning caused shaking and slurred speech, being known as 'Hatter's disease', which is believed to have inspired the character of the Mad Hatter in Lewis Carroll's *Alice in Wonderland*, a character made famous in Tenniel's drawing (below).

Illustration by John Tenniel of 'A Mad Tea Party' in the nineteenth-century print of Lewis Carroll's *Alice's Adventures in Wonderland* in which Alice meets the Mad Hatter, the March Hare and the Dormouse.

Another source of poisoning was the (mis)use of alkylmercury compounds used as fungicides on seeds. Between the World Wars, workers in fungicide-manufacturing plants developed mercury poisoning. In 1942, two young Canadian secretaries working in an office of a warehouse in Calgary, Canada, were fatally poisoned. The warehouse was storing diethylmercury. In the 1960s, Swedish farmers noticed birds flopping helplessly on the ground and then dying. The birds had been eating mercury-treated grain, or else the bodies of dead rodents that had consumed it.

More horrifying than this were epidemics of poisoning, caused by people eating treated seed grains. There was a serious epidemic in Iraq in 1956 and again in 1960, while the use of seed wheat (which had been treated with a mixture of C_2H_5HgCl and $C_6H_5HgOCOCH_3$) for food, caused the poisoning of about 100 people in West Pakistan in 1961. Another outbreak happened in Guatemala in 1965. The most serious was the disaster in Iraq in 1971–1972, when according to official figures, 459 people died. Grain had been treated with methylmercury compounds as a fungicide, and should have been planted. Instead, it was sold for milling and made into bread. It had been dyed red as a warning and also had warning labels in English and Spanish that no one could understand.

I'VE ALSO HEARD OF THE ACCIDENT AT MINIMATA

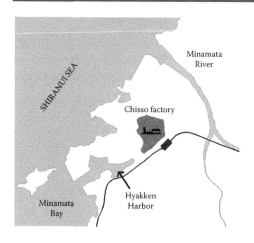

Map of Minimata in Japan, showing the Hyakken Harbor and Minimata River into which the methylmercury was discharged

'Accident' is debatable. Given what was known about the toxicity of mercury compounds at the time, people could have acted sooner, at the very least.

WHAT HAPPENED?

In the early 1950s, inhabitants of the seaside town of Minamata, on Kyushu Island in Japan, noticed strange behavior in animals. Cats would exhibit nervous tremors, dance, and scream. Within a few years, this was observed in other animals, and birds would drop out of the sky. Symptoms were also observed in fish, an important component of the local diet, especially for the poor. When human symptoms started to be noticed around 1956, an investigation began. Fishing was officially banned in 1957. It was found that the Chisso Corporation, a petrochemical company and maker of plastics such as polyvinyl chloride, had been discharging heavy-metal waste into the sea. They used mercury compounds as catalysts in their syntheses. It is believed that more than 1400 people were killed and perhaps, 20,000 people were poisoned to a lesser extent. The effects of mercury poisoning are still sometimes referred to as Minamata disease.

How Did This Happen?

Methylcobalamin, a coenzyme form of Vitamin B12, is capable of methylating 'inorganic' mercury compounds to form CH_3Hg^+(aq.), also by methylation of mercury itself. The actual mercury species present in solution may be CH_3HgOH. The CH_3Hg^+(aq.) ion is absorbed by plankton, which is in turn eaten by small fish. The fish eat so much of the contaminated plankton, and they excrete the mercury so slowly, that it gradually builds up in their systems. The small fish are then eaten by larger fish, and the concentration of mercury in the organism increases each time. Animals and humans eating these larger fish concentrate the mercury even more, so that the final concentration in animals higher up the food chain (such as humans) can be thousands or millions of times larger than was present in the original water. This process is known as 'biomagnification'.

So the World Has Learned from These Tragedies?

Regrettably not, if problems connected with the use of mercury in mining precious metals are anything to go by. Mercury forms mixtures with many metals, known as *amalgams*. It is used to extract metallic gold from gold ores; on account of this, the gold

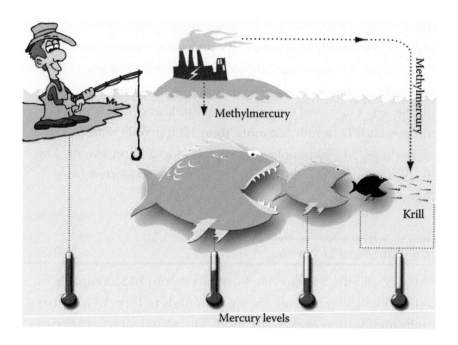

Biomagnification of mercury. Methylmercury compounds emitted from industry find their way to rivers and lakes where they are absorbed by krill and small aquatic organisms. These are then eaten by small fish, which in turn are eaten by larger fish. Since mercury is only slowly removed from the organism's body, the levels of mercury increase inside the cells of the organism, until at the top of the food chain – humans – the level is dangerously high

is being recovered by distilling off the mercury. Most mines do not recover the mercury, so that it finds its way into rivers, where it gets converted into the methylmercury ion, and this has been a problem in California as a legacy of the 1849 Gold Rush. Mercury was used to refine Mexican silver in the sixteenth century. The (mis)use of mercury in gold mining is currently an issue in the Amazon region of South America, West Africa and other places. Mercury is cheap and convenient to use, and that's all some people care about.

So Chemistry Is a Dangerous Subject?

Doing chemistry is safe, much safer than driving a car. A chemistry laboratory is about the safest place in a school or university, far safer than the sports field. But it is only by ceaseless vigilance and attention to safety that it remains so.

Chapter

19

DIMETHYLSULFIDE

(AND TRUFFLES)

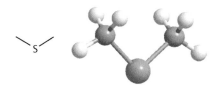

Dimethylsulfide

'Whoever says "truffle" utters a great word which arouses erotic and gastronomic memories among the skirted sex and memories gastronomic and erotic among the bearded sex'.

WHO SAID THAT?

A Frenchman called Jean Anthelme Brillat-Savarin.

WHO WAS HE?

He was born in Belley, to the East of Lyon in France, and studied medicine, law and chemistry (*naturellement*) in Dijon, before returning to Belley to become a magistrate, and then mayor in 1793. He had been a deputy (similar to a Congressman) in the

Jean anthelme Brillat-savarin
(1755–1826)

The cover of Savarin's book

pre-Revolutionary French government in 1789, but the advent of the Jacobins forced him to flee to the United States. Returning to France in 1796, he became a judge under Napoleon as counsellor to the Supreme Court of Appeal, where he remained until his death. This profession gave him the time to entertain his friends, be invited to the best tables in Paris, and write books, including his gastronomic memoirs *La Physiologie du Goût* (The Physiology of Taste) which appeared in bookshops on December 8th 1825, just two months before his death. Only 500 of this edition were printed; a first edition was on sale in 2005 at £5000. It has been in print ever since.

DID HE REALLY SAY THAT STUFF ABOUT TRUFFLES?

Well, he said it in French. What he actually said was:

> *'Qui dit truffe prononce un grand mot qui réveille des souvenirs érotiques et gourmands chez le sexe portant jupes, et des souvenirs gourmands et érotiques chez le sexe portant barbe'.*

SO HE MADE COOKING SEXY?

It seems so. He was certainly the father of gastronomy.

Black truffles ...

... and white truffles

WHERE DOES DIMETHYLSULFIDE COME IN?

To understand that, you have to start with truffles. They are
fungi that grow underground. There are white truffles and black
truffles, as Brillat-Savarin said (first talking about the Italian white
truffles):

> '*The highly esteemed white truffles are found in Piedmont: they
> have a hint of garlic, which does not detract from their perfection,
> as it does not repeat on you!*
>
> *The best French truffles come from Périgord and Haute
> Provence; they have their strongest taste around the month
> of January'.*

The most famous ones are the French black truffles, *Tuber
melanosporum*, and they have to be searched out.

HOW?

Traditionally, in Périgord, pigs, usually females, are used.

HOW DO PIGS FIND TRUFFLES?

At one time, it was believed that they were detecting the smell of
5α-androst-16-en-3α-ol, a steroid which has been identified in
truffles and is also present in boars' saliva. It was thought that the
pigs were conditioned to responding to a pheromone molecule.

Nowadays, dogs are used since they are easier to transport –
and less likely to eat the truffles! Truffle flies (*Suillia pallida*)
can be seen to hover above the ground where truffles lie buried
(the flies lay eggs above the fungi as the larvae use the truffle
for food).

5α-androst-16-en-3α-ol

A pig rooting for truffles

AND WAS IT A PHEROMONE?

Thierry Talou, a French chemist, made some synthetic truffle aroma (lacking 5α-androst-16-en-3α-ol) and separately buried samples of the synthetic aroma, real truffles and 5α-androst-16-en-3α-ol at different locations. Pigs ignored the 5α-androst-16-en-3α-ol but made for either the real truffles or the synthetic truffle aroma.

SO WHAT ARE THE MOLECULES THAT ARE ACTUALLY RESPONSIBLE FOR THE SMELL?

Truffles contain lots of small organic molecules, especially alcohols, aldehydes and ketones, such as 2-methylpropanal, butanone, 2-methylpropan-1-ol, 2-methylbutanal, 3-methylbutanal, 2-methylbutan-1-ol and 3-methylbutan-1-ol. However, the 'truffle' smell is due to sulfur-containing molecules, most notably CH_3SCH_3 (dimethylsulfide), as well as $CH_3CH_2CH_2SCH_3$ and $CH_3CH=CHSCH_3$.

Dimethylsulfide

Do These Smells Persist?

On keeping the truffle, the volatile sulfur compounds evaporate
faster than other molecules. After a while, the smell becomes
'mushroomy' and is now mainly due to 1-octen-3-ol, the principal
compound responsible for the smell of mushrooms (it is known as
'mushroom alcohol') as well as the related 1-octen-3-one, which
also has a mushroom smell (see p381).

Do White Truffles Smell the Same?

No, their smell is mainly due to $CH_3SCH_2SCH_3$, though also
including some CH_3SCH_3, $(CH_3)_2S_2$ and $(CH_3)_2S_3$.

The various other smelly molecules that contribute to the odor of white
truffles

What Sort of Smell Does Dimethylsulfide Have?

It has been described as extremely unpleasant, as it is a substance
added to odorless gases (like natural gas) as an odorizer, so
that leaks can be detected. Others say that the smell is at least
in part due to impurities, such as polysulfides, and that pure

dimethylsulfide has a more pleasant smell (a bit like 'sweetcorn'). The odor may also depend upon concentration.

Does It Crop Up Anywhere Else?

Recent research reveals dimethylsulfide to be one of the main odor components (along with H_2S and CH_3SH) of human flatus. It has been found to be an odor component of some beers (germinating barley produces S-methyl methionine, a source of $(CH_3)_2S$), and is responsible for the smell of cooked cabbage, possibly from bacterial metabolism of methionine or S-methylmethionine, present

"Ahhh, I love the aroma of zee truffles..."

in large quantities in brassicas. It has also been associated with the rotting smell of dead-*horse arum* florets, which fools flies into pollinating it by emitting a smell like a dead animal. It is produced by some sponges, though probably not as a protection against fish predators, but with other defensive functions (*e.g.* antimicrobial and antifouling). And you may smell it in the bracing air at the seaside!

Why Is That?

It is now recognized that dimethylsulfide is the volatile organic sulfur compound responsible for 75% of the global sulfur cycle.

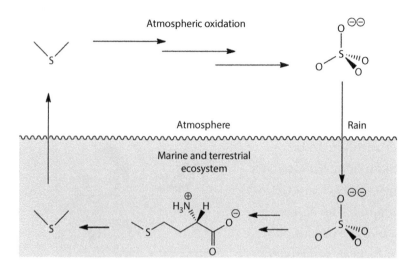

It has also been suggested that it has an important role in the climate, since its oxidation in the atmosphere leads to other sulfur compounds, including sulfuric acid, which act as cloud condensation nuclei, leading to decreased sunlight levels, and which also contribute to the acidity of rainwater. Dimethylsulfide is synthesized (by reduction) in various marine organisms and released to the atmosphere, largely from phytoplankton.

AND ARE TRUFFLES APHRODISIACS?

There is no evidence for it, but it makes a good story.

Chapter

20

DOPAMINE

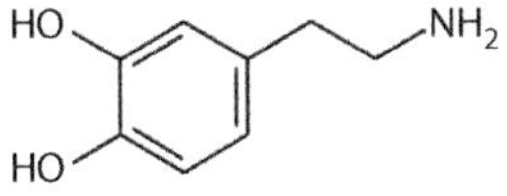

Dopamine

DOES IT MAKE YOU DOPEY?

The opposite in fact! Dopamine is a neurotransmitter found in the brain of all animals, and is closely related to adrenaline (see p9). In fact, it is the precursor to noradrenaline, which in turn is used to synthesize adrenaline in the body.

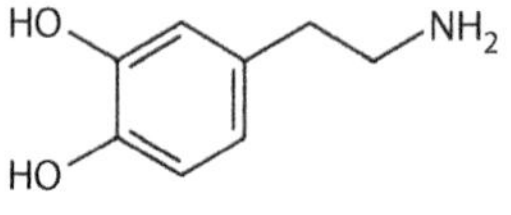

Dopamine

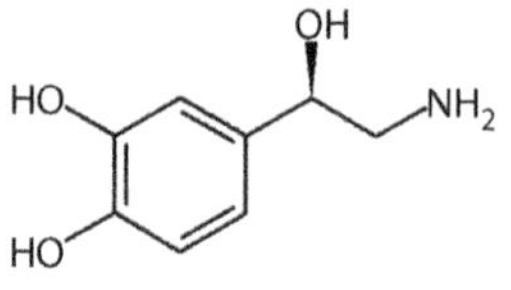

Noradrenaline
(Norepinephrine)

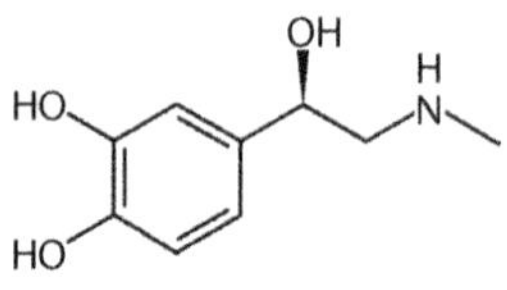

Adrenaline
(Epinephrine)

SO IT WAKES YOU UP?

Well, it keeps your brain active. It plays a major role in transmitting electrical signals through the brain and central nervous system. In a nerve cell, the electrical impulse travels down the length of the cell until it reaches one end. Here there is a short gap, called a synapse, before the next nerve cell begins. When the signal

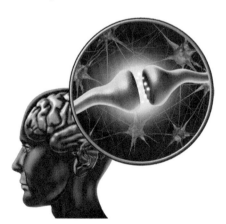

reaches the end of the first nerve cell, neurotransmitters such as dopamine are released and within a few milliseconds diffuse across the synaptic gap to lock on to a receptor on the surface of the next nerve cell. This generates a new electrical signal, which then travels down this nerve cell to the next synapse.

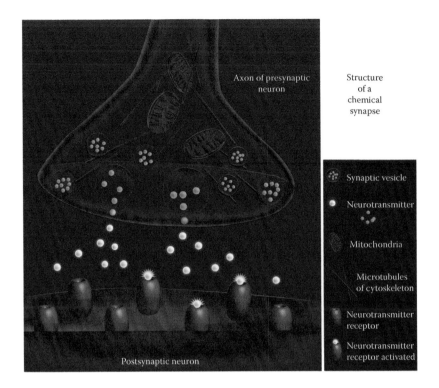

Axon of presynaptic neuron

Structure of a chemical synapse

Synaptic vesicle

Neurotransmitter

Mitochondria

Microtubules of cytoskeleton

Neurotransmitter receptor

Neurotransmitter receptor activated

Postsynaptic neuron

So Dopamine Allows the Brain to Function?

Yes, and many of the other nerve cells in the body. But as well as allowing nerves to operate, there are many receptors in the brain to which dopamine attaches, and when these are activated, different physiological effects occur. For example, it is believed that dopamine is responsible for a general feeling of well-being. It has also been linked with feelings of happiness, euphoria, excitement and positivity, as well as the eagerness to chase after goals or rewards. It plays a large role in reward-driven learning in animals, and every type of reward that has been studied (food, pain relief, sex) results in increased dopamine levels in the brain.

DOPAMINE MAKES YOU HAPPY!

Yes, it provides feelings of enjoyment and reinforcement which motivates someone to perform activities that are beneficial to the survival of the individual or species. That's why you enjoy eating food*, having sex and doing exercise. In some people, aggressive behavior also results in a dopamine 'high', which may explain our propensity to fight and our fascination with violent sports like football and boxing.

THERE MUST BE A DOWNSIDE … ?

Unfortunately, there is. Reward-driven learning like this can lead to addictions, such as over-eating, sex addiction or chronic violent behavior, which need to be combated, respectively, by dieting, rehab clinics or prison! Arguably, a more serious problem comes from the fact that the brain's dopamine receptors can be tricked by molecules with a similar chemical structure to that of dopamine. Molecules such as cocaine (see p105), amphetamine, methamphetamine and MDMA (see p309), bind to the dopamine receptors and produce enhanced feelings of well-being, even euphoria, in an artificial chemical 'high'. However, once the effects wear off, the user often suffers a deep crash, as the dopamine receptors return to normal. The receptors gradually become accustomed to these elevated levels of activity, so the user has to take higher and higher doses of the drug to get the same effect. All this leads to drug addiction, with the associated risk of overdoses. Nicotine (see p363) is another dopamine mimic that helps relieve feelings of anxiety, at the expense of being highly addictive.

---

* As well as dopamine receptors, cannabinoid receptors are also believed to be responsible for pleasurable feelings when eating food (see p497).

Methamphetamine   MDMA (Ecstasy)   Nicotine

AND ALL ANIMALS USE DOPAMINE AS A REWARD?

Nearly all vertebrates and invertebrates
use it, except for insects, which use
a related molecule, octapamine, as a
reward system. Curiously, in insects,
dopamine has the opposite effect –
it acts as a punishment signal helping to
form bad memories of events that should
be avoided in future.

Octopamine

CAN YOU HAVE TOO MUCH DOPAMINE?

Yes, having excess dopamine
in the brain doesn't simply
make you even happier! In
fact, overstimulation causes
the brain to malfunction,
leading to conditions
such as schizophrenia,
attention deficit
hyperactivity disorder
(ADHD) and restless leg
syndrome.

Excess dopamine in the brain can
cause schizophrenia, leading to
personality changes, hearing voices and
seeing things that are not there, and
violent tendencies

How Is It Made?

The usual starting material is L-tyrosine, which is an amino acid commonly obtained by ingestion of various types of high-protein food, including cheese, poultry, bananas, milk and peanuts. An enzyme then attaches a second OH group to the benzene ring forming levadopa (also known as L-dopa). A second enzyme then removes the carboxylic acid to give dopamine.

L-Tyrosine   L-Dopa   Dopamine

I've Heard of L-Dopa – It's a Treatment for Parkinson's, Isn't It?

Yes. Parkinson's disease is a neurological condition that progressively worsens. It gets its name from the London doctor James Parkinson who first described the disease in his *'Essay on the Shaking Palsy'* in 1817. It affects movements, such as writing, talking, walking and even swallowing, and is often characterized by slow movement and repetitive shaking. The actor Michael J. Fox and boxer Muhammad Ali are famous sufferers of Parkinson's disease.

Parkinson's disease is caused when the nerve cells in the brain that synthesize dopamine die. As a result, less dopamine is produced and this prevents certain parts of the brain from functioning correctly. Unfortunately, there is no cure for Parkinson's disease, but there are several medications available to control the symptoms. These drugs work by either boosting the production of dopamine in the brain, or by mimicking the effects of dopamine.

Muhammad Ali

Michael J. Fox

IS THIS WHERE L-DOPA COMES IN?

Yes, dopamine itself cannot be used because the molecule is too polar to cross the blood–brain barrier and so cannot enter the brain. Instead, L-dopa is used.

BUT L-DOPA IS ALSO POLAR ... ?

Yes, but the difference is that it's also an amino acid, which means that it can be recognized by proteins that carry amino acids across the blood–brain barrier. L-Dopa hitches a ride on one of these proteins and is safely ferried into the brain. Once there, it is converted to dopamine.

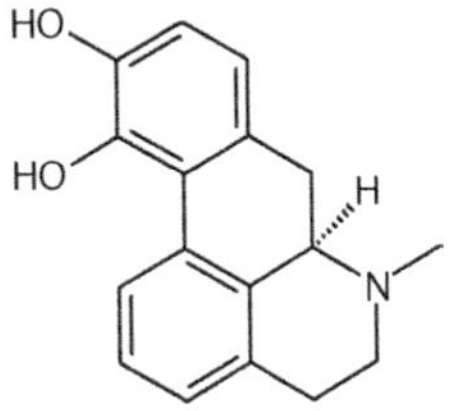

Apomorphine

Another drug that's used to treat Parkinson's is apomorphine, which gets its name from being a decomposition production of morphine (although the remaining structure is quite different from that of morphine – see p347). It is also a dopamine mimic, that, as well as helping with the symptoms of Parkinson's, has been used to treat erectile dysfunction (the increased dopamine response in the brain is linked with sexual response), and as a 'cure' for alcohol and opiate addiction.

Chapter

21

EPIBATIDINE

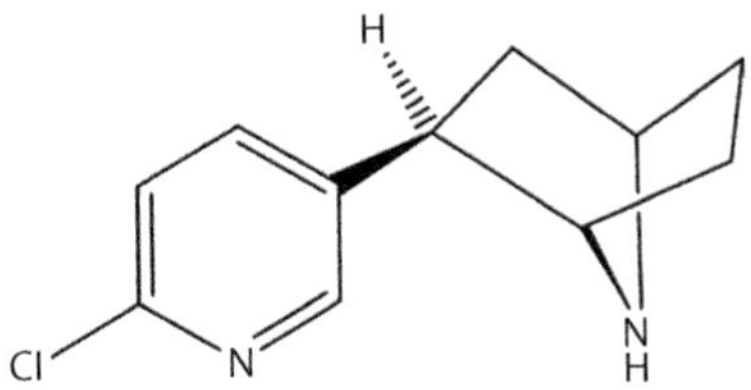

(+)-Epibatidine

It Can't Be a Natural Molecule ...

Why do you say that?

It Contains Chlorine

There are thousands of natural molecules that contain chlorine. Unfortunately, quite a lot of environmentalists don't realize that while some chlorine-containing compounds like DDT (see p135) can cause environmental problems, others are produced naturally by living systems.

Phantasmal poison frog
(*Epipedobates tricolor*)

So Where Does It Come From?

A frog.

Like Kermit?

Not any old frog, but a tiny, multi-colored one, from an Ecuadorean rainforest. These are one of a number of different families of frogs whose brightly colored skins warn predators not to mess about with them, as their skins exude a lethal poison. Colombian Indians put these poisons on their darts and arrows when hunting for food. This molecule was first isolated in 1974 from the rare 'phantasmal poison frog' *Epipedobates tricolor*. The researchers extracted all the chemical they could (from 750 frogs!) but only obtained 0.5 mg; however, the tools they had in their spectroscopy kit were simply not enough to work out its structure.

WHAT DID THEY DO NEXT?

They faced a number of problems. They only had a tiny amount of the poison; they hadn't a clue about the active agent and, even worse for the scientists, the frogs became a protected species. It wasn't until the early 1990s that improved techniques showed that the active agent was a compound they called epibatidine after the frog's Latin name. The scientists became really interested when the late John Daly (a leading researcher in poison dart frogs) injected it into mice and obtained a very strong Straub tail reaction (where the tail stands up and arches over the mouse's back), a sign of opiate behavior. This suggested that in small doses, this toxic and potentially lethal poison might be a useful painkiller.

SO IT WAS A PAINKILLER?

Yes, but an unusual one. It has been shown to be 200 times as strong as morphine in blocking pain in animals, but because epibatidine has a structure rather like nicotine, it operates through a different receptor in the brain than other opiate painkillers like morphine (see p347). This raised real hopes of it being a non-addictive painkiller.

SO DOCTORS ARE GOING TO USE EPIBATIDINE?

Chemists have found routes to synthesize it in laboratories, but it is too toxic to be safely used on humans (the lethal dose is only around 2 mg). Undeterred, chemists made several hundred similar molecules. They found that one of them, tebanicline, has a much lower toxicity than epibatidine, while remaining an effective painkiller. It went through clinical trials, though these were not successful enough to go further. Nevertheless, further research is still ongoing in this area and it is expected that studies of other related molecules will likely lead to new types of painkillers within the next few years.

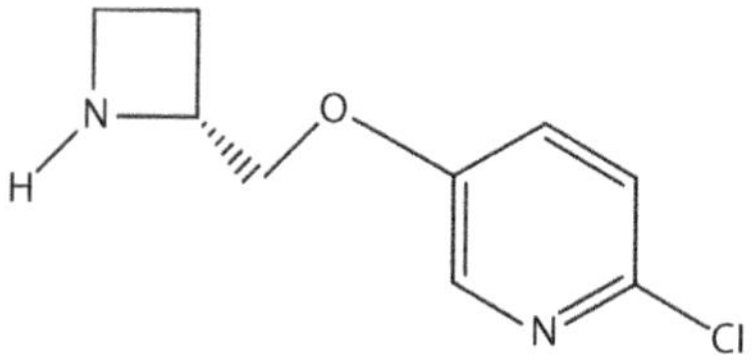

Tebanicline

ARE THERE ANY SIMILAR MOLECULES THAT OCCUR NATURALLY?

Indeed, recently a four-ring molecule named phantasmidine was discovered in *Epipedobates anthonyi* frogs; this also contains the unusual feature of a chloropyridine unit (the 6-membered ring containing a nitrogen and a chlorine side-group).

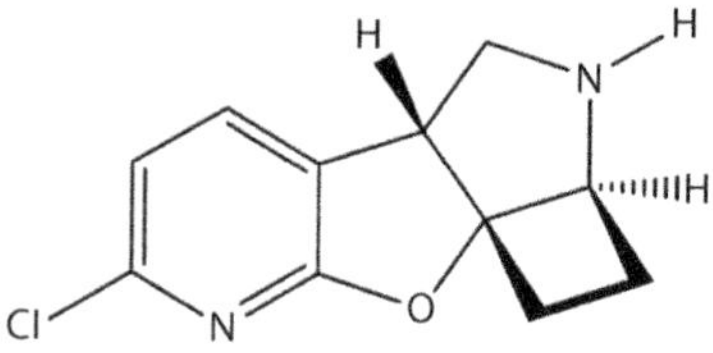

Phantasmidine

And there's batrachotoxin and homobatrachotoxin which also come from rainforest frogs (*Phyllobates terribilis*). They are among the most toxic substances known, more toxic than *curare* or tetrodotoxin (see p515) used by the puffer fish (itself over 1000 times more poisonous than cyanide). Around 136 µg is the lethal dose for a person weighing 150 pounds; that is, about two grains of table salt. On average one frog packs 1100 µg of batrachotoxin.

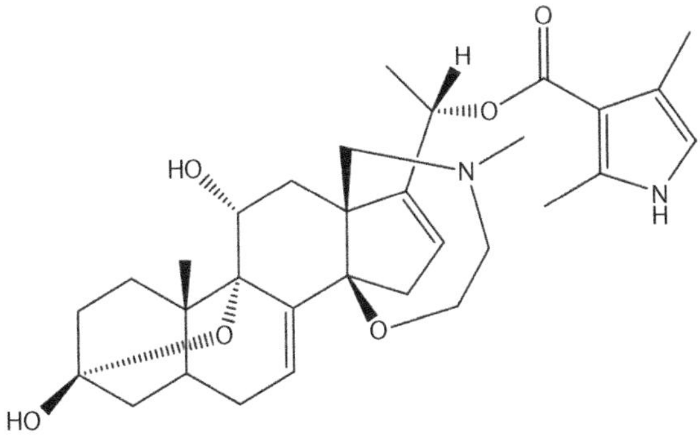

Batrachotoxin

WHERE DO THE FROGS GET THE POISON FROM?

For epibatidine, the frog's source of the poison is unknown. But scientists found that if the frogs are bred in captivity, they do not form epibatidine. One possibility is that the frogs eat something that contains epibatidine, such as insects and spiders on the forest floor. The insects may contain

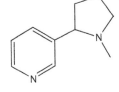

Nicotine

epibatidine from eating the vegetation – a possible clue is that the structure resembles nicotine, a plant product (see p363).

For batrachotoxin, the story is similar, but more complex. Captive-born *Phyllobates terribilis* don't have batrachotoxins in their skins. If poison frogs are caught in the wild, the amount of toxin in their skin often diminishes with keeping. As before, this suggests a dietary

The *Phyllobates terribilis* poison dart frog

origin for the toxin. In 1989, an American ornithologist called Jack Dumbacher was doing his graduate studies in Papua New Guinea. One day, he was freeing a Pitohui bird (a native songbird about the size of a jay) from a net, when his hand was bitten and scratched by the bird. Instinctively, he put his hand in his mouth and found it tingled and started to go numb. Intrigued by this, he eventually passed samples of the plumage to John W. Daly, the world expert on the chemistry of the substances secreted by poison dart frogs, who identified the presence of batrachotoxin. The natives of New Guinea have long known that these birds were inedible (in fact, they called them 'rubbish birds' as a result), and now the reason for this was known. More recently, batrachotoxins have been found in the feathers of another New Guinea bird, *Ifrita kowaldi*.

So What's the Link?

In 2004, it was reported that Melyrid beetles were found in the stomach of Pitohui birds in New Guinea. Some Melyrid beetles are now known to contain large amounts of batrachotoxin and varieties of this beetle are found all over the tropics, including Colombia and New Guinea, where they may form part of the diet of both the frogs and the birds. Whether the beetles make the batrachotoxin themselves or obtain it from their diet is not known at present. It would be unusual for a beetle to synthesize steroid molecules like batrachotoxin, so it may be that

A hooded Pitohui (*Pitohui dichrous*) from Varirata National Park, Papua New Guinea (Photo: Copyright Jack Dumbacher, California Academy of Science, with permission.)

the beetle gets it from small arthropods it eats, or maybe from plants.

However, it looks as if some other frogs can, indeed, make their own alkaloids. Australian myobatrachid frogs of the genus *Pseudophryne* contain two types of alkaloids in their skin extracts, some of which they get from their diets of formicine ants, but some they synthesis themselves.

AND THE OUTLOOK?

These poisonous frogs use a very wide range of substances of hitherto undreamed-of structures as venoms. Scientists are using these molecules to study the way in which nervous impulses are transmitted in animals. They are being studied for possible applications in areas like heart drugs and anesthetics. Who knows what other amazing chemicals lie in wait in the rainforests, each a potential medicine? However, the frogs are nearly all on the endangered list. For example, the phantasmal poison dart frogs currently can only be found in only seven locations in the Amazon rainforest, and are nearing extinction. With the rainforests being rapidly destroyed to make way for farmland, the world is in danger of losing valuable species like these forever – and with them go a vast pharmacy of undiscovered but potentially valuable medicines and drugs.

Chapter

22

ESTRADIOL
THE MAIN FEMALE HORMONE

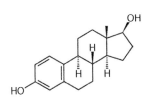

DON'T YOU MEAN ESTROGEN?

No, estrogen (sometimes spelt oestrogen) is actually the name given to a group of compounds, including estradiol, that are important in the female reproductive cycle. The name comes from the Greek *oistros* meaning 'producer of sexual desire'. Like the male equivalent, testosterone (see p489), estrogens are produced in all vertebrates and insects, which suggests that they appeared very early on in the evolution of life on Earth.

SO IT'S TO DO WITH SEX?

Yes, in both the gender and reproductive meaning of the word. The sex hormones determine whether an animal is male or female, and also the relative amounts of the different sex hormones determine the sexual characteristics of that animal, such as how masculine or feminine they appear and behave.

AND THE FEMALE SEX HORMONE IS ESTRADIOL?

That's the most important one, along with estrone and estriol, which are less potent. They are all based on the standard four-ring steroid structure, with just minor differences in side groups. And this includes testosterone too. It's quite amazing that all the differences between male and female come down to something as trivial as the side group of a molecule!

Estradiol. Two OH groups, hence the '-diol' suffix

Estriol. Three OH groups, hence '-triol'

Estrone. One OH group
and one ketone group,
hence '-one'

Testosterone – the male hormone.
Note the difference with estradiol
is only an extra methyl group, the
OH has been oxidized to a ketone,
and the phenyl ring reduced

Although it can be produced by many cells in the body, including
fat cells, in the brain and in artery walls (in males as well as
females), most estradiol is synthesized in the ovaries from
compounds derived from cholesterol (see p91).

What Does It Do?

It helps a woman prepare for pregnancy, by supporting the
reproductive organs, keeping the eggs healthy in the ovaries,
and instigating the monthly ovulation and menstrual cycle.
It is also responsible for the development of the female
secondary sex characteristics, which begin at puberty and
decline after the menopause. For example, estradiol initiates
the development of breasts, and alters the fat distribution
in a woman's body to make her more 'curvy'. It also helps
strengthen bones and joints.

Is That Why Post-Menopausal Women Have Brittle Bones?

Partly, yes. After menopause, women's estrogen levels drop
dramatically, and this results in their bones becoming weak

Look what HRT did for me!

and porous, a condition called *osteoporosis*. To combat this, post-menopausal women, or those who have had a hysterectomy, are often prescribed hormone replacement therapy (HRT), which usually involves ingesting a mixture of estrogen, progesterone and progestin to compensate for the amount they've lost. One common estrogen that's used in HRT has the trade name *Premarin*, and is isolated from the urine of pregnant mares.

The name comes from *pregnant mares' urine*, and is actually mostly composed of estrone sulfate. It may sound an odd thing to take, but this is converted directly into normal estradiol in the woman's body.

Estrone sulfate

DOES ESTRADIOL AFFECT MEN?

Exposure to any of the components of estrogen by males can cause anorexia, vomiting and feminization, such as breast growth, and also erectile dysfunction. In the developed world, sperm counts have been falling by about 1–2% every year for the past several decades. The reason for this is not entirely clear, but one possibility is that men are inadvertently ingesting small quantities of estrogens or estrogenic-like compounds, and these are damaging male fertility. Various chemicals have been blamed for possibly mimicking the effects of estrogen, including bisphenol A (BPA), polychlorinated biphenyls (PCBs) and phthalates.

But another route through which estrogens have made their way into the water supply is as a result of the disposal of unwanted or out-of-date steroid pills and contraceptive pills down the sink or toilet.

REALLY? HOW SO?

The female contraceptive pill contains the hormone progesterone (often called the pregnancy hormone) combined with an estrogen, usually a derivative of estradiol called ethynyl estradiol. High levels of progesterone such as those in the pill are usually only found when a woman is

Bisphenol A (BPA) is used to make polycarbonate plastic and the epoxy resins that are used as a lining in most food and beverage cans

Phthalates are plasticizers providing durability and flexibility to plastics such as PVC. High-molecular-weight phthalates are used in wall coverings, flooring and intravenous bags and tubing. Low-molecular-weight phthalates are found in cosmetics such as lotions and perfumes, and also varnishes and lacquers

Polychlorinated biphenyls (PCBs) are chemically stable, have low flammability and are electrical insulating, and their main use is as insulating fluids and coolants

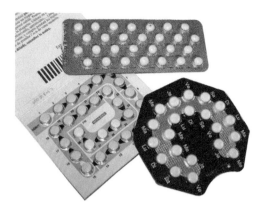

Different types of contraceptive pills

pregnant, so the pill works by tricking the body into thinking it's already pregnant. The female body's response to being pregnant is complex. But, among other effects, further ovulation is prevented (because no more eggs are needed if one has already been fertilized), the mucus in the cervix thickens making it difficult for additional sperm to enter, and the lining of the womb becomes thinner, so it is less likely to accept (another) fertilized egg. Other contraceptive pills, such as the progesterone-only pill (or mini-pill), contain only synthetic progestogens (compounds chemically related to progesterone but not progesterone itself, despite the name) and no estrogen. Nevertheless, all of these pills contain compounds that mimic estrogen and which can have hormone-disruptive effects on both wildlife and humans.

Progesterone – the pregnancy hormone. Note the remarkable structural similarity to testosterone, above

Ethynyl estradiol – the main difference between this and estradiol is the addition of the acetylene group

SHOULD WE BE WORRIED?

Well, yes. The male-to-female ratio in many animals and fish species have reportedly been altered by exposure to these chemicals. In some of these cases, the males have been 'feminized', and have partial female characteristics. And in humans, they are thought to be responsible for premature puberty in regions where high levels have been found in the local environment. Several countries have already started to ban these estrogenic compounds. For example, Canada, the United States and the European Union have recently (2012) banned the use of BPA in baby bottles. PCBs were banned in

The ultimate result of estrogen compounds getting into the environment?

1979 but do not degrade very effectively, so continue to persist in the environment. But there are many more natural and man-made estrogenic compounds all around us, the effects of which we know little or nothing about. Maybe everyone should all start to get used to wearing skirts and dresses!

Chapter

23

GLUCOSE

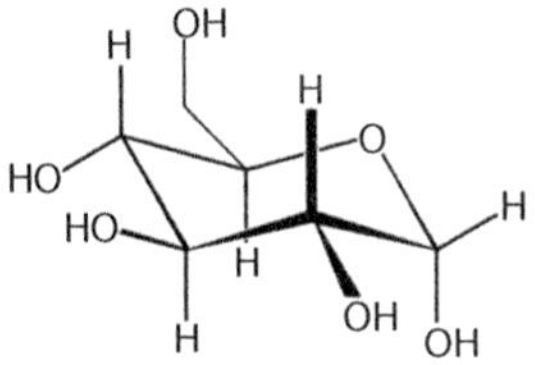

Glucose

'Do you like my grass and my buttercups? The grass you are standing on, my dear little ones, is made of a new kind of soft, minty sugar that I've just invented!'

Willy Wonka
Charlie & the Chocolate Factory, Roald Dahl

GRASS MADE OF SUGAR? THAT'S A BIT SILLY!

Actually, in a way, all grass is made of sugar. Well, a polymer of glucose …

GLUCOSE IS SUGAR?

It is, and it isn't. When most non-scientists use the term 'sugar' they mean the specific molecule sucrose (or 'table sugar', see p467); these are the white crystals we add to tea or coffee or sprinkle onto cakes to make them sweeter. But to a scientist, 'sugar' is a generic term that refers to a class of molecules called saccharides. Simple sugars (monosaccharides) contain only one sugar unit per molecule; a good example of this is glucose. Conversely, more complex sugars such as disaccharides contain two sugar units bonded together, such as in sucrose. And polysaccharides have many sugar units joined together to form a polymer; examples are cellulose or starch.

Glucose can be purchased as crystals, powder, tablets, a water solution or as glucose syrup which is made from corn starch

But Glucose Is the Simplest?

Not quite. Glucose contains six carbon atoms (and is called a hexose), but there are other monosaccharides such as fructose which are simpler because they only contain five carbon atoms. However, all sugars are carbohydrates (they contain only carbon, hydrogen and oxygen), and chemists use the suffix '-ose' when naming them. So the name glucose comes from the Greek for 'sweet' (*glukus*) +ose.

Who Discovered It?

The famous German chemist Emil Fischer determined the structure of glucose and two other sugars (fructose and mannose) over a period of about 10 years while working at the Universities of Erlangen, Würtzburg and Berlin, in the late 1800s. He was the first chemist to synthesize all three of these sugars starting from glycerol (see p207), and was awarded the Nobel Prize in Chemistry in 1902 for this work and his other work on protein structures.

Glucose   Fructose   Mannose

The next breakthrough came 35 years later when Norman Haworth of Birmingham University, who won the Nobel Prize in 1937, showed that glucose (and other sugars) existed in a ring-like structure, rather than a straight-chain arrangement, although the two forms interconvert when in solution and exist in equilibrium with each other.

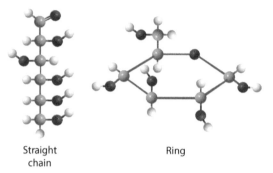

Straight
chain

Ring

The two ways of representing glucose, as a straight chain or as a ring

In fact, there are actually two different ring structures that can be formed depending upon how the glucose chains links up to form the ring. The isomer where the OH group on the first carbon is above the ring is called α-glucose while the isomer with the OH below the ring is β-glucose. The two forms interconvert (via the chain form) in water solution in a period of a few hours.

α-Glucose – the OH group on the right-hand side points down

β-Glucose – the OH group on the right-hand side points up

As well as the α and β isomers, glucose is chiral and has stereoisomers (mirror images). The one found almost exclusively in nature is the 'right-handed' version of glucose (so called because when polarized light is shone through a solution of it the polarization is rotated a few degrees to the right). This type of glucose is known as dextrorotary glucose (from the Greek *dextro* = right side), or D-glucose, sometimes shortened to 'dextrose' – which is the name commonly seen in the ingredients on packets of various foodstuffs.

The other (left-handed) isomer, L-glucose, is hardly ever found in nature but can be synthesized in labs. It is indistinguishable in taste from D-glucose, but it cannot be used as a source of energy by living organisms as the enzymes needed to metabolize it evolved to work only on the right-handed isomer. L-Glucose was once proposed as a low-calorie sweetener, particularly useful for patients with diabetes, but it was never marketed due to excessive manufacturing costs.

L-Glucose

And, of course, there are α and β isomers of L-glucose too.

WHY DOES IT MATTER IF GLUCOSE IS A RING OR A CHAIN?

Well, it's one possible answer to a question that bugged biologists for years: why is glucose, and not any of the other sugars, so abundant in nature and act as the universal energy source for plants and animals? Scientists recently found that sugars in their chain form react quite readily with the amino groups of proteins and enzymes. Sometimes, this reaction can destroy the protein or prevent an enzyme from functioning. Many of the long-term problems of diabetes (blindness, kidney failure, *etc.*) are believed to be due to these unwanted reactions of excess glucose in the blood attacking and damaging proteins, enzymes or lipids in organs around the body (see p255). But because the ring form of glucose is so stable, glucose prefers to remain in its benign ring form most of the time. Thus, it is nowhere near as damaging as other sugars, which spend longer in their reactive chain forms. As a result it's believed that animals and plants have evolved to use glucose as an energy source as it's less harmful than other sugars.

SO GLUCOSE IS AN ENERGY SOURCE?

It's actually the main energy source used by nearly all living things during respiration. All the food we eat is broken down and converted ultimately to form glucose. Because it is water soluble, the glucose can then be transported in the bloodstream (where it is referred to as 'blood sugar' by medics and diabetics) to the cells, whereupon it reacts with inhaled oxygen that has been carried there from the lungs by another molecule in the blood, hemoglobin (see p227). The oxidation of glucose releases a lot of energy, and produces carbon dioxide and water as waste products. These are removed from the body either by exhalation or urination, or for plants by evaporation from the pores in their leaves. This whole process is actually quite a complex metabolic process known as glycolysis, but it can be simplified to

Physical exercise requires the energy to be released from glucose to power muscles

$$C_6H_{12}O_6 + 6O_2 \rightarrow 6CO_2 + 6H_2O + \text{Energy}$$

This is almost the exact opposite of the reaction that occurs during photosynthesis (see p81). The energy that's released is used to convert ADP to ATP (see p1), which is a molecule that stores energy locally within the cell and can be easily broken apart to release the energy to power metabolic processes.

Starch is widely used in laundry to stiffen collars, sleeves, and petticoats, and also helps with cleaning. Sweat and dirt preferentially sticks to starch rather than to the fibers of the clothing. This then washes away along with the starch during laundry, and then more starch would be reapplied for the next wearing

IF IT'S WATER SOLUBLE, HOW IS GLUCOSE STORED?

Good point. Glucose itself would simply dissolve in the animal or plant fluids and rapidly be excreted from the organism. So the trick is to convert the glucose into an insoluble form, which will stay where it's put. This is done by polymerizing the glucose using enzymes, and the resulting polysaccharides have very different properties depending upon which form of glucose is used.

If α-glucose is polymerized, the result is either starch or glycogen. Starch is the polysaccharide energy storage molecule used by plants, and is found in potatoes, rice, maize, corn and cassava, among many others. It is the most common carbohydrate in the human diet.

Rice – a good source of starch …   … as are potatoes

The polysaccharide that makes up starch comes in two types, an unbranched version called amylose which makes up 20–30% of the total, and a branched version called amylopectin which makes up the rest. Amylose is made up of about 300–3000 glucose units bonded together in a linear or helical structure and is insoluble in water. This compact form is easy to pack

Amylose – the glucose units are joined together in a line. The resulting polymer chain can be linear, or can twist around on itself, eventually forming a helix shape. The chains can pack together to form compact semi-crystalline granules of up to 100 μm in size, as shown for both the linear and helical form on the right together into a small space and so makes amylose an efficient energy store for plants.

Amylopectin – the glucose units are joined in a line, as in amylose, except every so often there's a side branch which starts a new parallel side line

Amylopectin chains – the chains branch and then pack together

In contrast, amylopectin is highly branched, with branches every 24–30 glucose units. It is also much larger being composed of up to 20,000 glucose units. The large number of polar end groups makes it soluble in water, and also more easily digested than amylose.

How Is This Stored Energy Released?

Glucose molecules in both the forms of starch are bound together by relatively weak bonds. These can be broken by enzymes called amylases, which are found in both plants and animals. In humans, saliva has a high concentration of amylase, and allows us to digest foodstuffs high in starch. It's known that people who eat a high-starch diet have more amylase enzymes than those who eat a low-starch diet. And chimpanzees, who eat virtually no starch, have very low amounts of amylase in their saliva. This has led scientists to propose that the evolution of amylase enzymes was a crucial moment in human evolution, as it allowed proto-humans to eat roots and grains that were previously indigestible. It may have coincided with the discovery of fire, since cooking starch makes it much more digestible. Arguably, the ability to digest starch enabled early humans to make the transition from being hunter-gatherers to agricultural farmers, which kick-started the road to civilization.

WHAT ABOUT IN ANIMALS?

Animals (and, surprisingly, fungi) use another polysaccharide called glycogen as their way of storing glucose. Glycogen has almost the same structure as amylopectin (and is often called 'animal starch' for this reason) except it is even more highly branched, with branches occurring every 8–12 glucose units. Glycogen is found in the liver and muscles, and forms an energy reserve that can be quickly released to meet a sudden need for glucose (see p9). However, it is less compact than the energy stored in triglycerides (body fat, see p207) which are used for long-term, slow-release energy reserves.

WHAT ABOUT THE POLYMERS OF β-GLUCOSE?

The bonds which join together β-glucose units are much less prone to hydrolysis reaction than those of α-glucose. This unreactivity makes β-glucose polymers very strong and chemically stable; as a result, plants use them to make structural components like cell walls. An example of one of these is cellulose, which is the most abundant organic polymer on Earth since it makes up between 50% and 90% of the dry content of all plant matter.

INCLUDING GRASS?

Yes, of course – so Willy Wonka wasn't too far from the truth. Unlike starch, cellulose is a straight-chain polymer, with no coiling or branching. The molecules adopt extended, stiff, rod-like conformations. The multiple OH groups from one rod form hydrogen bonds with oxygen atoms on an adjacent rod, which hold the rods firmly together side-by-side. Many of these rods, locked together in this way, form fibers with high tensile strength. This is what gives cell walls their strength and plants their rigidity.

IS THAT WHY HUMANS CAN'T DIGEST CELLULOSE?

Yes. Humans can only partially digest cellulose; most of it simply passes through our digestive tract unchanged, although it does act as roughage. Cows and other ruminant animals are able to digest cellulose with the help of bacteria that live in their guts, and a multi-compartment stomach. The plant material, such as grass, is eaten and then stored in a compartment in the animal's stomach for several days while bacteria break down the cellulose into smaller molecules (mostly small fatty acids like acetic acid, propanoic acid and β-hydroxybutyric acid). The animal then regurgitates the contents (the 'cud'), rechews it to break it down further, before swallowing it into a different compartment of the stomach for further digestion.

IF WE CAN'T EAT CELLULOSE, WHAT CAN WE DO WITH IT?

Well, we can burn it to provide heat. A wood fire is mostly burning cellulose. And we can wear it. Textiles made from linen, cotton and other plant fibers are just cellulose, and another textile fiber, rayon, is made from processed cellulose. And we can write on it – cellulose is the main component of paper, and of cardboard boxes. And we can even blow things up with it – nitrocellulose (guncotton) is used in smokeless gunpowder, which is made by simply treating cotton with strong acids under icy cold conditions. We can watch movies on it – until the 1930s, all movies were made using 'celluloid' film. This was made using nitrocellulose, which is why so many old movies used to spontaneously catch fire in storage. Nitrocellulose was also briefly used as an alternative for ivory to make billiard balls. However, not only were the balls very flammable, but sometimes they would explode upon impact. An owner of a billiard saloon in Colorado in 1869 wrote to the inventor of these balls (John Wesley Hyatt) complaining that he personally didn't mind the balls exploding, except for the fact that when they did so every man in his saloon immediately pulled a gun at the sound!

ANY OTHER GLUCOSE POLYMERS?

Well, there's lignin, the second-most abundant organic polymer in the world. It comprises perhaps 25–30% of the weight of dry wood. Unlike cellulose and some other polysaccharides used in building plant cell walls, lignin is hydrophobic, and so impermeable to water. Thus, many of the tubes (xylem) within plants that carry water and nutrients around the plant are made from lignin or use lignin to line them. Without the lignin, water would leak out through the walls. As a biopolymer, lignin is quite unusual in that it doesn't have a defined structure. It comprises around 50 glucose units bonded together in a seemingly random arrangement, together with various types of substructures that appear to repeat in a haphazard manner. There are different types of lignin found in different plants and trees, which differ according to the degree of polymerization, the arrangement of the glucose units, the hydrogen bonding between substructures, *etc.* When wood or charcoal is burned, the lignin is oxidized to guaiacol and syringol and their derivatives. These two molecules impart the characteristic smell and taste to smoked and barbecued foods.

Guaiacol   Syringol

Another important glucose-based polymer found in nature is chitin, which is a version of cellulose where an OH group on each

… while the exoskeletons
of insects …

Tree bark is composed of lignin …

… and the shells of crustaceans
are made of chitin

glucose unit is replaced with an acetylamine group (CH_3CONH). This increases the hydrogen bonding between adjacent polymer chains, giving the chitin polymer matrix increased strength. As a result, chitin is used for the hard exoskeletons of insects, and of crabs, lobsters and other crustaceans, and the internal shells and beaks of squids.

So, you might say that as well as being Nature's universal energy source, glucose is also one of Nature's most versatile building materials. Not only is the grass made of sugar, but also a large part of the biological world!

Chapter

24

GLYCEROL

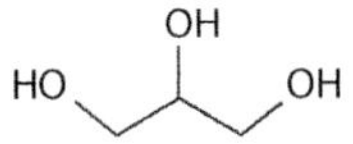

Glycerol

Is That the Same as Glycerine?

Yes, it's just a different name. Many people have heard of glycerine due to it being used in glycerine soaps.

And Nitroglycerine, Of Course?

Yes, but we will come on to that later.

Ok. So It's a Component of Soap?

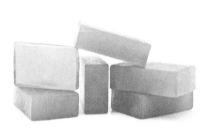

Bars of glycerine soap

Not normally – the exception being the translucent glycerine soaps which have glycerol added to them to change their properties. Normal soap is formed when fats and oils are converted into soap (in a process called saponification) by reaction with a base like sodium hydroxide, and glycerol is the waste product.

So It's a Component of Fat, Then?

Yes, and quite an important part. Fats are made from long-chained hydrocarbons with a carboxylic acid group at one end. These are called fatty acids, and the sodium salts of these are often used as soaps and detergents, because one part of the molecule is soluble in water while the other part is soluble in oil and greases (see p271).

WHAT DOES THE GLYCEROL DO?

The glycerol has three hydroxyl groups which can link to the carboxylic acid group of three different fatty acids, producing a fat. Depending on which fatty acids are used, we can get different types of fat with different properties. If the long chains have several double bonds in them (see p287), then we call it a poly-unsaturated fat. Alternatively, if the long chains are mainly single C–C bonds, then we get saturated fats, like stearic and lauric acids.

An example of a triglyceride made by joining glycerol on the left to three different fatty acids. In this case, the acids are (from top to bottom): palmitic acid, oleic acid and linoleic acid. This triglyceride is typical of what constitutes the body fat of humans

THOSE ARE BAD FOR YOU, RIGHT?

Eating too much saturated fats causes an increase in cholesterol in the bloodstream, which can lead to blockage of arteries and cause heart problems (see p91). And unsaturated fats have also been linked with a variety of illnesses, such as heart disease, cancer and diabetes, especially the *trans*-fats (see p287).

SO EATING ANY FAT IS BAD FOR YOU?

Eating too much of any type of fat is bad for you. But many fats contain fatty acids that are essential for the smooth operation of the body, but which the body cannot synthesize itself, and therefore we have to obtain them by eating them in our diet. For example, some fatty acids are converted into biologically active molecules such as prostaglandins (see p419) which control inflammation and affect the central nervous system. These are called essential fatty acids (EFAs) because if we didn't eat them, there would be dire health implications.

Also, fats are a very important way for the body to store energy for long time periods, so that it can survive periods of when there is little food or even famine. Typically, women have around 25–31% fat in their bodies whereas men have 18–24%. However, in some obese people, more than 50% of their body weight can be fat.

What Is Body Fat Composed of Exactly?

Body fat is a triglyceride composed of glycerol bonded to three fatty acids. In naturally occurring triglycerides, the lengths of the chains in each of the fatty acids vary. Because they are synthesized using the molecule acetyl CoA, which adds two carbons at a time, natural fatty acids found in plants and animals are usually only made from even numbers of carbon atoms, such as 16, 18, or 20. However, some bacteria can synthesize odd- and branched-chain fatty acids. As a result, the fat from many ruminant animals, such as cows, contains odd-numbered fatty acids, such as 15, due to the action of bacteria breaking down vegetation in the animal's gut.

'I'm not obese ... this is just my internal collection of triglycerides ...'

In the early 1800s, scientists believed that human fat was composed of three fatty acids: stearic acid, oleic acid and the recently discovered (1813) margaric acid. The latter was probably named after the Greek word *margarís*, meaning 'palm-tree', from whose oil it was first isolated. Later analysis showed that margaric acid was actually a mixture of stearic acid and the previously unknown palmitic acid (also named after the palm tree). Even though it never really existed, the name margaric acid is the origin of the word 'margarine'.

Nowadays, we know that most human fat is mostly composed of oleic (47–52%), palmitic (22–25%) and linoleic (11–13%) acids, with smaller contributions from stearic (4–8%), myristic (2–3%) and palmitoleic (4–8%) acids. (A typical example of a triglyceride found in human fat is shown in the figure on p209.) However, the exact amounts of each acid depend upon a number of factors, including gender, race, diet and which type of body fat you mean. In essence, you are what you eat – in that the meat you eat has a fat composition that is very similar to the fat which makes up your body.

CAN HUMAN FAT BE USED FOR ANYTHING?

In many cultures, human fat was often believed to have magical healing properties, and was used as an ointment for treatment for anything from toothache to gout.

HOW DID THEY OBTAIN IT?

From executions. Many executioners would make a tidy side profit by recovering the fat from their executed victims and selling it on to medics and pharmacists. In the late nineteenth century, human fat was produced and offered in Germany under the trade

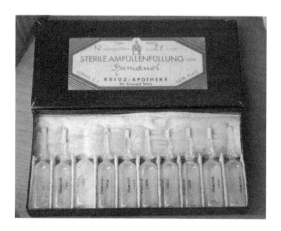

Ten ampules of humanol, a sterile preparation of human adipose fat,
in the Apotheken Museum, Heidelberg, Germany

name *Humanol* as a sterile, liquified preparation for injections.
In 1909, it was introduced for surgical treatment of scars and
wound disinfection. But in the 1920s it went out of fashion after
the low cure rates proved it didn't actually work! However, even
as late as the 1960s various manufacturers supplied human fat in
preparations for external use as anti-wrinkle creams. The fat here
was obtained from placentas collected by midwives and obstetric
departments.

And if you remember the Brad Pitt movie *Fight Club*, human fat was
used to make soap!

YUCK, YES. ANYWAY, LESS ABOUT FAT – TELL ME ABOUT NITROGLYCERINE

If you take the three hydroxyl groups of glycerol and bond it to
three NO_2 molecules, you get nitroglycerine, which is really an
organic nitrate rather than a true nitro compound – but the old
name has stuck. It's been used since the 1860s as an explosive, and

since the late 1800s as a treatment for angina and heart failure. Many heart patients carry around a GTN spray which they administer into their throat and nose if they get angina pain. GTN stands for glyceryl trinitrate, and is the medical name for nitroglycerine; apparently it's called this to avoid alarming patients. In the body, nitroglycerine is converted into nitric oxide (NO), causing the cells that line blood vessels to relax, making the vessels dilate (widen), which improves the flow of blood to the heart. The process is brought about by the enzyme mitochondrial aldehyde dehydrogenase.

Nitroglycerine

So That's Good

But it's not quite that simple. Nitroglycerine is also being tested as a treatment for male erectile dysfunction, as the NO released from the nitroglycerine will improve blood flow in the way that *Viagra* does. There is one type of person who should not use *Viagra*; that is someone who is already taking nitroglycerine, as a combination of the two molecules can lead to a fatal heart attack.

An Explosive Chemical Helps Heart Patients?

Yes, although the main use of nitroglycerine is still as an explosive. In fact, it is also part of the story of the Nobel prizes?

How So?

Nitroglycerine contains carbon (a fuel source), a supply of nearby oxygen (an oxidizer) and nitrogen, which prefers to bond to itself

Alfred Nobel

as N_2 molecules. Thus, the molecule readily falls apart with the release of a lot of energy and a lot of gas.

$$4C_3H_5N_3O_9(s) \rightarrow 6N_2(g) + 12CO(g) + 10H_2O(g) + 7O_2(g)$$

This explosive release of gas in a very short time period is called a detonation, and nitroglycerine is so unstable that it can be set off by heat, flame or even a slight shock. This sensitivity made it difficult to use in practice, and many people were accidentally killed in mines, quarries and battlefields simply from mishandling it. The situation changed when Alfred Nobel discovered that if nitroglycerine were absorbed into an inert clay called *kieselguhr*, it became much more stable and hence much safer. Such sticks of clay, wrapped in greased water-proof paper, became known as dynamite, which Nobel patented in 1867. Dynamite revolutionized mining, quarrying, tunneling, and many other engineering applications. However, it also revolutionized warfare, allowing far more deadly explosive weapons to be developed.

In 1888, a French newspaper wrongly wrote an obituary for Alfred, mistaking him for his brother who had recently died in Cannes. Alfred read his own obituary, and was shocked at what it said:

> *'The merchant of death is dead. Dr. Alfred Nobel, who became rich by finding ways to kill more people faster than ever before, died yesterday'.*

Apparently, Nobel was so worried that this would be his legacy to the world that he decided to do something positive to improve his reputation. He used the vast wealth he'd accumulated from his dynamite business to set up the Nobel Foundation to give annual prizes for science, notably Chemistry, Medicine and Physics. Later, Nobel Prizes for Literature, Economics and Peace were added to the list.

How Is Nitroglycerine Made?

Very carefully! A 1:1 mixture of concentrated nitric acid and sulfuric acid is used to create protonated nitric acid ($H_2NO_3^+$), which then reacts with the hydroxyl groups on the glycerol. This effectively adds an NO_2 group to each OH site in turn, releasing water. The problem is that this process is exothermic, and the heat given off speeds up the reaction further. This positive feedback can lead to a runaway reaction, ending in overheating and detonation. To prevent this, the glycerol is added slowly to the acid mixture, and the whole container is continually cooled to ~22°C. Even so, there is usually a trapdoor in the bottom of the reaction vessel which can open if the temperature rises above a certain critical threshold (~30°C), dumping the mixture into a large pool of very cold water – a process called 'drowning the charge'. Nowadays this is computerized, but in previous years a worker used to watch the temperature dial carefully, with his finger on the trap-door button. Allegedly the worker was given a one-legged stool to prevent him falling asleep on the job.

Chapter

25

HEAVY WATER
DEUTERIUM OXIDE, D₂O

'I wanted *heavy* water
but I only got halfway
there …'

WHAT IS HEAVY WATER?

It is just like ordinary water, H_2O, but the two hydrogen atoms are the 'heavy hydrogen' isotope, deuterium (D), which has a mass twice that of an ordinary hydrogen atom. Around 1912, the concept of isotopes – atoms of the same element having different neutron numbers – was conceived independently by the Polish chemist Kazimierz Fajans and by the British chemist Frederick Soddy.

WHO INVENTED HEAVY WATER?

No one invented it. D_2O molecules were already there, as a tiny fraction of the molecules in any sample of water (0.0156%); you can't invent it but you can discover it.

OK, WHO DISCOVERED HEAVY WATER?

During the 1920s, Lord Ernest Rutherford had suggested that a heavy hydrogen isotope could exist, and there was some evidence for this idea. The American scientist Harold Clayton Urey found that if liquid hydrogen was fractionally distilled, he could obtain a fraction richer in the heavier isotope, and when he examined the atomic spectrum of these samples, he detected lines that were not due to 'ordinary' hydrogen, thus were caused by the presence of small amounts of deuterium (D).

WHAT'S THAT GOT TO DO WITH HEAVY WATER?

Well, dealing with a liquid boiling at around 100°C is easier than handling liquid hydrogen boiling around –250°C. The great American chemist Gilbert Newton Lewis (1875–1946) set about isolating heavy water. There were two ways of doing this. One was electrolysis, using an electric current to split up water. Because heavy water molecules are split up more slowly than H_2O, as a sample of water is electrolyzed, it becomes enriched in D_2O. Another way is to use fractional distillation, as D_2O has a higher boiling point than H_2O, so Lewis and his student Ronald T. MacDonald set about this too, using a 72-feet-high distillation column. So they obtained D_2O, or DOD, if you prefer it that way.

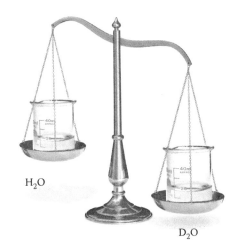

Heavy water is about 10% heavier than normal water

WHAT DID THEY DO WITH IT?

Armed with his supply of deuterium oxide, Lewis set out to investigate its properties – not just the obvious ones, like melting and boiling points – but also whether it was toxic.

He obtained three young white mice *'of respectable ancestry'*, used two as controls and fed the third with his precious supply of heavy water. He wrote: *'During the course of three hours the mouse received, in three doses, a total of 0.54 g of 87% and 0.26 g of 71% heavy water, containing altogether 0.66 g [of pure D_2O]. This would be equivalent, weight for weight, to a consumption of 4 or 5 liters of heavy water by an adult human being. The mouse survived and on the following day and thereafter seemed perfectly normal. Nevertheless, during the experiment he showed marked signs of intoxication'.*

So It Is Not Toxic?

As with most poisons, the dosage matters. Small amounts of heavy water are not toxic, and some think that it would need perhaps 20–25% of the water content of the body to be heavy water for serious poisoning to occur; this would mean the victim drinking neat heavy water for several days on end.

What Difference Does Being Heavier Make to the Properties of Heavy Water?

There are small but measurable differences, as you can see from the table.

	H_2O	D_2O
Boiling point (°C)	100.00	101.42
Freezing point (°C)	0.00	3.81
Density of liquid (g/cm³)	0.9999 (277 K)	1.1056 (293 K)
Density of solid at m.p. (g/cm³)	0.917	1.018
Temperature of maximum density (°C)	3.98	11.2
pH (pD) (298 K)	7.00	7.43

One consequence of the D_2O molecule being significantly heavier than H_2O is that an ice cube made from heavy water will sink if placed in liquid H_2O, whereas normal ice floats (see p557).

How Different Are the Actual Molecules?

Looking at isolated H_2O and D_2O molecules by microwave and infrared spectroscopy shows a shortening in the bond length in

D_2O, but only by 0.0037 Å. For an isolated H_2O molecule, O–H is 0.9724 Å, and H–O–H is 104.50° while for an isolated D_2O molecule, O–D is 0.9687 Å, and D–O–D is 104.35°. A 2008 study of liquid H_2O and D_2O by X-ray and neutron diffraction techniques indicates that the two molecules have very similar, but not identical, structures. This found that the O–H bond in H_2O is 1.01 Å, while the O–D bond in D_2O is 0.98 Å, shorter by 0.03 Å. On the other hand, the intermolecular bond in D_2O (the 'hydrogen bond') is longer by 0.07 Å.

AND AFTER ALL THIS EFFORT, I SUPPOSE UREY AND LEWIS SHARED A NOBEL PRIZE?

Urey was awarded the 1934 Nobel Prize in Chemistry 'for his discovery of heavy hydrogen'. Lewis – who had been Urey's PhD supervisor – won nothing! At this, Lewis stopped work on heavy water. Despite his work on heavy water, not to mention inventing the covalent bond (plus dot-and-cross diagrams), coming up with

Harold Urey

Gilbert Lewis
(© Sciencephoto com. With permission.)

the concept of acids and bases as electron-pair acceptors and donors, respectively, developing a theory of electrolytes and also formulating thermodynamics for chemists (no mean feat), Lewis was never to win a Nobel prize.

THAT'S NOT FAIR! WHAT HAPPENED NEXT?

Research continued on the properties of D_2O. Obtaining large quantities was very difficult, as it required large amounts of electricity for the electrolysis process. At that time, the only place this was carried out was at Norsk Hydro's plant at Rjukan, near Telemark in Norway, where they first obtained really pure heavy water in early 1935, and started supplying it to researchers. Usually scientists only asked for quantities up to 10 grams or so, but in late 1939, Rjukan was receiving orders from the German chemical giant I.G. Farben for up to 100 kg of D_2O a month.

Vemork hydroelectric power station at the Rjukan waterfall in 1935

Why the Sudden Need?

In late 1938, the German scientists Otto Hahn and Friedrich 'Fritz' Strassmann found that uranium atoms could be split by slow neutrons, releasing energy, and more (fast) neutrons. They had discovered nuclear fission. Other scientists, including the Frenchman Jean Frédéric Joliot-curie started working on this as well. But it was realized that if the fast neutrons that were released by fission could be slowed down, they would then go on to split another uranium atom, and another, to form a chain reaction. Slowing down fast neutrons can be achieved by letting them collide with a suitable molecule (called a 'moderator') that can absorb some of their kinetic energy. Either graphite or heavy water were suitable, but at the time the German scientists – notably the great theoretician Werner Heisenberg – did not think graphite would work, so they needed D_2O.

What Did the Norwegians Do?

To begin with, they stalled for time. By early 1940, French intelligence became aware of what was going on and in early March that year Lieutenant Jacques Allier, a *Deuxième Bureau* agent, managed to spirit out of the country the entire Rjukan stock of heavy water, 26 five-liter containers, just before German troops invaded Norway. The heavy water stocks went first to France, then to England, and Joliot-Curie was instrumental in helping find it a safe home.

What Did the Germans Do Next?

They now controlled Rjukan, and started to produce heavy water themselves, and ship it back to Germany. As the British and the Americans became aware of the potential of nuclear fission, they decided to destroy the plant. An initial raid on November 19th

DVD cover of the movie '*The Heroes of Telemark*'

1942 (*Operation Freshman*) failed; over 30 commandos were killed when their gliders crashed and the men were executed by the Gestapo. In the next attempt (*Operation Gunnerside*) on the night of February 27–28 1943, Norwegian commandos badly damaged the plant, and finally a Norwegian resistance party sank the ferry *Hydro* when it was transporting heavy water to Germany on February 20th 1944. The 1966 film '*The Heroes of Telemark*', starring Kirk Douglas and Richard Harris, is loosely based upon these events. This helped to curtail the German research project and prevented the Nazis from developing a nuclear bomb. Ironically, in the American project to develop an atom bomb, the *Manhattan project*, they used graphite as a moderator, not heavy water.

IF HEAVY WATER ISN'T USED IN NUCLEAR POWER STATIONS, WHAT USE IS D_2O?

In fact, Canada did use heavy water as the moderator in nuclear power plants, which used natural-abundance uranium. These days heavy water is obtained using the Girdler process, which employs an exchange reaction for which the equilibrium constant is very favorable ($K \sim 1.01$):

$$HOH(l) + HSD(g) \rightleftharpoons HOD(l) + HSH(g)$$

In the laboratory, chemists use deuteration to help study the mechanisms of reactions, and also in nuclear magnetic resonance

(NMR) spectra, as D has different nuclear properties to H, and so does not give resonances in the proton-NMR range. It can also be used to assign the signals due to labile (weakly bonded) N–H and O–H groups in compounds such as alcohols.

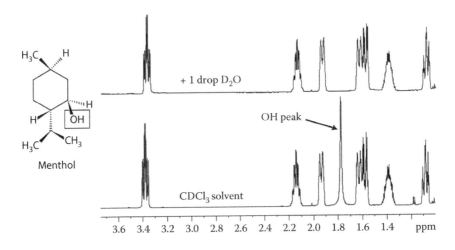

Adding a drop of D_2O to a non-aqueous solution of a molecule (such as menthol) allows you to easily identify the protons which can exchange using ^1H NMR. In the above example, the peak in the lower spectrum due to H from the OH group in menthol disappears when D_2O is added (top spectrum) because the Ds in solution replace only the Hs in the OH group. The other Hs remain firmly bonded to the molecule whether in the presence of D_2O or not, and so their peaks do not change (Reproduced with permission of Glenn A. Facey, University of Ottawa NMR Facility BLOG, Canada, 2007.)

CAN YOU HAVE SEMI-HEAVY WATER?

Of course, but it's not called that. It's called partially deuterated water, with the formula DOH, or HOD, depending on your preference. In fact, you can replace the Hs in almost any organic molecule with Ds to get fully or partially deuterated versions of the molecule. An example is the solvent used in the NMR experiment in the figure which uses deuterated chloroform ($CDCl_3$). One interesting application that is developing is the concept of deuterated drugs. Molecules involving deuterium are generally more stable than the analogues with the lighter H atoms, so are harder to metabolize in the body. This might permit lower doses and fewer side effects.

Chapter

26

HEME

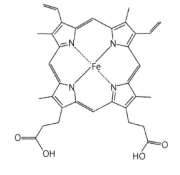

Heme

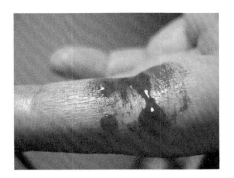

Is That the Same as Hemoglobin?

Not quite, but they're related. Heme is the small molecule that's responsible for the reversible binding of oxygen and CO_2 in the bloodstream. Globin is the name given to the large, globular-shaped protein that wraps around heme protecting it from attack from all sides, except through a small opening or crevice. The combination of globin + heme is called myoglobin, and four myoglobin units joined together form hemoglobin (sometimes spelled haemoglobin).

Why Do We Need Hemoglobin?

Most living organisms perform respiration, the breakdown of food products to release energy in the presence of oxygen (see p387). However, oxygen is not very soluble in water, and so in order for animals to convey enough oxygen from the lungs (or gills) to the muscles, they require an efficient oxygen-carrying molecule. In vertebrates, this molecule is hemoglobin.

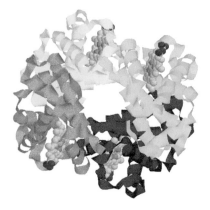

Low-resolution representation of hemoglobin, showing the four globular myoglobin units

Hemoglobin showing the positions of the four heme groups within each myoglobin unit, and with the protein backbone now being shown as a ribbon

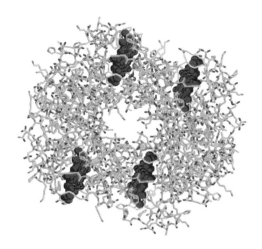

Ball-and-stick representation of hemoglobin, again with the heme groups highlighted, showing the huge complexity of the 3D structure

Hemoglobin (frequently abbreviated as Hb), makes up 97% of the dry content of red blood cells, and serves as the oxygen carrier in blood. The presence of hemoglobin within the red blood cells increases the oxygen carrying ability of a liter of blood from 5 to 250 ml. Each hemoglobin molecule can transport four oxygen molecules, one attached to the heme groups inside each myoglobin subunit.

WHAT ABOUT MYOGLOBIN?

Well, apart from being a component of hemoglobin, myoglobin (frequently abbreviated as Mb) is found in muscles. It serves as a reserve supply of oxygen and also facilitates the movement of O_2 within the muscle. The higher the concentration of myoglobin in the muscle cells, the longer an organism can hold its breath. Whales, seals and other diving animals have muscles with very high levels of myoglobin. Myoglobin (or rather the heme group it contains) is responsible for the red color of meat.

SO WHAT IS HEME?

Although the hemoglobin and myoglobin molecules are very large, complex proteins, the active site is actually a non-protein group called heme. The heme consists of a flat organic ring surrounding an iron atom. The organic part is a porphyrin ring based on porphin, and is the basis of a number of other important biological molecules, such as chlorophyll (see p81) and cytochrome.

Porphin   Heme

The ring contains a large number of conjugated double bonds, which allows the molecule to absorb light in the visible part of the spectrum. The iron atom, the attached protein chain, and any molecules bonded to the iron, modify the wavelength of the absorption and give heme, myoglobin and hemoglobin their characteristic color. Oxygenated hemoglobin found in blood from arteries is bright red, but without oxygen present (as in blood from veins), hemoglobin turns a darker red. Venous blood is often depicted (incorrectly) as blue in color in medical diagrams, probably because veins sometimes look blue when seen through the skin. (The phrase 'blue blooded' is often used to refer the aristocracy, and this is because unlike the peasants who were usually deeply suntanned from working in the field (or covered in dirt), the upper classes had very white, pasty skin, through which their blue veins were very prominent.) The appearance of venous blood as dark blue is caused by the reflection of blue light away from the outside of venous tissue if the vein is 0.05 cm deep or more. Take the blood out of the vein and it will initially be dark red, which will rapidly turn bright red on exposure to air.

Two blood drops, bright-red oxygenated blood on the left, and darker-red deoxygenated blood on the right

Hemoglobin-containing red blood cells flowing in an artery

So How Does Heme Bind Oxygen?

The key is the iron atom in the middle of the heme ring. This iron atom binds to the four nitrogen atoms in the center of the porphyrin ring, but this leaves two free bonding sites for the iron, one above and one below the plane of the heme ring. The heme group is located in an inert crevice in the myoglobin molecule, but with two reactive histidine groups above and below it.

One of the free bonding sites of iron is joined to the lower one of these histidines, leaving the final bonding site on the upper side of the ring available to bond with oxygen (or another gas molecule). The second histidine group hangs around nearby, and serves several purposes. It modifies the shape of the crevice so that only small

(Protein chain)

H   Histidine

N

N   H

H

O_2

Heme plane   Fe

H   N   H

N

(Protein chain)

Schematic diagram of the binding site in myoglobin

molecules can get in to react with the iron atom, and it also helps to make the reaction *reversible*, such that the oxygen can be released when required by nearby tissues. It is amazing to realize that the whole complex three-dimensional structure of the large myoglobin protein is designed purely to produce exactly the correct shaped crevice, with the correct two histidine groups in exactly the right positions to facilitate this reversible oxygen uptake.

WHAT ABOUT CO_2?

The waste product of respiration is carbon dioxide (see p61), and this has to be transported in the blood back to the lungs to be removed. Hemoglobin does this job too. When we breathe, oxygen in the lungs passes through the thin-walled blood vessels and into the red blood cells, where it binds to the hemoglobin, turning it into the bright red oxy-hemoglobin. The blood then passes around the body until it reaches cells and tissues which require oxygen to sustain their processes. These cells are rich in waste CO_2, which displaces the weakly bound O_2, bonding to hemoglobin to form dark-red carbamino-hemoglobin. This then travels in the veins back around to the lungs where the CO_2 is again displaced by fresh oxygen, and the CO_2 is exhaled.

Are O₂ and CO₂ the Only Molecules That Can Bind to Hemoglobin?

The crevice in the myoglobin is so small and so cleverly shaped that only small gas molecules can enter and reach the iron. Gases in the air, such as nitrogen and the noble gases (argon, xenon, *etc.*), can reach the active site, but are inert and so do not bond to the iron. Nitrous oxide (N_2O, laughing gas, see p371), also binds reversibly to hemoglobin, which transports it to the brain where it acts as a sedative. This allows N_2O to be used as a breathable painkiller or anesthetic, particularly by dentists, often mixed with air (in a 50:50 mixture called *Entenox*) to prevent the patient from asphyxiating.

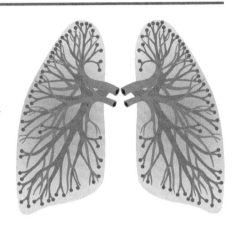

Lungs contain a branched network of blood vessels which get narrower and smaller, allowing O_2 to enter the hemoglobin and CO_2 to leave it

Certain other molecules, like carbon monoxide (CO), are small enough to fit into the protein crevice but form such strong bonds with the iron that the process is now *irreversible*. Thus, high concentrations of CO rapidly use up the body's limited supply of hemoglobin molecules, and prevent them from binding to oxygen. This is why CO is poisonous – the affected person rapidly dies of asphyxiation because their blood is no longer able to carry enough oxygen to keep the tissues and brain supplied. When hemoglobin combines with CO, it forms a very bright-red compound called carboxy-hemoglobin. Indeed, the characteristic signature of CO poisoning is that the victim displays bright-red lips. Hemoglobin's binding affinity for CO is 200 times greater than its affinity for oxygen, meaning that even small amounts of CO dramatically reduce hemoglobin's ability to transport oxygen. When inspired air contains CO levels as low as 0.02%, headache and nausea occur. If the CO concentration is increased to 0.1%, unconsciousness will follow. In heavy smokers, up to 20% of the oxygen active sites can be blocked by CO.

Another poisonous molecule that binds to hemoglobin is hydrogen cyanide (HCN). Once cyanide is taken into the bloodstream, the majority (92–99%) is found bound to hemoglobin in red blood cells. From there it is taken to the body's tissues where it binds to an enzyme called cytochrome oxidase and stops cells from being able to use oxygen. Other small molecules such as NO, H_2S and SO_2 also bind irreversibly to hemoglobin, and so are poisonous.

SO THE CENTRAL METAL ATOM IS THE KEY?

Yes, this is where all the action happens. Iron-based hemoglobin is used in almost all vertebrates, but other creatures use other metals. Most mollusks and some crabs have blood based upon a molecule called hemocyanin, which has copper at its core. Their blood is gray-white to pale yellow in color, but it turns dark blue when they bleed and it's exposed to the oxygen in the air. Some sea cucumbers or sea squirts contain a vanadium-based blood protein called hemovanadin, which turns a mustard-yellow color when exposed to air. Although plants don't have a bloodstream, they contain chlorophyll, which has a structure that is remarkably similar to that of heme, except with magnesium replacing the central iron (see p81). In plants, this molecule is used for photosynthesis – but it's amazing to see how the chemistry of almost identical molecules can be used by plants and animals for very different purposes.

Did Dracula Have a Hemoglobin Deficiency?

Vampires are, of course, fictional, and are based on the original novel by Bram Stoker. But vampire bats do exist in South America, and survive almost exclusively on a diet of blood, usually from cattle or livestock, not humans. Blood is a rich source of protein and iron, and even forms part of traditional dishes in many countries. In the United Kingdom, there is 'black pudding' which is fried pig blood mixed with oatmeal, and in Germany, there is *Blutwurst* (blood sausage) which is made from pig's blood mixed with fillers such as barley. The Masai tribe of Tanzania drink the blood  of cattle directly from the neck of the live animal. And in various oriental countries, snake blood is taken as part of traditional medicine and believed by the locals to cure a variety of ailments including farsightedness and hair loss, as well as increasing sexual performance. But a diet consisting solely of blood would be deficient in vitamins, essential fatty acids and minerals, leading to a variety of ailments. The absence of roughage would also cause chronic constipation. No wonder Dracula always looked so miserable!

Chapter

27

HEXENAL
(AND 'GREEN GRASS' SMELL)

WHAT'S THE CONNECTION WITH GRASS?

(Z)-3-Hexenal (or *cis*-3-hexenal) is the key aroma substance emitted when grass is cut and other types of vegetation are damaged. It has a very low odor threshold of 0.25 parts per billion.

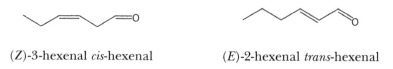

(Z)-3-hexenal *cis*-hexenal   (E)-2-hexenal *trans*-hexenal

WHAT'S AN ODOR THRESHOLD?

The odor threshold is the lowest concentration of a vapor in air which can be detected by smell.

CAN YOU BOTTLE IT?

Unfortunately, (Z)-3-hexenal is unstable and tends to rearrange to the more stable (E)-2-hexenal ('leaf aldehyde'), which has an odor threshold of 17 ppb. The related molecule (Z)-3-hexen-1-ol ('leaf alcohol') has an odor threshold of 70 ppb – it's used in perfumes.

HOW IS IT FORMED?

When grass is cut, the breakdown of fats and phospholipids leads to long-chained fatty acids such as α-linolenic acid and linoleic acid (see p287). These are oxidized by enzymes, and then the long carbon chain is chopped up by another enzyme, hydroperoxide lyase, usually into C_6 and C_{12} fragments. The initial C_6 product is (Z)-3-hexenal, but others are made too.

These volatile compounds, mainly C_6-aldehydes and C_6-alcohols, including (Z)-3-hexenal, (E)-2-hexenal and (Z)-3-hexenol, make

Synthesis of various hexenal isomers by the action of enzymes on linolenic acid in plants. I.F. is another enzyme, called the Isomerization Factor, which converts one isomer of hexenal into another

up the 'green odor' from cut leaves, and are collectively known as GLVs, 'Green Leaf Volatiles'.

WHAT ROLE DO THESE MOLECULES HAVE?

One may be in protecting the plant from bacteria, for example, allowing the cut ends to heal. (E)-2-Hexenal is one of a number of aldehydes found in coriander, widely used in cooking. Spices have long been associated with preserving food, and both (E)-2-hexenal and (E)-2-decenal have been found to be active against a range of bacteria, including *Salmonella choleraesuis* (the most common cause of septicemia). Alkenals from olives have similar antibacterial properties.

It also appears that emissions of molecules like (E)-3-hexenal from damaged plants prime other nearby plants to activate their defense mechanisms, such as the formation of protease inhibitors and release of molecules that attract pest predators. So they may act like warning signals, letting other plants know that danger is near.

WHY DO THESE SMELLS HAVE SUCH AN IMPACT?

Marcel Proust *ca.*
1900

There is no doubt that smell taps into our memories and into subconscious emotions. The classic example linking smell with memory occurs in the masterpiece of the French novelist Marcel Proust (1871–1922, left), '*A la Recherche du Temps Perdu*' ('*In Search of Lost Time*'). Very early on, in the first book ('Swann's Way'), the protagonist Charles Swann found that the smell from a piece of a small madeleine cake soaked in tea acted as the trigger for a whole raft of memories from his childhood. On Sunday mornings at Illiers, near Chartres (Combray in the novel), Proust would go to the bedroom of his aunt Elizabeth (Léonie, in the novel) to say good morning before Mass, and she would give him a small piece of madeleine that had been dipped in her cup of lime-blossom tea.

Players of the traditional English summer game, cricket, are familiar with the smell of cut grass. It is very potent. Alan Dixon played first-class cricket for Kent. He was an all-rounder, both a middle-order batsman and a bowler at medium pace and of off-spinners. He decided to retire in the autumn of 1957 at the age of 23 as he was not sure of a place in the First XI. He went off to become a traveling salesman. He later wrote:

> '*But it came round to the spring [1958], and I stopped the car by a cricket field where they were mowing the grass. And the smell of that new-mown grass meant so much. I got out of my car and rang Les Ames [the Secretary of Kent County Cricket Club]. "Leslie," I said, "it looks as if I've made a mistake".*'

Cricket is usually played on newly mown grass pitches

Dixon played for another 13 seasons and did not retire until 1970, when he was a member of the Kent team that won the County Championship.

How Does It Work?

We are hot-wired for memory. Olfaction is closely linked with the limbic system, of which the hippocampus and amygdala are a part, and which is responsible for emotions and memory. The sense of smell has a direct link to the cerebral cortex in the brain; messages involving other senses like touch or taste have a more circuitous route. How the message gets through is another matter, with the traditional 'lock and key' theory and the 'vibrational' theory linking molecules and smell both currently in vogue.

And, of Course, People Sing Songs about Grass …

Tom Jones for one, in 'Green Green Grass of Home' and The Inkspots, with 'Whispering Grass'.

Tom Jones

WHERE ELSE DO THESE MOLECULES OCCUR?

No single molecule has a 'fresh tomato' smell, but (Z)-3-hexenal makes the greatest molecular contribution to the tomato volatiles. (E)-2-hexenal and hexyl acetate are chief among molecules responsible for the freshness of apple juice flavor, and thus hexenals can be used as additives to make flavors have a greener 'edge' to them.

Insects use them too.

Hexenal is the unlikely connection between fresh tomatoes ...

... apple juice ...

... and bed bugs

REALLY?

Bedbugs use a mixture of (*E*)-2-hexenal and (*E*)-2-octanal to deter unwanted advances during mating! Stink bugs, important agricultural predators, use (*E*)-2-hexenal as an attractant and pheromone, and it is thus a potential weapon to protect crops.

Some ants use (*E*)-2-hexenal as an alarm pheromone, while the Japanese scarab beetle, *Phyllopertha diversa*, has olfactory receptor neurons that are specific for green-leaf volatiles, namely, (*E*)-2-hexenal, (*Z*)-3-hexenyl acetate and (*Z*)-3-hexenol.

Green leaf volatiles, including (*E*)-2-hexenal, disrupt responses by the spruce beetle, *Dendroctonus rufipennis*, and the western pine beetle, *Dendroctonus brevicomis*, to attractant-baited traps.

AND GREEN GRASS SMELL REALLY DOES SEEM TO BE GOOD FOR YOU …

Behavioral studies in monkeys involving positron emission tomography (PET) have shown that the 'green odor' mixture of (*E*)-2-hexenal and (*Z*)-3-hexen-1-al has a healing effect on the psychological damage caused by stress. The mixture activates regional blood flow in the parts of the brain called the primary olfactory cortex (like other smelly molecules) and also the anterior cingulated gyrus, the latter being linked with the 'healing effect'. It has similarly been found that a mixture of (*Z*)-3-hexenol

and (E)-2-hexenal reduces stress in rats. 'Green odor' has likewise been found to reduce pain sensation in humans. So next time you mow the lawn, do your body a favor and inhale those lovely healing hexanals.

Chapter

28

HYDROGEN PEROXIDE

Marilyn Monroe in a
scene from the movie
Gentlemen Prefer Blondes

WHO WAS THE FIRST FAMOUS PEROXIDE BLONDE?

This was Jean Harlow (1911–1937). A natural ash-blonde, she became
a 'platinum blonde' through regular use of hydrogen peroxide,
ammonia and other chemicals, including Clorox bleach (see p443).
Her film *Platinum Blonde* (1931) popularized the expression. She was
followed by Marilyn Monroe (1926–1963), who starred in *Gentlemen
Prefer Blondes* (1953) and others, up to Reese Witherspoon of *Legally
Blonde* (2001). Not forgetting Debbie Harry of *Blondie*.

Jean Harlow – the
original peroxide
blonde

Debbie Harry of *Blondie*
(© Press Association
Images. With permission.)

HOW DOES PEROXIDE BLEACH HAIR?

The hydrogen peroxide oxidizes the melanin pigment inside the
hairs to a colorless substance. Using very slightly alkaline hydrogen
peroxide solution softens the cuticle and makes the hair more
permeable to the H_2O_2, so the peroxide can reach the melanin.

HYDROGEN PEROXIDE'S UNSTABLE, ISN'T IT?

It depends on what you mean by 'unstable'. It is stable with respect
to the elements, hydrogen and oxygen, of which it's composed:

$$H_2(g) + O_2(g) \rightarrow H_2O_2(l)$$

However, its decomposition products, water and (di)oxygen gas are even more stable, so its decomposition is exothermic.

$$H_2O_2(l) \rightarrow H_2O(l) + \tfrac{1}{2}O_2(g)$$

As a result, concentrated H_2O_2 will gradually turn into water, releasing oxygen gas. To slow down this reaction, it's usually kept in a refrigerator at 4°C. Since some glasses can speed its decomposition, it is often kept in bottles made of Teflon or polythene. Stabilizers are usually added to hold back decomposition.

However, many substances catalyze its decomposition. The best catalyst is the enzyme *catalase* in the human liver which does the important job of removing toxic H_2O_2 (a by-product of human metabolism) from the body. Even dust can act as a catalyst. The catalysts work by reducing the activation energy for its decomposition.

So What Does Hydrogen Peroxide Look Like?

When it's pure, hydrogen peroxide is an almost colorless (very pale blue) substance that resembles water (and mixes with it in all proportions), although it has a slightly lower freezing point (−0.41°C) and higher boiling point (150.2°C) than water. Unlike ice which floats on liquid water, solid H_2O_2 sinks when placed onto its liquid form.

In terms of its structure, it is skewed with a dihedral angle of 111.5° (gas phase), which is a balance between maximizing bonding interactions and minimizing repulsion between the electron lone pairs and the O–H bond pairs.

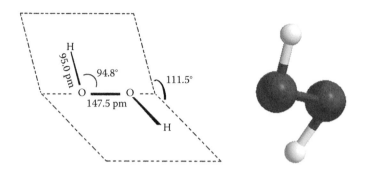

AND IT'S A BLEACH?

Yes, it is an excellent oxidizing agent, and can be thought of as 'water with added O'. The oxygen can attack the double bonds in dyes and pigments, reducing them and removing the color. Around 2 million tonnes of hydrogen peroxide are made each year. About half of that is used to bleach wood pulp or paper. It is seen as an environmentally friendly alternative to chlorine-based bleaches. In domestic uses, it is used as a mixture with sodium hydroxide to bleach wooden surfaces. It is also used to bleach textiles.

If you get it on your skin, the skin is whitened. Medically, hydrogen peroxide is used to disinfect skin, as it can kill bacteria by oxidizing them to death. When it comes into contact with a cut, it fizzes as oxygen is released, probably due to the action of *catalase* in the blood. Dilute H_2O_2 solution (around 3%) is used as a mouthwash, and also for cleaning and disinfecting contact lenses. It also bleaches teeth. In the aerospace industry, it is used to disinfect satellites and space probes, both before they go into space (otherwise the various spacecrafts that are searching for signs of life on other planets, such as the *Curiosity* rover on Mars, might detect a false reading from hitchhiking Earth bacteria), and when they return to Earth in case they've picked up any unwanted 'space-bugs' (we've all seen enough science fiction films to know what these might do!).

A bombardier beetle spraying its enemy (in this case an irritating pair of tweezers) (From T. Eisner and D.J. Aneshansley, *Proc. Natl. Acad. Sci. USA.* 1999, 96, 9705. © (1999) National Academy of Sciences, USA. With permission.)

Controversially, it has been suggested that the use of hydrogen peroxide solution – for example, in very low concentrations intravenously – can be used as a therapy for cancer ('oxygen therapy'). However, the American Cancer Society says that there is nothing to support this, and that it may well be harmful.

AND THAT'S IT?

Definitely not. Hydrogen peroxide has found all sorts of uses, often in weapons. Starting with the bombardier beetle …

WHAT'S THAT?

When assaulted, bombardier beetles spray a hot mixture of irritating *p*-benzoquinones and hydrogen peroxide from the end of their abdomen. They can aim this spray accurately in any direction. Usually in the wild they are aiming at predators like ants.

HOW DO THEY DO THAT?

The bombardier beetle stores a mixture of hydrogen peroxide and hydroquinones in a sac in its body. At the moment, the defense is needed, the mixture is pumped into a reaction chamber, where it is mixed with enzyme catalysts (*catalases* and *peroxidases*) which speed up the reaction, decomposing the peroxide to oxygen. This oxidizes the hydroquinones to quinones, reactions which release a lot of heat. The oxygen gas then propels the hot spray out of the jet on the abdomen.

SO HOW ABOUT MAN-MADE WEAPONS?

A Me163 fueled up and ready for take off ... or explosion, pilots could never be quite sure which would happen first ...

Well, first there was the Messerschmitt 163 rocket aircraft ('Komet'). This wasn't a conventional jet aircraft, it was actually the only rocket aircraft to ever fly in operational service in World War II. It was fitted with a rocket motor that made it far faster than any other fighter in the air, but the fuel used made it somewhat 'unsafe'.

The two fuel components were called '*C Stoff*' (57% methanol, 30% hydrazine, 13% water) and an oxidizer, '*T Stoff*' (80% H_2O_2, 20% water), from the German '*Stoff*' meaning substance, stuff, or fuel. They were kept in separate tankers at opposite ends of the airfield and were always at least half a mile apart. It is said that any organic matter (including humans) could spontaneously combust in contact with the *T Stoff*. The fuels were loaded separately, with the aircraft and crew washed down carefully after the first fueling, and again after the second one. A catalyst $(Ca(MnO_4)_2 + K_2CrO_4)$ was mixed with some peroxide, generating a mixture of steam and oxygen that drove the pumps feeding the two fuels to the combustion chamber. The explosive reaction produced a mixture of hot gases, steam and nitrogen, which would exhaust out the back of the plane, propelling it forward:

$$2H_2O_2(l) + N_2H_4(l) \rightarrow 4H_2O(g) + N_2(g)$$

The plane would be doing 500 mph by the edge of the airfield, climb vertically at 440 mph, then approach American bombers at over 550 mph. No Allied fighter could keep pace with it, but the Me163's only shot down nine bombers, and lost far more of their own in accidents. If a Me163 crash landed and somehow did not explode (they usually did), fuel lines could fracture, and the pilot would be dissolved alive by the *T Stoff*!

Amazing! And There's More?

Hellmuth Walter, who designed the Me163, also designed a peroxide-powered U-boat (type XVIIB). It ran on very concentrated (95%) H_2O_2 ('*Perhydrol*') which, on decomposition (using a catalyst), gave out superheated steam to drive a turbine. The oxygen formed was used for life support for the crew. These U-boats were fast underwater (25 knots) but there were too many technical problems for them to be successful (such as using unstable 95% peroxide in an enclosed environment).

A Soviet 'Oscar'-class submarine, similar to the *Kursk*

In the past, hydrogen peroxide was used, sometimes fatally, as a torpedo fuel. On August 12th 2000, the Russian nuclear attack-submarine *Kursk* was taking part in a naval exercise off the Kola Peninsula near Murmansk. This was a huge vessel, 155 m long, twice the length of a Boeing 747. At 11:28 local time, there was an explosion on board, equivalent to about 100 kg of TNT (see p535), which seems to have been caused by a leak of hydrogen peroxide fuel from a torpedo. This may have caused a fire, which detonated several other torpedoes, leading to a much bigger explosion 135 s later, reportedly with a force of around 7 tonnes of TNT which was recorded by seismic stations 5000 km away, and measured 3.5 on the Richter scale! The submarine sank in 108 m of water and the entire crew of 118 men was lost.

Don't Terrorists Use It Too?

Yes, unfortunately. Hydrogen peroxide was the main ingredient in the bombings that killed 52 passengers on the London Underground and bus network on July 7th 2005. The actual explosive was tricyclic acetone peroxide (TCAP, a.k.a. triacetone triperoxide, TATP), which is made by simply mixing cold hydrogen peroxide, acetone and sulfuric acid. Because of the low cost and the ease with which these common ingredients can be obtained, TCAP has recently become the terrorists' weapon of choice. This is despite the fact that making it without using proper laboratory equipment and procedures is extremely dangerous. TCAP is notoriously sensitive to shock, heat and friction, and can often explode for no apparent reason. It's no wonder that some of terrorist groups that

The London bus that was destroyed by a terrorist TCAP bomb in 2005 (© Press Association Images. With permission.)

use it have nicknamed it 'Mother of Satan' due to the number of casualties it caused among the would-be bomb-makers.

Tricyclic acetone peroxide (TCAP)

Hexamethylene triperoxide (HMTD)

Another peroxide-based explosive is hexamethylene triperoxide (HMTD), which, again, can be made simply and cheaply from common liquids (hydrogen peroxide, hexamine and citric acid). It, too, is extremely sensitive to shocks, heat and friction, and can detonate when put into contact with most common metals, which makes it very dangerous to make under non-lab conditions. HMTD (along with TCAP) were implicated in the 2006 transatlantic aircraft bomb plot, in which terrorists aimed to blow up 10 airliners traveling between the United Kingdom and the United States/Canada. Luckily, the plot was foiled before it could take place, but these liquid explosives are the reason that passengers on airplane flights now have to declare all their liquids and carry them in a transparent bag for inspection by security.

At the inquests for the London bombings at the Royal Courts of Justice in 2011, Lady Justice Hallett made the point: '*So you get cross-examined by the chemist [pharmacist] if you want to buy too many aspirin, but you can buy as much hydrogen peroxide as you like at the market?*' Most governments have since acted on this problem, and now keep a watchful eye on sales of hydrogen peroxide from pharmacies and chemical suppliers.

Chapter

29

INSULIN

A monomer of insulin

THAT'S GOT SOMETHING TO DO WITH BLOOD SUGAR LEVELS AND DIABETES, RIGHT?

That's correct. When you eat food, it's broken down during digestion to produce smaller molecules and sugars, especially glucose (see p193). The glucose dissolves in the bloodstream and travels around the body waiting for the call to action. But if the glucose isn't used up, its levels can build up to a point at which it becomes toxic.

SO HOW DOES THE BODY DETECT THIS?

Cells in the pancreas sense the increasing concentration of glucose in the blood, and send signals to other specialized cells, also in the pancreas, called the 'islets of Langerhans'.

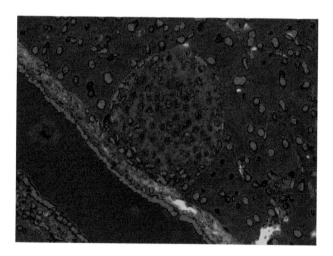

A spherical islet of Langerhans from the pancreas of a mouse, situated next to a blood vessel. It has been stained to visualize the insulin, which is found in the central spherical region (Taken from Paul Langerhans Institute, Dresden)

SOUNDS LIKE A PLACE OUT OF *THE LORD OF THE RINGS* ...

They are named after the German anatomist Paul Langerhans who discovered these cells in 1869 while studying sections of a pancreas through a microscope. He didn't know the function of 'these little heaps of cells', but it was suggested by later workers that they might be involved in digestion. To prove this, experiments were performed whereby the pancreas was removed from dogs to see what effect

Paul Langerhans

it would have on their digestion. All the dogs developed the symptoms of diabetes within a few days of the operation. In one experiment in 1889, two medics, Oscar Minkowski and Joseph von Mering noticed a swarm of flies feeding on the urine of a dog that had had its pancreas removed a few days before. The urine was tested and found to contain a lot of sugar, the prime diagnostic for diabetes. Indeed, the name *diabetes mellitus* means 'flowing through of honey' or more literally 'sweet urine'. Further experiments narrowed down the link between diabetes and the pancreas to the islets of Langerhans, and the search was on to find what chemical they secreted.

AND THAT WAS INSULIN?

Yes, but it took teams of scientists nearly 30 years to isolate and identify it. The major breakthrough came at the University of Toronto in 1920, when a team led by Canadian Frederick Banting and his assistant Charles Best, under the direction of J.J.R. Mcleod, successfully isolated the secretion, which they initially called *isletin*, but was later renamed *insulin* (from the Latin for island: *insula*). They showed that diabetic dogs (created by removing their pancreas) could be kept alive almost indefinitely by simply injecting them with isletin.

DID IT WORK ON HUMANS?

Not at first. The extracts they had at that time were so impure that they often failed to work, or caused severe allergic reactions. Eventually, aided by biochemist James Collip, they found a method to purify it sufficiently and began to have very promising results with human diabetic patients. At a local hospital in 1922, there was a ward of about 50 comatose children, all dying from ketosis brought on by advanced diabetes. Banting, Best and Collip went from bed to bed, injecting each child with their newly purified insulin, until they'd done the entire ward. In a dramatic moment almost worthy of Hollywood, even before they'd reached the last dying child, the first few children were waking up from their comas, while their parents and relatives shouted in joy. Although insulin hadn't cured the diabetes, it prevented it from being fatal and relieved all the symptoms, so long as it was taken regularly. The drug company Eli Lilly then took up the challenge, and worked out how to make large quantities of it, and improved the purity, so that it became widely available.

Charles Best and Frederick Banting *ca.* 1924

J.J.R. Mcleod *ca.* 1928

James Collip *ca.* 1930

What Happened to the Researchers?

This was a bit controversial. Banting was awarded the 1923 Nobel Prize in Medicine, jointly with the lab director McLeod, but the assistants who had done much of the work, Best and Collip, were awarded nothing. Banting was furious that Best hadn't been acknowledged, and in a very generous gesture shared his prize with Best. Not to be outdone, Mcleod shared his part of the prize with Collip. In fact, insulin has been responsible for several more Nobel Prizes. The 1958 Nobel Prize for Chemistry was awarded to Frederick Sanger, who determined the primary structure of insulin, the first protein to have its structure determined. In 1964, Dorothy Hodgkin was awarded the Nobel Prize in Chemistry for developing the technique of X-ray crystallography, and 5 years later insulin was one of the first proteins to have its large-scale three-dimensional structure to be determined by her this way. And in 1977, Rosalyn Sussman Yalow received the Nobel Prize for Medicine for developing a radioimmunoassay that could detect and measure insulin in a blood sample.

But How Exactly Does Insulin Work?

Once it's released into the bloodstream, the insulin molecules begin their job of seeking out cells that require energy, and bind to receptors on the surface. This 'opens the door', allowing glucose molecules to be ferried (by a special 'taxi' molecule called Glut-4) across the cell membrane into the cell. There, they react with oxygen also delivered by the blood (see p227), to produce CO_2 as a waste product which diffuses back out into the bloodstream. The energy released in this reaction is used temporarily to make ATP (see p1), which is then used later by the cell as a local energy source. Alternatively, if cells don't require energy immediately, then insulin goes into plan B – it triggers enzymes which stimulate the liver to convert glucose into glycogen for storage (see p193) and (adipose) fat cells to absorb fatty acids from the blood and turn them into triglycerides to store as body fat (see p271).

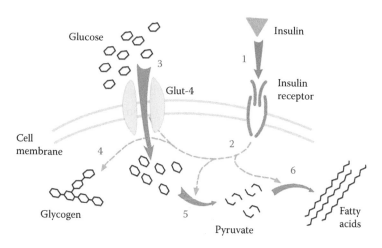

The mechanism of action of insulin. (1) Insulin binds to its receptor on the cell membrane. (2) This triggers a host of activity within the cell. (3) The Glut-4 protein activates and ferries glucose from the blood across the cell membrane into the cell interior. Once inside, the glucose is either (4) converted into glycogen for storage or (5) metabolized into smaller pyruvate molecules releasing energy for use by the cell. The pyruvates are either broken down further in the presence of oxygen to make CO_2, or they are converted into fatty acids for storage as body fat

So Insulin Acts a Bit Like a Doorman or Butler Ushering Guests into a Private Dinner Party?

I suppose so. If the guests (glucose) arrive and the butler (insulin) is not ready to greet them, they can't enter the house (cell). But normally, the butler is alerted to the arrival of the guests by the doorbell ringing (pancreas sensors), and so goes to open the door (Glut-4) to let them in. But only if they are the *right* guests, and only if the party is not already full.

Ok, Enough of the Analogies. What Actually Is Diabetes?

Diabetes is a condition whereby the body has too much sugar in the bloodstream. This can be a genetic illness (Type 1 diabetes) whereby the body fails to produce insulin, most probably because the islets of Langerhans don't work properly. Or it can appear later in life when body cells fail to properly utilize the insulin that's present (Type 2 diabetes). Type 2, which is also known as insulin resistance, can result from a variety of causes, such as poor exercise, a diet rich in saturated fats, and of course, consumption of too much sugar, especially in the form of sugary drinks (see p467). As well as causing a whole bunch of unpleasant symptoms, such as frequent urination, blurred vision and itchiness, diabetes can lead to far more severe conditions such as blindness, heart disease, stroke, lower-limb amputation, kidney failure, and is estimated to knock 10 years off your life expectancy.

Wow! What Can Be Done about It?

Diabetes can be controlled with regular injections of insulin and by diets in which the sugar intake is carefully controlled. Diabetics often need to take blood samples several times a day to check their blood sugar levels. This involves pricking a finger and putting a drop of

Diabetics need to test a sample of their blood several times a day to check their blood sugar levels

blood onto a detector. This is obviously very inconvenient, not to mention painful, so there are hopes that in future diabetics will be fitted with an embedded sensor which will continuously read their blood glucose level and send this wirelessly to an external receiver.

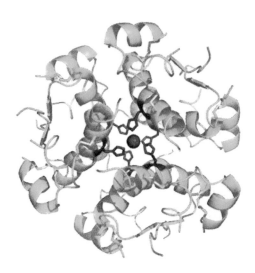

Simulation of the structure of the insulin hexamer consisting of
six insulin monomers bound together. The amino acid chains are
depicted as ribbons, except for the six histidine residues which
point toward the center of the molecule. At the very center, bonded
to the histidines and locking the whole structure together, is a
spherical zinc ion

For many years, it was difficult to get hold of enough human
insulin to treat all the diabetics in the world. But animals also
produce versions of insulin as well, and these can often be used
instead. Insulin from cows differs from human insulin by only
three amino acid groups, while that from pigs differs by only one
amino acid group. Even some species of fish possess insulin that
is sufficiently similar to that from humans that it can be clinically
effective. Indeed, until recently, nearly all the medical insulin used
by diabetics was actually obtained from pigs. Nowadays, nearly all
medical insulin is made commercially using genetically engineered
yeast or *Escherichia coli* bacteria whose DNA has been modified to
trick them into producing human insulin.

WHAT DOES INSULIN LOOK LIKE?

Well, it's a protein, so it has a rather large complicated structure involving many thousands of atoms. It is composed of two chains made up of amino acids (51 in total), linked together by disulfide bonds. These chains then contort and wrap around themselves to form a complex three-dimensional shape. This is a monomer of insulin, but in this form it's too reactive to be stored in the body. Instead, six of these monomers join together to form a hexamer, which is much less reactive. The hexamer can be stored for long periods, but readily broken apart to release the reactive monomers when required. The hexamer is locked together by a zinc ion, which sits at the center of four histidine residues, in a similar way to the iron atom that sits at the center of the heme molecule (see p227) or the magnesium at the center of chlorophyll (see p81).

I BET THE SCIENTISTS WHO DISCOVERED INSULIN MADE A FORTUNE, DIDN'T THEY?

Well, maybe, but only indirectly, and later on. But, whether through bad planning or sheer generosity, they sold the patent for insulin in 1922 to the University of Toronto for … one half dollar!

Chapter

30

KISSPEPTIN

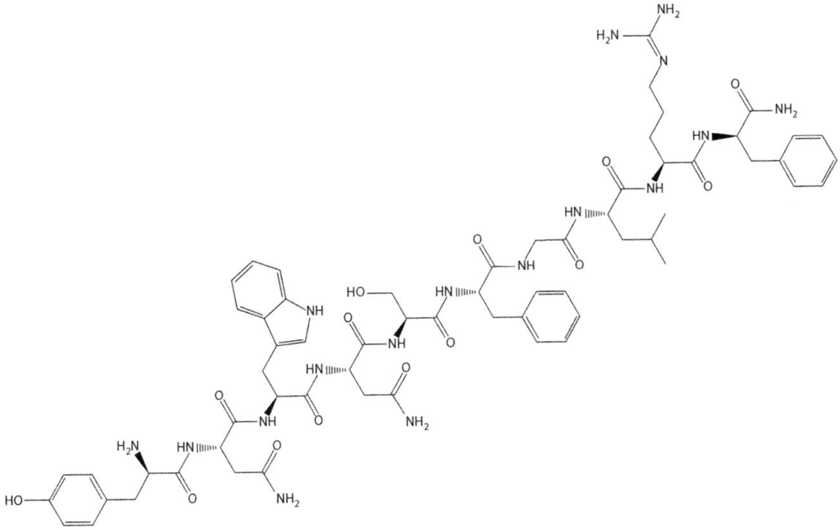

Kisspeptin-10

Surely Nothing to Do with Gene Simmons and Co.?

The rock band KISS in concert
in 2004

No, not the rock band Kiss, although it has similarities that it also sends teenagers wild. Kisspeptin is a protein that, when released from the brain, triggers a cascade of biochemical changes that leads to puberty, turning children into adults. Puberty is still one of the great unsolved mysteries in biology. Scientists have known for years that puberty begins when a child's brain releases hormones that cascade through the body, leading to the maturing of testes in boys and ovaries in girls. However, the exact nature of the hormones wasn't fully understood until a few years ago, when kisspeptin was discovered. Adolescents who lack a functioning kisspeptin system fail to achieve puberty. We now know that, almost literally, 'puberty begins with a kiss'!

So, It Started with a Kiss ...?

Kisspeptins are actually a group of proteins with similar structure, and are products of the *KiSS-1* gene. This gene was discovered in 1996 when scientists were searching for a gene that prevents cancer from spreading. This groundbreaking work took place in research labs in the town of Hershey, Pennsylvania, the location also famous for the Hershey's chocolate factory. Since Hershey's most popular product is a chocolate confectionary called a Hershey's *Kiss*, the discoverers of the gene named it *KiSS-1* after this chocolate. However, the nomenclature also has a scientific grounding as the inclusion of SS in the name also indicates that the gene is a suppressor sequence, *i.e.* it prevents some biological process from occurring (such as cancer spreading). Kisspeptin was first given the name metastin, because it was thought to play a role in tumor metastasis, but later named kisspeptin after its gene precursor. The receptor for kisspeptin in the brain was identified only recently and was given the rather less whimsical name GPR54 (G-protein coupled receptor 54).

The *KiSS-1* gene is found on the long arm of chromosome 1. The protein it makes is a peptide containing 145 amino acids, which are then cleaved into smaller, 54-amino-acid chunks. These may also be further truncated down to 14-, 13- or even 10-amino-acids fragments with carboxylic acid terminations. These N-terminally truncated peptides are known as the family of kisspeptins. Kisspeptin-10 is shown in the figure on p265.

... NEVER THOUGHT IT WOULD COME TO THIS

A single injection of kisspeptin in animals has been found to stimulate a huge increase in the secretion of gonadotropins (hormones that stimulate the production of testosterone (see p489) in men and estradiol (p185) in women). Repeated injections into immature rats have been shown to advance the age of puberty.

Very recently, a group of Japanese scientists delivered an antibody directed against kisspeptin into the brain of female rats. This stopped the rats' reproductive cycle, demonstrating that inhibiting the effect of kisspeptin, even after puberty, still blocks the reproductive function. Therefore, as well as being vital to initiate puberty, kisspeptin is necessary for reproductive function to continue later in life. Later, Spanish scientists discovered that administering kisspeptin in food-restricted rats still stimulated the release of gonadotropins. This is a remarkable discovery, since normally when a mammal is facing starvation, its reproductive system becomes dormant to conserve the body's stored food supply. Kisspeptin injections over-ride this defense mechanism, and can kick-start normally dormant reproductive systems.

(sigh) ...and I was looking forward to a nice quiet winter...

It has also been discovered that kisspeptin plays a role in seasonal-breeding animals. Many mammals only become fertile during the annual breeding season, and this is controlled by the duration of daylight per day. During short winter days, these animals have a reduced amount of kisspeptin in certain parts of the brain, and sexual activity is switched off (or greatly reduced). But in the long summer days, kisspeptin amounts increase, stimulating the animals to breed. Artificially injecting animals (such as ewes) with kisspeptin during the winter months causes ovulation even in the non-breeding season. This has implications for farmers, who may soon be able to choose when their livestock reproduce rather than be restricted by the natural annual cycle.

THE KISS OF LIFE?

Because of its link with the *KiSS-1* gene, kisspeptin has been suggested as a possible treatment for some forms of cancer. In particular, when breast and prostate cancers develop, they are often nurtured by the sex hormones oestrogen and testosterone (see p489 and p185). If the production of these hormones could be switched off, the tumors should shrivel and die. One way to do this might be to block the kisspeptin receptor in the brain. Scientists are now actively looking for molecules that can block this receptor.

Kisspeptin's role in switching on sexual hormones could be vital in controlling the timing of puberty when this goes wrong. There are a number of conditions where children reach puberty much too early (five or six years old), and these could possibly be treated by simply taking pills containing kisspeptin blockers. Conversely, for young people in whom puberty does not begin as normal in their teenage years, it may be possible to administer kisspeptin to kick-start the process.

Early-onset or late-onset puberty may be treated using kisspeptin

It has recently been found that kisspeptin can restart the female reproductive system in women who have stopped ovulating due to an imbalance in their sex hormones. Kisspeptin is therefore a potential basis for a new infertility treatment – the kiss of life indeed!

Chapter

31

LAURIC ACID

(THE FATTY CONSTITUENT OF COCONUT OIL)

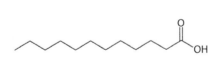

THE FLESH OF A COCONUT IS MAINLY FAT?

Yes, in fact around 50% of the content of coconut flesh is lauric acid (or dodecanoic acid), a fatty acid.

Lauric acid

DOES THAT MEAN THAT EATING COCONUTS IS BAD FOR YOUR HEART?

Yes, and no. Fatty acids are known to increase cholesterol in the bloodstream, which can lead to blocking of the blood vessels in the heart, producing heart attacks. And lauric acid is also known to produce the largest increase in total cholesterol of all the fatty acids. But this isn't as bad as it seems, because most of this increase is due to an increase in the so-called 'good' cholesterol. Medics have classified cholesterol into two types: high-density lipoprotein (HDL) and low-density lipoprotein (LDL), depending on the density of the lipoproteins (see p91). LDLs are very bad for you, as they greatly increase the chance of getting arterial blockages, whereas HDLs are not such a problem. As a result, medics believe that lauric acid lowers the ratio of total cholesterol to HDL cholesterol more than any other fatty acid, either saturated or unsaturated. So eating coconuts may actually decrease the risk of getting heart disease, although that has yet to be properly proven.

IS IT FOUND ANYWHERE ELSE?

It comprises 6.2% of the total fat in human breast milk, 2.9% of the total fat in cow's milk and 3.1% of total fat in goat's milk.

Does It Have Any Other Uses?

Lauric acid is also known to the pharmaceutical industry for its excellent antimicrobial properties, and the monoglyceride derivative of lauric acid, monolaurin, is known to have even more potent antimicrobial properties against lipid-coated RNA and DNA viruses, numerous pathogenic gram-positive bacteria, and various pathogenic protozoa.

What about Coconut Oil?

As well as foodstuffs, coconut oil can be used as a feedstock for biodiesel fuel. Various tropical island countries are using coconut oil as a fuel source to power trucks, automobiles, and buses, and for electrical generators. Before electrical lighting became commonplace, coconut oil was used for lighting in India, and was exported under the name *cochin oil*. Coconut oil can be used as a skin moisturizer, as an engine lubricant and a transformer oil, and acids derived from coconut oil can be used as a herbicide. Coconut oil is used by movie theater chains to pop popcorn!

What Happens When You Eat Lauric Acid?

Much of it gets converted into monolaurin (or glyceryl laurate), which may help protect against bacterial infection. Because of this, monolaurin is commonly used in deodorants. Finally, monolaurin gets converted into the HDL version of cholesterol.

Monolaurin, the monoglyceride formed by joining one lauric acid
molecule to glycerol (see p207)

But Its Main Use Is in Soaps and Detergents, Right?

Yes, because it is relatively cheap, has a long shelf-life, is non-toxic and safe to handle, it is often used for the production of soaps and cosmetics. The lauric acid is extracted, then neutralized with NaOH to give sodium laurate, which is a soap. This can be further processed to give sodium lauryl sulfate (SLS), a detergent.

SLS (or SDS, if you prefer its other common name sodium dodecyl sulfate) is the main component of most soap-based products, and if you were to look in your bathroom, you're guaranteed to find at least one product containing it, for example, your shampoo or toothpaste. In industry, it is found in products such as engine degreasers or carpet cleaners. SLS is inexpensive and an excellent foaming agent. It has a high pH as it is an alkali substance and has the appearance of a white powder.

How Does a Detergent Work?

SLS is a surfactant, which means a molecule that is active at the surface of a liquid. The molecule has two parts, an ionic water-soluble head group and a long carbon chain, or tail group, which is insoluble in water. In the case of SLS, the sulfate head group has a negative charge, so SDS is called an anionic surfactant.

Sodium lauryl sulfate (SLS)

Hydrophobic tail   Hydrophilic head

Na⁺

Dirt that is hydrophilic (or soluble in water) is easily removed by rinsing with water, but dirt that is hydrophobic (insoluble in water), *e.g.* oils, fats and grease, are much harder to remove from their substrate (the surface that they are coating). Surfactants have many roles in detergency. First, they promote wetting of the substrate. At the substrate, a liquid droplet will form to have the least surface tension possible and thus the least surface area in contact with the substrate. Surfactants enable the surface area of contact between the droplet and the substrate to be larger, and thus increase wetting of the substrate.

How a detergent works. (a) The long-chained tails of the detergent molecules do not like to be surrounded by polar water molecules and so are attracted to the oil/fatty deposit on the substrate. (b) Agitation causes the hydrophobic tails to wrap around and solvate the fatty deposits. They eventually pull the deposits free of the surface and encapsulate them in a sphere of surfactant molecules called a micelle, where they can be rinsed away. At the same time, the substrate is covered with surfactant molecule to prevent re-deposition.

Second, the hydrophobic tails of the SLS will dissolve fatty acids or grease deposits that are attached to the surface of a solid and hold them in suspension, moving them into the liquid bulk phase and thus making it easier to wash them away. This process is known as emulsification, because a dispersion of a chemically different liquid within another liquid, such as water with oil particles in its bulk phase, is known as an emulsion. Once the dirt is surrounded by the surfactant, it will also be prevented from re-depositing upon the substrate. The surfactant will also coat the substrate, as mentioned before, and this will also prevent re-deposition of dirt.

Are All Detergents Made from SLS?

No, but it is one of the most common ones. The salt of any long-chained fatty acid would act as a detergent. The one used in most household soaps is the sodium salt of stearic acid, which has the same structure as lauric acid except with a chain of 18 carbons rather than 12.

Is It the SLS Which Makes Detergents Foam?

SLS makes toothpaste and bubble-bath foam

Yes. SLS is an excellent foaming agent, and this is one of the reasons it's included in many personal care products, such as toothpaste and washing-up liquid. However, its ability to foam has a very little effect on the functional performance of the product. Foaming properties are actually added due to consumer demand. The amount and quality of foam produced is associated by consumers as an indicator as to whether the product is working. This myth is propagated by advertising companies as it is a visual, tangible feature of SLS, and it would be hard to show the cleaning process otherwise.

For foam to exist, there must be a substituent with the bulk of the liquid to lower surface tension. If one were to shake a bottle of water, air bubbles would be trapped briefly but they would be short-lived due to the high surface tension and instability of the bubble. Hence, on the addition of the surfactant SLS, the surface tension of the bubble is lowered and thus has more stability and a longer life-span.

SLS's foaming properties do have a use in dental care besides consumer satisfaction, although its performance does not rely upon it heavily. The foaming action allows the polishing agent in toothpaste to be suspended and the detergency properties to reach otherwise inaccessible areas and cavities in the mouth. SLS also shows antimicrobial effects on bacterial flora in the mouth and hence is the most commonly chosen surfactant for toothpastes.

Is It Safe?

You'll often see warnings on the back of many beauty products, such as shampoo, stating 'if this gets in contact with your eyes wash thoroughly with water, if irritancy persists, see a doctor'. This is because SLS in high doses can be toxic, especially if swallowed, and is an irritant to skin, eyes, and the respiratory system if inhaled. However, it has been deemed safe by many investigatory pharmaceutical associations in small doses, and so the warning messages are probably just there to prevent the manufacturers being sued. There are many unfounded rumors on the Internet about SLS being a carcinogen, usually proposed by 'health food' suppliers or environmental groups, but no scientific link has ever been established between SLS and cancer.

Sodium lauryl ether sulfate and ammonium lauryl ether sulfate are common analogues that are also used regularly as an alternative, the former having less health consequences. These compounds are prepared by the addition of ethylene oxide and are also structurally related compounds found in many beauty products of a lower pH and have a higher compatibility with other surfactants.

It May Be Safe for Us, But What about the Rainforests?

A palm oil plantation in Malaysia

Good point. SLS is a by-product of palm oil, which is extracted from palm trees. These are mainly grown in Indonesia and Malaysia and accounts for a large percentage of income for much of the local population. However, in order to make way for palm oil plantations, mass deforestation has occurred in these countries. Deforestation on such a large scale has severe implications for the climate of the whole planet. When cutting down ancient rainforests, great quantities of carbon are released into the atmosphere from peat bogs, adding hugely to global warming (see p61).

The loss of trees which carry out photosynthesis (the process of converting carbon dioxide to oxygen) also greatly adds to the global warming problem, since the palm oil trees which replace them are not as large, as densely packed or as efficient at photosynthesis as those in the rainforest. Many environmental groups have claimed that the deforestation being carried out in order to make room for plantations is actually more harmful to the Earth's climate than the benefits that could be gained by using palm oil as a biofuel.

"Hmm, I may not have any trees left, but at least my hair is clean!"

Chapter

32

LIMONENE

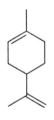

Limonene

THE NAME SOUNDS LIKE LEMONS

Yes, it does. In fact, limonene makes up 60–80% of the essential oil from lemon peel.

IS IT TRUE THAT LIMONENE SMELLS OF ORANGES AND LEMONS?

You mean, something along the lines of *'Oranges and lemons smell the molecules of limonene'*, like the nursery rhyme *'Oranges and lemons, say the bells of St Clement's*'*?

It's a good question; the answer takes a little explaining. First, there are two slightly different limonene molecules which are optical isomers. (*R*)-Limonene is the form found in lemons and oranges, and it also occurs in a lot of the essential oils from other citrus fruits, as shown in the table below.

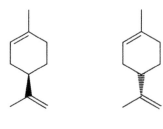

(+)-(*R*)-limonene   (+)-(*S*)-limonene

Essential Oil	% Limonene
Lemon	60–80
Sweet orange	88–97
Bitter orange	75–86
Mandarin	80–94
Grapefruit	86–95
Bergamot	26
Lime	50–64

WELL, WHAT ABOUT THE S-ISOMER?

A paper published in 1971 said that (*R*)-limonene is responsible for the smell of oranges and that (*S*)-limonene causes the smell of

---

* Thought to refer to the church of Saint Clement Danes in the City of London.

lemons. Because this appeared in one of the most prestigious scientific journals (*Science*), people didn't question it for many years. In fact, this turned out to be incorrect. The version of limonene found in citrus fruits – including lemons – is (+)-(*R*)-limonene and if you buy it from a commercial supplier, it indeed has a 'citrus' smell,

Zesting an orange

described by one expert as 'fresh citrus, orange-like'. However, if you buy commercial (*S*)-limonene, which is obtained from turpentine – itself obtained from sources like pine trees – the predominant smell is 'pine' (not lemons!).

So That's It?

Well, probably not. Charles Sell, probably the best perfume chemist in the United Kingdom, says in an article he wrote for the scientific journal *Flavor and Fragrance Research* that the more that (+)-limonene is purified, the less smell it has, and that its orange smell is due to small amounts (~1%) of certain aldehydes, many of which like octanal (see p381) have a very intense orange smell. Certainly, lemons contain up to 13% of intensively smelling aldehydes like citral (a natural 4:1 mixture of geranial and neral).

Geranial   Neral

ISN'T LIMONENE A TYPE OF TERPENE?

Yes, limonene is actually a monoterpene, built up from two isoprene units, though that isn't how it is put together in plants (which involves a complex biosynthetic pathway). However, pyrolysis (heating it until it breaks apart) of limonene does afford isoprene.

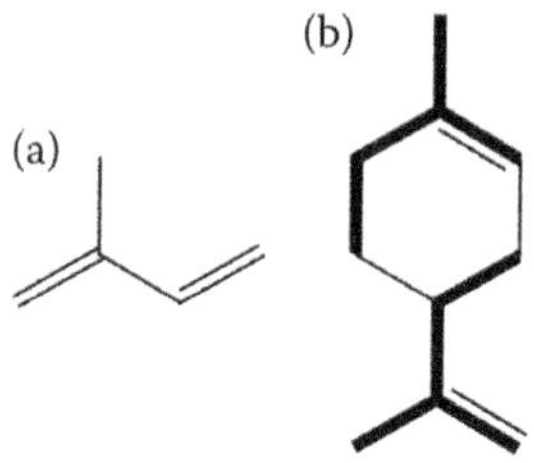

Isoprene (a) and how two of these units combine together to make limonene (b).

SO WHAT USE IS LIMONENE?

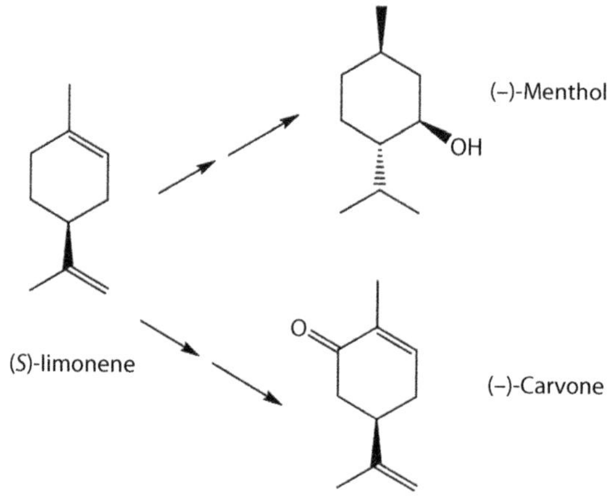

It repels many insects, so it is possible that plants make it as an insecticide, as it is active against species like ants, cockroaches and fleas. Specific plants use it as a starting material to make other important molecules, including (–)-menthol in peppermint, (–)-carvone in spearmint and (+)-carvone in caraway and dill.

AND IF YOU'RE NOT A PLANT?

Limonene is used to make all sorts of things from floor waxes to perfumes. As a liquid hydrocarbon, it is good at dissolving non-polar molecule, so it is a non-toxic 'green' solvent, often used for degreasing, and has the advantage of being biodegradable and being sourced from the waste products of the citrus

Orange Guard household cleaner contains limonene (Courtesy of Orange Guard Inc. With permission.)

juicing industry. It's also used to make polymers and adhesives, and it is found in insecticides, like the *Orange Guard* products.

Chapter

33

LINOLEIC ACID

IS IT USED TO MAKE LINOLEUM FLOORS?

No, the word *linoleic* comes from the Greek word *linon* (flax), and *oleic* meaning relating to or derived from oil. Linoleic acid, (also called *cis,cis,*-9,12-octadecadienoic acid) is the main fatty acid found in vegetable oils such as soybean oil, corn oil and rapeseed oil. It is used for manufacturing margarine, shortening, and salad and cooking oils, as well as soaps, emulsifiers, and quick-drying oils. It is an example of a *poly-unsaturated* fatty acid.

I'VE HEARD OF POLY-UNSATURATED FATTY ACIDS, BUT WHAT DOES THAT MEAN?

Fats are typically straight long-chained hydrocarbons, and fatty acids are fats which have a carboxylic acid group at one end. 'Unsaturated' means it contains a carbon–carbon double bond somewhere in its structure, while 'poly' means it contains more than one of them. In fact, linoleic acid contains two C=C double

Linoleic acid has 18 carbons in its chain. The carbon next to the carbonyl is labeled α (alpha) and the others are labeled sequentially by Greek letters. By convention, the last carbon in the chain (whatever the chain length) is named after the last Greek letter, ω (omega). The bonds are numbered from the ω end, and if there is a double bond at the 6th bond, as in linoleic acid, it's called a ω-6 fatty acid bonds, as shown in the figure. Linoleic acid belongs to one of the two classes of Essential Fatty Acids (EFAs) that humans require.

Essential? For What?

These acids are called 'essential' because they are needed by the human body but cannot be synthesized from smaller components. Therefore they must be obtained by eating appropriate food. EFAs (especially arachidonic acid which is a 20-carbon chain) are converted by enzymes into the body into a variety of biologically active molecules which control inflammation and the central nervous system, including prostaglandins which help regulate body temperature, heal wounds and fight infections (see p419). If a person does not eat sufficient amounts of these EFAs (*i.e.* at least a tablespoon day), they may start to suffer symptoms such as dry hair, hair loss and poor wound healing.

Are These the Same as the Omega Fats I've Heard About?

Yes, except that the fatty acids become 'fats' when the carboxylic acid groups in three fatty acids are bonded to glycerol to make a triglyceride (see p207). The two families of EFAs are called ω-3 (or omega-3 or n-3) which comes from fish oils, and ω-6 (omega-6, n-6) which comes from vegetable oils (linoleic acid is one of these). The above figure shows how the naming convention works. When they were discovered to be essential nutrients in 1923, the two families of EFAs were designated as 'vitamin F'. But around 1930, it was realized that they are better classified with the fats than with the vitamins, and so the name vitamin F was dropped.

AND WE MUST GET OUR EFAS FROM EATING VEGETABLE OIL?

Yes, and this is one of the reasons why the so-called 'Mediterranean diet' is supposedly so healthy. For non-Mediterraneans, one of the most common sources of EFAs is in margarine.

BUT MARGARINE ISN'T AN OIL?

Margarine is a specially treated vegetable oil that has been solidified to resemble and be a substitute for butter. In order to convert the liquid linoleic oil (and its triglyceride) into soft solid margarine, hydrogen is bubbled through the oil in the presence of a nickel catalyst under fairly mild conditions (175–190°C, 20–40 p.s.i.). Hydrogenation in this way does a number of things. Firstly, hydrogen attaches to some of the double-bonded carbons, increasing the saturation level. In doing so, the molecules lose some of the rigidity associated with double bonds and so are able to flex. This allows them to pack closer together, raising the melting point, and turning the oil into a solid fat. The removal of some of the reactive double bonds in this way also reduces the chances of attack by oxygen, so that the fat becomes rancid much less readily, increasing its shelf-life. Superheated steam is then passed through the molten fat to remove any impurities (especially bad-smelling acids and aldehydes). But

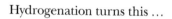

Hydrogenation turns this ...   ... into this

since this process also removes the coloration from the fat, artificial
coloring agents made from carotenes (see p73) of various kinds
are added to make it appear yellow and buttery. Other additives
include butanedione (to make it smell like butter), vitamins A and D,
emulsifiers (to sharpen the flavor) and binding agents (lecithins) to
hold the whole thing together.

But Does All This Chemical Processing Have Any Unwanted Effects?

Yes, the hydrogenation can lead to isomerization of the double
bond, from *cis* to *trans*. An example of this is shown in the diagram
below, in the two forms of 9-octadecenoic acid (another component
of vegetable oil), the *cis* form (oleic acid) and the *trans* form (elaidic
acid). Since the *trans* fats are straighter than their bent *cis* isomers,
they can pack together easier and so have a higher melting point.

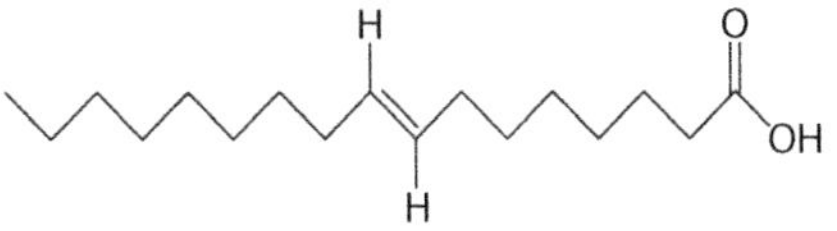

Elaidic acid (*trans*-9-octadecanoic acid)

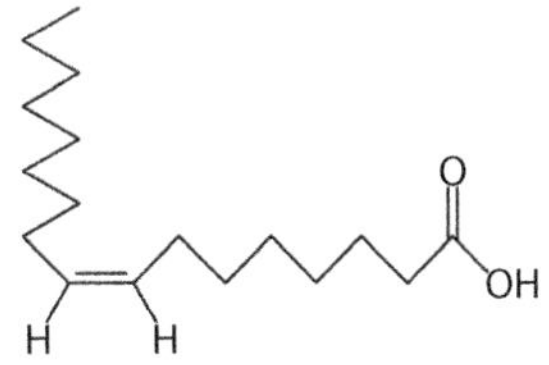

Oleic acid (*cis*-9-octadecanoic acid)

By selecting a particular hydrogenation catalyst, temperature,
stirring speed and pressure, manufacturers can control the precise
composition of the margarine (*i.e.* ratio of *cis* to *trans*) to create a
particular 'mouth feel', melting range and stability.

And the *Trans* Fats Are Bad for Your Health?

The primary source of *trans*-fatty acids (TFAs) in the Western diet
is hardened vegetable oils (*i.e.* margarine), which typically contain

35–50% of TFAs, compared to less than 5% in butter. However, unlike the *cis*-fatty acids, which occur naturally, the TFAs are artificial and so cannot be metabolized in the human body as efficiently as their isomers. Over the past three decades, TFAs have been linked to cancer, diabetes, heart disease, low birth rate and obesity. However, these results are still controversial, as a number of other studies (funded by the US soybean industry) have concluded that TFAs are as safe as their *cis* counterparts.

But What about *Saturated* Fats?

Saturated fats are made from saturated fatty acids, which are also long-chained hydrocarbons with a carboxylic acid group at one end – the difference is that these do not contain any double bonds, an example being lauric acid (see p271). Such saturated fatty acids are known to increase cholesterol in the bloodstream, which can lead to blocking of the blood vessels in the heart, producing heart attacks (see p91). Vegetable-based margarines contain no cholesterol, whereas 100 g of butter contains 178 mg of cholesterol.

So, Which Is Best, Butter or Margarine?

In the question of which is the healthiest to spread on our toast – butter, containing saturated fats which have been linked to coronary disease and strokes, or margarine, with the possible risks associated from TFAs – well, the jury is still out … But best of all, just use pure vegetable oil on your bread, like the Mediterraneans do!

Chapter

34

LYSERGIC ACID DIETHYLAMIDE (LSD)

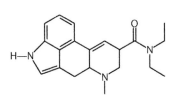

LSD – That's 'Acid', Right?

Yes, technically it's lysergic acid diethylamide, or just 'acid'. It's also known as LSD_{25}, or colloquially as cubes, pearly gates, heavenly blue, royal blue, wedding bells, sugar, sugar lump, big D, blue acid, the chief, the hawk, instant Zen, *etc.*

I'll Stick to LSD. How Long Have People Known about It?

Rye infected by ergot. The ergot kernel is the darker husk drooping down from the ear of rye

Well, the story starts with a fungus called *Claviceps purpurea* which can grow parasitically on some cereals such as rye, particularly in wet summers. It forms fungal growths known as *ergot*. These growths resemble cereal grain and contain ergot alkaloids, including lysergic acid, from which LSD is derived.

In the Middle Ages, the flour used to make bread was sometimes contaminated with this infected grain. Many people died from ergot poisoning, some 40,000 during an outbreak in 994 AD; some got gangrene, and others had epileptic type fits and convulsions known as St Anthony's fire, or experienced hallucinations. Ergotism may also explain the hysterical attacks in Salem, Massachusetts in 1692 that led to the Salem witch trials, though this is disputed, and it may even be one of the factors that led to the French Revolution in 1789. More recently, there was an outbreak in some parts of Southern Russia in 1926–1927, and in August 1951 there was a mysterious outbreak in the Provencal village of Pont Saint Esprit.

SO ERGOT'S NO USE?

Since the sixteenth century, ergot has had medicinal uses, originally to induce contractions and bring about childbirth. The early 1930s brought a new era in ergot research, beginning with the determination of the chemical structure of the main chemically active agents, the ergot alkaloids. Finally, W.A. Jacobs and L.C. Craig of the Rockefeller Institute of New York succeeded in isolating and characterizing the structure common to all ergot alkaloids. They named it *lysergic acid* after the process of *lysis* – the process of breaking down of cells that releases their contents.

In 1938, the Swiss chemist Albert Hofmann, made an ergotamide drug to be used to stop bleeding during childbirth. It was given the code name LSD-25 as it was the 25th lysergic acid derivative synthesized by Hofmann. In April 1943, he accidentally ingested some of it (probably via his hands) and discovered its propensity to cause hallucinations. Three days later, on April 19th 1943, he deliberately ingested 0.25 mg of LSD, and discovered that it was much more potent than other ergot alkaloids. His diary, written later, continues the story:

Albert Hofmann

'*Here the notes in my laboratory journal cease. I was able to write the last words only with great effort. By now it was already clear to me that LSD had been the cause of the remarkable experience of the previous*

*Friday, for the altered perceptions were of the same type
as before, only much more intense. I had to struggle
to speak intelligibly. I asked my laboratory assistant,
who was informed of the self-experiment, to escort me home.
We went by bicycle, no automobile being available because
of wartime restrictions on their use. On the way home, my
condition began to assume threatening forms. Everything in
my field of vision wavered and was distorted as if seen in a
curved mirror. I also had the sensation of being unable to
move from the spot. Nevertheless, my assistant later told me
that we had traveled very rapidly. Finally, we arrived at home
safe and sound, and I was just barely capable of asking my
companion to summon our family doctor and request milk from
the neighbors'.*

A. Hofmann
LSD: My Problem Child[*]

Hofmann survived his LSD trip and woke up next morning feeling
very clear-headed; in fact, he survived until he died on April 29th
2008, aged 102.

So LSD Became a Useful Drug?

In the post-war period, it became a research chemical, investigated
by psychologists and psychiatrists looking for applications in
psychoanalysis. The CIA tested it on servicemen and on members
of the public as a possible mind-control agent. In 2008, it was
alleged that the outbreak of 'ergotism' in the peaceful French
village of Pont Saint Esprit in 1951 was nothing of the sort, but

---

[*] Courtesy of The Multidisciplinary Association for Psychedelic Studies
(MAPS), The Beckley Foundation, and Albert Hofmann.

due to food spiked with LSD in a CIA experiment.

Timothy Leary, a Harvard academic, had been involved in testing LSD and became consumed by its possibilities; President Richard M. Nixon once called him 'the most dangerous man in America'. As the 1950s moved into the 1960s, Leary became an advocate for LSD, which entered the counter-culture as a recreational drug and *entheogen* (a drug which supposedly makes the user feel more spiritually aware), with claims that its users accessed higher states of

Timothy Leary being arrested in 1972

consciousness and helped them search for religious enlightenment. Aldous Huxley took it on his deathbed on November 22nd 1963, the same day that John F. Kennedy was assassinated. Many of the great 1960s rock groups took LSD and there were overtones of it in both lifestyle and lyrics. Just think of The Byrds, The Beach Boys, Jefferson Airplane and the Grateful Dead, while the Beatles created one song that was immediately linked with LSD in the public consciousness due to its psychedelic lyrics and chorus that had the initials LSD.

> *'Picture yourself in a boat on a river with tangerine trees
>   and marmalade sky,*
> *Somebody calls you, you answer quite slowly, a girl with
>   kaleidoscope eyes.*
> *Cellophane flowers of yellow and green towering over
>   your head.*
> *Look for the girl with the sun in her eyes and she's gone'.*
>
> *'Lucy in the Sky with Diamonds'*, The Beatles

John Lennon, however, always insisted that the song wasn't about
LSD but was inspired by the vivid images in a picture his son,
Julian, drew while at nursery school, insisting, 'I had no idea it spelt
LSD'. Paul McCartney too, backs up this story, saying that he and
John had spent an evening writing lyrics about 'cellophane flowers'
and 'newspaper taxis', and 'We never noticed the LSD initial until
it was pointed out later – by which point people didn't believe us'.
Nearly 50 years on, the controversy over this song still rages…

This era reached its climax in the 1967 *Summer of Love* in the
Haight-Ashbury district of San Francisco, but it went downhill
from there. The following year, Haight-Ashbury was plagued by
intravenous use of methamphetamine (see p309) while concern
spread that there were psychological risks to some LSD users.

In the 1980s, LSD made a comeback, accompanying the new
recreational drug Ecstasy (MDMA, see p309) in the rave and
dance club scene, in what became known as 'acid house'. Much of
this was supplied to Europe from a lab in the United States run by
William Leonard Pickard and Clyde Apperson, both of whom are
now serving long sentences in prison. At the peak of production,
their lab was making 1 kg of LSD every 5 weeks. After the lab
was raided and shut down in 2000, it is believed that LSD supply
dropped by 90%, although it has gradually risen again over the
successive years.

IS IT SAFE?

Although LSD is generally thought to be non-toxic and non-
addictive, various governments around the world outlawed it
after a number of fatal accidents were reported. Some accidents
allegedly involved people under the influence of LSD jumping to
their deaths off high buildings thinking they could fly. Research

in the 1960s and the 1970s showed that there was also a considerable psychological risk with the drug and that high doses, especially in inappropriate settings, often caused panic reactions. The effects not only depended upon the individual, but upon their setting. A blue sky could bring on euphoria but a passing cloud might

LSD is often absorbed in paper strips or tabs, and then placed into the mouth

result in a massive 'down'. For individuals who have a low threshold for psychosis, a bad LSD trip could be the triggering event for the onset of full-blown psychosis. Years after people have stopped taking LSD, powerful flashbacks can occur. Research on potential therapeutic uses of LSD was abandoned for political reasons in the mid-1970s.

What Dose Is Needed?

Doses as low as 20–25 µg are capable of producing marked effects. LSD is, in fact, a very powerful psychoactive drug, as even 10 µg can produce symptoms, though a 'normal' dose would be in the range of 50–100 µg. Because the dose is so low, LSD is usually administered either as microdots or small squares of blotting paper that have been infused with a drop of LSD solution.

How Does LSD Work?

LSD has two chiral centers, but only (+)-LSD, with the absolute configuration (5R,8R), is psychoactive. LSD is structurally related to mescaline and to 2-phenylethylamine, which is a parent of a group of drugs known as the amphetamines that act on the central nervous system carrying messages to the brain (see p309). While this area is still open for research, some scientists think that LSD may compete with and block the action of the neurotransmitter molecule serotonin (see p451). It's interesting to spot that serotonin and LSD share the same core structure, an indoleamine group. The belief is that because LSD binds to the nerve receptor more strongly than serotonin, it effectively blocks it and prevents further signals passing along that nerve. Subsequent impulses therefore have to use different nerve pathways, which pass through lesser used parts of the brain. It's like when a main road into a city gets blocked by a traffic jam, the cars then have to by-pass the road block and take side roads through the suburbs. In the case of the brain, impulses traveling through the 'suburbs' of the brain create all sorts of unusual effects, which are experienced as hallucinations, mood changes, altered states of consciousness, *etc.*

LSD   Serotonin   Indoleamine

WHY DO PEOPLE TAKE LSD?

Good question. It alters perception, produces distorted images, intense light and vivid colors; stationary objects seem to move. The person may seem to be removed from time and space; feelings of ecstasy and terrified doom are both possible. The effects vary unpredictably from one person to another and depend upon the user's mood at the time.

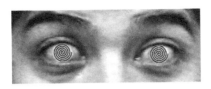

If suppose people take LSD for the same reason they go on roller-coasters or go bungee-jumping – for the excitement of trying something new or gaining new experiences – and this impulse overrides the potential dangers.

Chapter

35

MEDROXYPROGESTERONE ACETATE

ACETATE

THE DRUG USED FOR

CHEMICAL CASTRATION

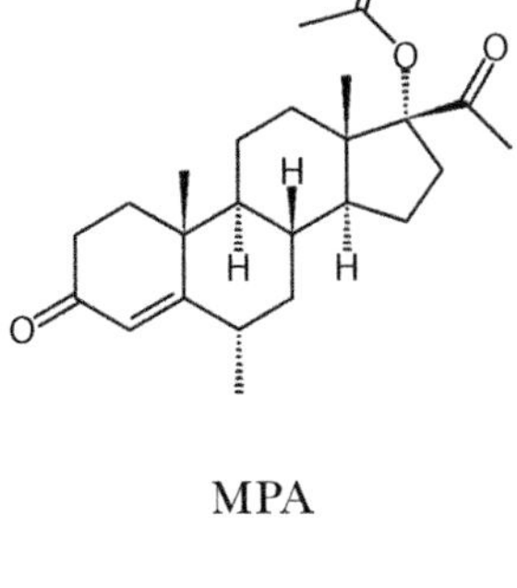

MPA

THAT SOUNDS PAINFUL!

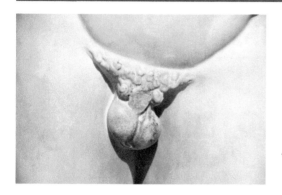

Actually it isn't. Chemical castration is administering a drug designed to reduce sexual activity and libido, and usually just involves a simple injection.

In 1857, Pope Pius IX thought that the 'accurate representation of the male form' of the statues might threaten the religious life of the Vatican. The story goes that he grabbed a mallet and smashed the genitalia off every male sculpture inside the museum to prevent 'lust in the Vatican'. This was allegedly known as *The Great Castration*

WHAT'S IT USED FOR, THEN?

It is often used in many countries as a way of treating male sex offenders, particularly child molesters, to prevent them having the urges to reoffend.

SO, 'CASTRATION' IS NOT ACTUALLY INVOLVED?

That's right. It doesn't involve surgery – nor the removal of any body parts! And the process is generally considered reversible after treatment is discontinued. However, there are reports of some permanent side effects such as loss of bone density or even feminization!

HOW DOES CHEMICAL CASTRATION WORK?

Medroxyprogesterone acetate (MPA) is an anti-androgen drug which is closely related to the human hormone progesterone.

As such it has a whole series of effects on the human nervous, endocrine and reproductive systems. In women, MPA is used safely as a contraceptive, for hormone replacement therapy, or to treat menstrual problems. But on men it counteracts the effects of androgen (the 'male' hormone), leading to a reduction in their sex drive, their compulsive sexual fantasies, and their capacity for sexual arousal.

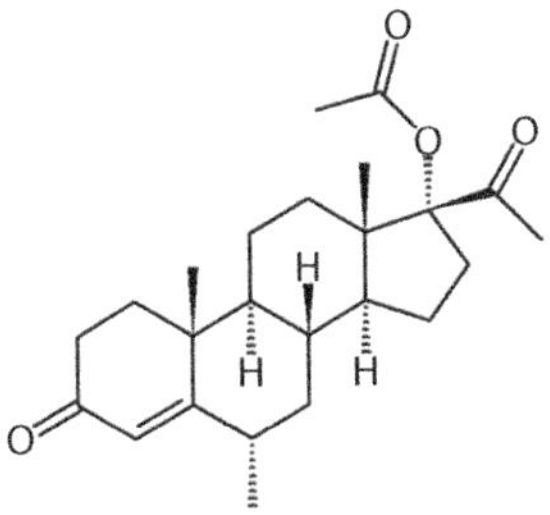

MPA 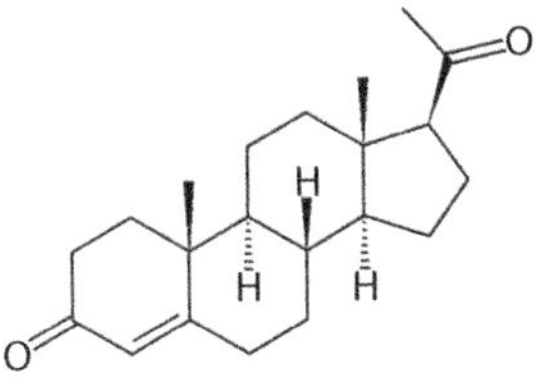   Progesterone

ISN'T IT CONTROVERSIAL?

Yes. There are a number of civil rights groups, such as the American Civil Rights Union, that object to chemical castration on the grounds that forcibly administering *any* drug is a 'cruel and unusual punishment' prohibited by the 8th Amendment. Another objection is that it only works on males, and until a similar treatment is found for women, this type of punishment is unequal. Despite these objections, many countries in the world now accept chemical castration as a humane alternative to lifelong imprisonment or surgical castration for dangerous sexual predators.

ARE THERE ANY FAMOUS CASES?

One of the most infamous cases (and arguably the most unjust) was that of the famous British scientist Alan Turing, who helped

Statue of Alan Turing made in slate at Bletchley Park (where he worked as a codebreaker in WWII), with a photo of him on the wall behind

break the German *Enigma* code in World War II, and went on to become the founder of modern computer science. He also created the famous 'Turing test' which is still used to decide whether a computer has artificial intelligence. However, Turing was a homosexual, which was illegal in the United Kingdom in 1952. Rather than face a prison sentence, Turing opted instead to undergo chemical castration, to 'cure' him of his homosexual urges. Unfortunately, he suffered from some of the unwanted side effects of the drug, developing enlarged breasts and bloating. He committed suicide 2 years later. In 2009 – following an online petition with 30,000 signatures – the then UK Prime Minister Gordon Brown issued a public apology on behalf of the British Government for the 'appalling actions' done to Turing in the 1950s.

More recently, in 2010 a man called Ryan Yates was jailed for attempting to murder a 60-year-old woman in an Aberdeen park during an attempt to rape her two young granddaughters. He was jailed for 10 years, and later agreed to be chemically castrated in order 'to try anything to overcome his problems which have blighted his life'.

CAN ANY OTHER CHEMICALS BE USED FOR CASTRATION?

Although MPA is most commonly used in Europe and the United Kingdom, in the United States, a variant of this is used as an aqueous suspension, called *Depo Provera* (DMPA, which is also used by women in the United States as a 'birth control shot'). This method allows the drug to be administered by an injection into the muscle a few times a year. Other drugs that are used include Cyproterone (and its acetate) and Benperidol, which is often prescribed to sex offenders as a condition of their parole.

Cyproterone-acetate   Benperidol

Chapter

36

METHAMPHETAMINE

(OR METHEDRINE, SPEED, CRANK, METH, ICE, GLASS, CRYSTAL METH, ETC.)

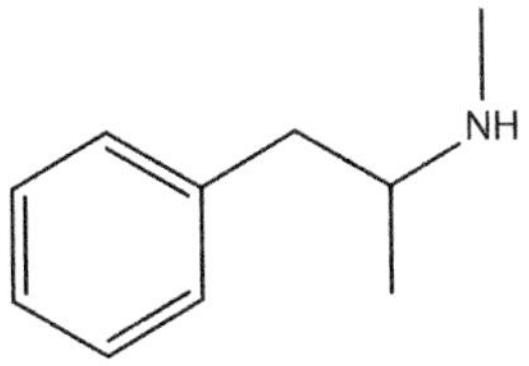

Methamphetamine

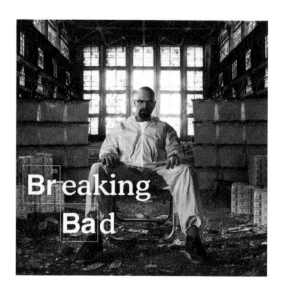

The hit US TV series *Breaking Bad* is about a high-school chemistry
teacher who turns to making and selling crystal meth after learning
he has terminal cancer (From 'Breaking Bad' © 2013 Sony Pictures
Television, All Rights Reserved. With permission.)

WHY IS METHAMPHETAMINE SO BA^{56}D?

It is a dangerously addictive drug, a Schedule II controlled substance
in the United States and a Class-A drug in the United Kingdom.

WHY IS IT REGARDED AS DANGEROUS?

Methamphetamine is very addictive. Its use, even in small amounts,
has been linked with damage to the dopamine-producing cells in
the brain; it also affects brain development in the fetus of pregnant
women who are meth-users. At low levels, methamphetamine is a
stimulant. It promotes the release of above-normal amounts of the
important neurotransmitter dopamine (see p169), which stimulates
regions of the brain linked with vigilance, as well as the action of
the heart.

As a result, for a short while, the user feels sharper, stronger and more energetic. For this reason, amphetamines were used widely by American, British, German and Japanese forces in World War II; they were used by watch-keepers on ships to stay alert, and bomber crews used amphetamines to maintain alertness on long night raids. In 1950s America, long-distance truck drivers used amphetamines to help them stay awake, and students knew that amphetamine 'pep pills' helped them stay alert when revising, and helped

Frequent use of amphetamines turns a casual user into an addict.

them score better marks. Under the influence of amphetamines, Jack Kerouac produced '*On the Road*' in a continual frenzy of typing in three weeks of April 1951 (though many of its ideas had already been published). As late as 2002, USAF bomber pilots on amphetamine 'go pills' mistakenly bombed Canadian forces in Afghanistan, killing four and wounding eight. At high-level exposure, methamphetamine is linked with heart disease, psychosis and an increased risk of Parkinson's disease, as well as social and family problems, including risky sexual behavior.

Is It a New Drug?

Both methamphetamine and its close relative, amphetamine, are structurally derived from phenylethylamine. Unlike the amphetamines, phenylethylamine occurs naturally in the human body, where it has stimulant effects on the nervous system.

Phenylethylamine   Amphetamine   Methamphetamine

Amphetamine itself was first made in 1887, and
methamphetamine in 1919. Under the nomenclature system
of the time, amphetamine was originally called alpha-
methylphenylethylamine, which was shortened to 'amphetamine'.
It was not until 1930 that it was discovered that amphetamine
raised blood pressure. In 1932, after it was found to shrink
mucous membranes, Smith, Kline and French marketed a nasal
inhaler containing amphetamine to treat nasal congestion using
the trade name *Benzedrine*. Within a short period of time, people
found that amphetamine could be extracted from the wadding
in these inhalers and used for its 'high'. Amphetamine abuse had
begun. To counter this, the authorities made amphetamines a
prescription medicine.

WHY IS IT SO POTENT?

The human body possesses an enzyme (monoamine oxidase)
which has evolved to digest the amines that we ingest in foods like
cheese, as well as phenylethylamine that occurs naturally in the
body. But the presence of the methyl group on the side chain in
both amphetamine and methamphetamine means that they no
longer fit the active site of this enzyme. Amphetamines have only
been around for about 100 years, so the human body hasn't yet
developed enzymes to break them down, and so they linger in the
body, producing a long-lasting effect.

How Is Methamphetamine Made?

Two methods that have been described both involve the reduction of ephedrine (itself readily available) using either lithium in liquid ammonia as the reducing agent, or else red phosphorus and iodine to generate hydrogen iodide for the reduction. Ephedrine has a chiral carbon atom, and is found in nature as a single isomer, so the methamphetamine made from it is also just the one isomer. As it happens, this is the isomer associated with stimulant properties.

Ephedrine   Phenyl-2-propanone

A different route involves the reaction of phenyl-2-propanone (P2P; phenylacetone), with methylamine. This produces a mixture of the two isomers.

Laboratories in many places, notably throughout the United States, are engaged in clandestine manufacture of methamphetamine. The chemistry is well understood, and widely published, notably on the web, though not easy to achieve. Much of this synthetic chemistry is difficult and dangerous outside a proper laboratory. Fires and explosions, as well as chemical burns to individuals, are reportedly common in D.I.Y. meth-labs. Some of the chemicals involved, like phosphine (PH_3), are very toxic. Don't try these at home, folks. It is very dangerous, as well as being illegal.

WHY DO THE ISOMERS OF METHAMPHETAMINE HAVE DIFFERENT EFFECTS IN THE BODY?

Methamphetamine contains a 'chiral' carbon, a carbon atom with four different groups attached, and therefore exists as two optical isomers, which are non-superimposable mirror images of each other. The isomers have identical chemical reactions, solubilities and melting points. They differ in their opposite rotations of plane-polarized light and, because they have different 3D structures, they fit protein receptors differently, and therefore have different effects upon the body. L-Methamphetamine is simply a decongestant, and has no stimulant activity. But its optical isomer, D-methamphetamine, is the stimulant commonly known as 'speed'.

(S)-(+)-Methamphetamine or D-methamphetamine or 'Dexedrine'

(R)-(−)-Methamphetamine or L-methamphetamine

SO JUST THE ONE FORM OF METHAMPHETAMINE IS ILLEGAL?

Sports drug legislation does not distinguish between the two isomers, and this cost a British sportsman an Olympic medal. On February 23rd 2002, the Scot Alain Baxter finished third in the slalom at the Winter Olympics at Salt Lake City. He became the first Briton ever to win a medal in alpine skiing. After that, things rapidly went downhill for Baxter (no pun intended). Within days, tests on a urine sample revealed traces of methamphetamine. After a 2-day hearing, the International Olympic Commission announced that Baxter must return his bronze medal. His third place would be removed from the Olympic record books. Baxter's subsequent appeal failed, but while the Court of Arbitration for Sport (CAS) upheld the International Olympic Committee's decision to strip him of the medal, they accepted that Baxter had acted honestly throughout.

Why Did Baxter Take Methamphetamine?

Alain Baxter had a longstanding medical condition, nasal congestion. Back in the United Kingdom, he regularly used British *Vicks* nasal inhalers, which use menthol, camphor and methyl salicylate (wintergreen) to shrink inflamed membranes. Baxter had the same problem at Salt Lake City, so when he saw a *Vicks* inhaler in a shop that appeared identical to the ones he used in the United Kingdom, he bought it. Sadly, although it looks the

Alain Baxter on his way to winning the bronze medal for Great Britain's first ever Olympic alpine skiing medal at the Salt Lake 2002 Winter Olympic Games in Deer Valley February 23rd 2002. Jean-Pierre Vidal of France eventually won the Olympic Gold in the slalom, ahead of silver medallist Sebastien Amiez of France (© Press Association Images. With permission)

same, American *Vicks* contains compounds not found in the British version, including l-methamphetamine, which is a decongestant, but without stimulant activity.

If the l-Methamphetamine Baxter Took Was Not a Stimulant, Why Did He Lose His Medal?

IOC regulations simply forbid the use of 'methamphetamine'. They do not distinguish between the two isomers. Neither did the IOC testing procedures as used at the 2002 Winter Olympics. The optical isomers of methamphetamine can be separated (and distinguished) by high-pressure liquid chromatography (HPLC) using a chiral column. Baxter requested that the IOC test his urine samples to confirm which isomer was present, a request that was declined.

But If the IOC Forbids Methamphetamine, Sportsmen Must Have Abused It?

The IOC banned amphetamines years before steroids. At the 1952 Oslo Winter Olympics, several speed skaters who had been taking amphetamines needed medical assistance. As others have remarked, this gave a whole new meaning to the expression 'speed skating'. However, at the 1960 Olympics held in Rome, the Danish cyclist Kurt Jensen collapsed and died of a heart attack from an amphetamine overdose, while on July 13th 1967 Tom Simpson, 1965 World Champion and the first Briton to wear the famous yellow jersey, collapsed from heat exhaustion on Mont Ventoux during the Tour de France, before dying of heart failure. They found amphetamines in his jersey, amphetamines which allowed him to go beyond his pain threshold, but also caused blood to flow away from the skin and thus increased the risk of heat stroke.

And Other People Abused Amphetamines, Too

After World War II, they became prescription drugs and were widely prescribed as pick-me-ups. They were once used as a slimming aid, as they reduced appetite. Although their unlicensed possession became illegal both in the United Kingdom (1964) and the United States (1970), people continued to take them.

Amphetamines also 'stimulated' much of 1960s British youth culture, particularly the Mods. This was epitomized by the Mod group '*The Small Faces*', with their 1967 speed anthem 'Here Come the Nice':

> *Here comes the nice, looking so good,*
> *He makes me feel like no-one else could.*
> *He knows what I want, he's got what I need,*
> *He's always there, if I need some speed.*

Dexy's Midnight Runners, the well-known 1980s UK band, were named after *Dexedrine*, a brand of dextroamphetamine popularly used as a recreational drug among Northern Soul fans at the time. The 'midnight runners' referred to the energy the *Dexedrine* gave, enabling the band/fans to run/dance into the midnight hours.

Among the many celebrated victims of methamphetamine abuse, Johnny Cash, the American singer, survived. He found God. Others were not so lucky. Amphetamines contributed to the deaths of Judy Garland and Edie Sedgwick, Andy Warhol's muse.

Where Did the Slogan 'Speed Kills' Originate?

Methamphetamine started to be known as 'speed' in the early 1960s, possibly a consequence of its use in 'speedballs' with heroin, or possibly because of its go-faster characteristics. In the Haight-Ashbury region of San Francisco, the 'Summer of Love' in 1967 turned sour the following year when many residents had moved on from marijuana and psychedelics like LSD (see p293) to injecting methamphetamine intravenously. The resulting epidemic of methamphetamine abuse and consequent illnesses (and crime) led to the neighborhood changing to a less friendly, indeed dangerous, one, and its motto changed from 'peace and love' to 'speed kills'.

WHAT'S THIS 'CRYSTAL METH' STUFF?

In the late 1980s, 'crystal meth' or 'ice' became available in Hawaii, produced in Far Eastern laboratories by careful recrystallization of methamphetamine hydrochloride, resulting in quite large, clear, crystals. Though a salt, it is volatile enough to vaporize unchanged (unlike amphetamine sulfate). Its use spread to the American mainland. Methamphetamine abuse has become a real problem right across the country, becoming the

Crystal meth is actually colorless, not blue as depicted in *Breaking Bad*. The blue color was invented by the scriptwriters to allow them to differentiate Walter White's supposedly extremely pure crystal meth from inferior versions

leading illicit drug in rural United States. Its use has become endemic in Japan (controlled by *yakuza* gangs) and Southeast Asia, including countries like Cambodia and Thailand, where it is known as '*yaa baa*' (*yaa* = medicine, *baa* = crazy), after its propensity to make people behave in crazy, uncontrolled ways.

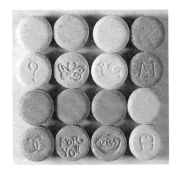

Ecstasy pills

AND ECSTASY?

There have been many variations of the methamphetamine structure used as recreational drugs. Perhaps the most notorious recently is 3,4-methylenedioxy-*N*-methylamphetmine (MDMA), which is known as 'Ecstasy' for

MDMA (Ecstasy)

the feelings of peace and euphoria it supposedly brings. MDMA became fashionable among the 'rave' culture in the 1980–1990s for its apparent ability to allow the user to robotically repeat the same motions over and over again in all-night dance music events. It too made popular music culture, when the UK techno band *The Shamen* released a single in 1992 called 'Ebeneezer Goode' which had the controversial chorus: ''Eezer Goode, 'Eezer Goode, He's Ebeneezer Goode', the first part of which sounds identical to, 'E's are good' – 'E' being common slang for Ecstasy. Although it is not addictive, and causes far fewer fatalities than other methamphetamines, MDMA is banned or severely restricted in most countries.

Chapter

37

METHANE

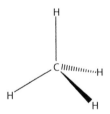

Me-Thane – Sounds Like a Selfish Molecule?

It's meth-ane, not 'me'-thane. It is the smallest hydrocarbon and the lightest stable organic molecule. A methane molecule has a regular tetrahedral structure; its H–C–H angles are all 109½°.

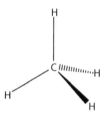

Structure of methane   Space-fill model of methane

It is not very soluble in water on account of its non-polar nature. It is a colorless, odorless gas at room temperature (m.p. −183°C, b.p. −164°C) and is less dense than air.

If It Is Odorless, How Come I Can Smell It When I Turn the Gas On?

You're not smelling methane – because methane is odorless. Natural gas supplied for domestic use has a trace of an obnoxious organic sulfur compound such as Me_3CSH or Me–S–Me (see p161) deliberately added for ease of detecting any leak.

And It's a Fuel, Right?

The combustion of methane is a very exothermic reaction and so methane is widely used as a fuel, not least because the volatile and non-toxic products make it 'clean', though one product, carbon dioxide, is a 'greenhouse gas'.

$$CH_4(g) + 2O_2(g) \rightarrow CO_2(g) + 2H_2O(g)$$

Incomplete combustion yields carbon (soot) or toxic carbon monoxide.

Natural gas (methane) burning in a gas cooker

Methane is the main component of 'natural gas', which contains between 70% and 95% methane, often over 90%. This can be purified to remove other low-boiling components like noble gases and slightly larger alkanes, then condensed at around −160°C as liquefied natural gas (LNG). This is a much more compact way of storing methane, and when pipelines are not available (such as across oceans or in remote rural areas), it can be transported in cryogenic tankers by sea or road.

How's It Formed?

Billions of years ago, volcanic activity was responsible for large amounts of methane in the Earth's atmosphere. Subsequently, the first primitive bacteria synthesized methane from CO_2 and hydrogen, and methane levels did not fall until the arrival of the first organisms to photosynthesize, which generated oxygen (see p81). The concentration of methane in the Earth's atmosphere was around 1800 ppb in 2010, more than double the pre-1750 value.

Nowadays, the main source of methane is 'natural gas', which originates in the same way as crude oil, by the anaerobic decomposition of microscopic sea animals under the pressure of layers of silt and mud over a period of millions of years. This can also occur in swamps and marshes, when plant matter decays ('marsh gas'), and in landfill sites; in digesters breaking down sewage sludge; and by enteric fermentation in livestock (*e.g.*

ruminants such as cattle and sheep), as well as being formed when termites digest the cellulose in the wood they have eaten.

Is That Why It's Found in Coal Mines?

A replica Davy lamp with apertures for measuring flame height

Yes, bituminous coal under pressure exudes methane, and this can collect in pockets in the coal and adjacent strata. Together with other gases like CO and CO_2 this mixture is known to miners as 'firedamp', or if it also contains hydrogen sulfide, 'stinkdamp'. If the pocket of gas is highly pressurized, it is known as a 'bag of foulness'. These mixtures are highly explosive when mixed with ~10% oxygen, and have been responsible for many deadly mining disasters. In olden times, miners used to take canaries down into the mines, since these birds were much more sensitive to toxic gases (CO, CO_2 and methane) than humans. Signs of distress from the bird were an indicator that conditions were unsafe, and that the miners should vacate that area rapidly. Canaries were used rather than flame detectors due to the explosion hazard.

In 1815, Sir Humphry Davy invented the miners' safety lamp, which consisted of a wick lamp with the flame enclosed behind a mesh screen which acted as a flame arrestor. If flammable gas mixtures were present, the flame burned higher and with a blue tinge. Miners could measure the height of the flame using a metal gauge on the side of the lamp. Whether through tradition or superstition, Davy lamps, and the later electric lamps, did not completely replace canaries for many years. Indeed, canaries were still in use in UK coal mines as late as 1987!

WHICH COUNTRIES HAVE THE MOST METHANE?

As you might expect, countries rich in oil and coal reserves also produce the most natural gas, with Russia producing the most by quite a large margin, followed by the United States, Canada, the European Union, and then the various Arabian Gulf counties.

A North Sea platform used for extraction of oil and natural gas

SO, NORTH SEA GAS IS LARGELY METHANE

Yes, and this forms one of the main supplies of gas for Western Europe. North Sea oil (and gas) was discovered in the early 1960s, with first supplies being piped ashore in the United Kingdom in 1975. The resource is shared between the United Kingdom, Norway, Denmark, Germany and the Netherlands, mainly the first two. Production peaked in 1999. This enabled the United Kingdom to become – briefly – an exporter of energy and also enabled the Conservative Government of the 1980s and early 1990s to move away from dependence upon coal-fired power stations toward the use of gas as a fuel (the 'Dash for Gas'). The United Kingdom is now a net importer, looking to obtain supplies from other sources, including Russia.

WHAT ABOUT THE UNITED STATES?

Natural gas supplies around 20% of the energy requirements of the United States, with most of it being produced in the oilfields of the Gulf of Mexico and Alaska and piped to mainland United States via a complex system of interstate pipelines. Because of rising demand and falling supplies, the United States has recently led the world in developing the new and controversial technology of 'fracking'.

FRACKING?

Short for hydraulic fracturing. This involves pumping pressurized liquid deep underground to fracture layers of rock. Any oil or natural gas trapped in the rock can then seep through the fractures back to the well-head, where it is collected. Shale (a fine-grained sedimentary rock) is particularly rich in oil and gas, and the gas collected (mostly methane) is often called shale gas. However, as with all energy production methods, fracking is controversial. There are concerns that fracturing the rocks contaminates underground water supplies, causes pollution, and contributes to Global Warming by creating yet more carbon dioxide to be released into the atmosphere (see p61). There have even been reports of minor earthquakes that may have been caused by fracking.

"The drill's broken? Oh Frack!"

SO POLITICS MATTERS?

Currently, only 15 countries account for 84% of the worldwide extraction of natural gas. This means that access to this vital resource has become an important issue in international politics, and countries compete for control of sources and pipelines. In the early 2000s, Ukraine and Belarus argued with the state-owned energy company in Russia (Gazprom) over the price of natural gas that it was supplying to them. Gas supplies for some of these regions were then cut-off for long periods until the dispute was resolved. Since much of the gas supply for Europe comes via these same pipelines, there are concerns that withholding such gas deliveries could be used by Russia as an economic weapon in future political disputes. Indeed, the dispute between Ukraine and Russia in early 2014 focused the World's attention on Russia's dominance of natural gas supplies to Europe, and was one alleged reason why many European countries were unwilling to support tougher sanctions against Russia.

APART FROM A FUEL, IS METHANE ANY USE?

Its reaction with steam in the presence of a hot nickel catalyst affords the mixture of CO and H_2 known as 'synthesis gas', widely used to make chemicals such as hydrogen, methanol and ethanoic acid.

$$CH_4(g) + H_2O(g) \rightarrow CO(g) + 3H_2(g)$$

AND METHANE IS A 'GREENHOUSE GAS'?

Methane is classed as a greenhouse gas. Compared with CO_2, it is shorter-lived in the atmosphere, but each methane molecule has 25 times the direct Global Warming Potential of a CO_2 molecule (calculated over a 100-year timescale). Because of this, there is concern about methane emissions from diverse natural sources, such as mud volcanoes and ruminants, together with the possibility of losses from methane hydrates (see p328), released as the oceans warm up. The 2010 estimates put methane emissions at around 566 million metric tonnes a year, double the pre-industrial value, with natural sources making up 37% of this.

It has been suggested that trees and other green plants in aerobic environments are responsible for adding large amounts of methane to the Earth's atmosphere, but research using ^{13}C-labeling to distinguish emissions from plants from ^{12}CH$_4$ 'background' in the atmosphere has found no evidence for this, and no biochemical pathway is known that would enable this to happen. Fungi have been found to produce it, but in negligible amounts on a global scale.

How Does Methane Make Flaming Snowballs?

When they are under pressure, such as under cold sediments at the bottom of oceans, methane molecules get locked up inside cavities within the structure of ice, forming what is known as methane clathrate (other alkane molecules are larger, and cannot fit into the cavities). This methane is largely formed by anaerobic bacterial breakdown of organic matter. Since methane clathrates have been estimated to store more energy than the total fossil fuel

reserves of the whole world, there has been interest in harvesting this methane as an energy resource, but also concern at the prospect of oceanic warming, liberating the methane causing rapid global warming. In 2013, it was reported that methane had been extracted successfully from frozen methane clathrate off the coast of Japan.

Burning 'ice'. The ice is frozen methane clathrate, which can be set on fire

And It's Found on Other Planets?

There is evidence for methane being present in several planets in the solar system, particularly Mars and Titan, Saturn's largest moon. Three groups of researchers detected methane in spectroscopic studies of the atmosphere of Mars in 2003–2004. Scientists debate whether it has a bacterial source, or if it originates in iron and magnesium containing silicate rocks, known to be capable of producing hydrogen, which could then react with carbon or its compounds forming methane.

In January 2005, the European Space Agency's *Huygens* probe sampled the atmosphere of Titan by GC–MS, detecting methane, but this lacked the ^{12}C enrichment associated with living organisms, making it likely that the methane originates from geological processes somewhere in the interior. There is also liquid methane on the cold (–180°C) surface of Titan, and this has apparently carved river channels upon the surface. As a result of this, several space agencies are now looking into the possibility of making rocket engines powered by liquid methane. The idea is that one day these rockets might 'refuel' in outer space by sucking methane from the atmospheres of distance planets and moons (like a planetary gas station), greatly extending their range.

Titan's surface image taken by *Huygens*. This is the only image from the surface of a planetary body outside the inner Solar System. Much of the atmosphere is composed of methane (Courtesy NASA/JPL-Caltech)

WHAT ABOUT COWS AND GLOBAL WARMING?

Cows and other ruminants have methanogenic bacteria in their rumen which can turn digested carbon compounds into methane, emitted as flatus, some 200–300 liters of gas a day. Most of the flatus is nitrogen, but cases have been reported of cows' flatus being ignited, though not explosively. It has been suggested that nearly half the methane emissions come from agriculture, with livestock and cultivation of rice prime causes (60–80 million metric tonnes a year each). In New Zealand, where sheep and cows do cause half the nation's methane emissions, a 'flatulence tax' has been suggested (for the animals not the population!). It's been suggested that small changes to Chinese farming practices have sharply cut methane emissions from paddy fields.

AND THE TERMITES?

Termites use micro-organisms – either protozoa or bacteria, depending upon the termite – in the gut to help digest plant material (cellulose). An individual termite produces around 0.5 micrograms of CH_4 per day. On scaling this up, worldwide emissions of up to 20 million metric tonnes a year from the 'cathedral' nest mounds produced by the termite colonies have been estimated.

Chapter

38

2-METHYLUNDECANAL

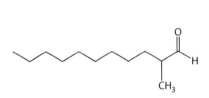

'Alone among fragrances known to me, it gives the irresistible impression of a smooth, continuously curved, gold-colored volume that stretches deliciously, like a sleepy panther, from top note to drydown'.

Luca Turin

Quatre-vingt ans plus tard, Chanel no.5 resplendit toujours d'un éclat unique, fulgurant. [(*Eighty years later, Chanel no.5 still shines with a unique, dazzling brightness.*)]

L'abcdaire du parfum [(*The ABCs of perfume*)]

Is That the Molecule That You Smell in Chanel No. 5?

It is one of a number.

Five?

More than that. All perfumes are mixtures of chemicals, and are normally analyzed in terms of three components called 'notes', as an analogy with the notes that make up a musical harmony.

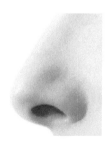

Top note (*head note*), the initial impact of a fragrance, which predominates for the first few minutes, leading to the …

Middle note (*heart note*), which contains the main fragrances and personality of the perfume, and lasts for several hours. Eventually, this is replaced by the …

Bottom note (*base note, end note* or *drydown*), which is the residual smell that persists when all else has fled, and remains with you until the perfume has totally evaporated. These molecules also help to reduce the volatility of the others.

Each chemical makes a different contribution to the overall effect.

So Coco Chanel Invented It?

She opened her first shop in 1909 in Paris and steadily developed her line in fashions, but when in 1919, she looked at introducing her first perfume, she went to an expert, Ernest Beaux.

Who's He?

Ernest Beaux (1881–1961) was a Russian-born son of a French father, chief perfumer to A. Rallet & Co., a Moscow soap and perfume manufacturer. He eventually became the senior perfumer at Rallet in 1907. His first major perfumes were *Bouquet de Napoleon* (1912) and *Rallet No. 1* (1914). After the Russian Revolution (when he fought with the White Russians), he left for France in 1919. Chanel's lover, the Grand Duke Dmitri Pavolivch, introduced the two, reportedly on the beach at Cannes.

Ernest Beaux

So What's Special about No. 5?

Until late into the nineteenth century, perfumes tended to be made from floral extracts. During the 1880s, perfumers started to use synthetic materials, the first example being man-made coumarin, in *Fougère Royale* (1882), then synthetic vanillin was used in *Jicky* (1887). So *Chanel No. 5* wasn't the first to use synthetics, but it was one of the first to make major use of aldehydes.

Not the First?

In 1904, the French chemist Georges Darzens devised a reaction – which now bears his name. The Darzens reaction is relatively simple, and involves reacting a long-chain ketone with ethylchloroacetate, followed by hydrolysis of the product to the corresponding acid, which readily undergoes decarboxylation. This produces long-chain aldehydes, where the length and composition of the chain can be varied depending upon the starting ketone. The important thing is that this simple reaction made these aldehydes readily available. They are important molecules because of the smells that they possess. Straight-chain aldehydes with 10–13 carbon atoms in the chain have citrus-orange smells.

Undecanal (*undeca* = 11 carbons, *al* = aldehyde). Green, fatty, floral-orange smell

Introducing a methyl side chain at carbon 2 makes the odor stronger and also more pleasant, making the smell less fatty, so 2-methylundecanal (2-MNA) became a useful perfume ingredient.

2-Methylundecanal (2-MNA). The methyl side group is on carbon-2. Herbaceous-orange-ambergris-like smell

WHY 2-MNA?

It is an abbreviation for the old, non-systematic name, methylnonylacetaldehyde.

AND DARZENS USED THESE ALDEHYDES IN PERFUMES?

Darzens may have supplied it to Piver for use in *Floramye* (1905), but we know that it was the perfumer Robert Bienaimé who used 2-MNA in his groundbreaking perfume *Quelques Fleurs* (Houbigant, 1912). Beaux was hot on his heels in using aldehydes, however. An original unopened bottle of Beaux's perfume *Rallet No. 1* (1914) was recently analyzed. It showed that Beaux used the aldehydes undecanal, dodecanal and undec-10-enal, along with neroli (spicy sweet floral) and ylang-ylang (*i.e.* floral), for the top note; jasmine absolute, *rose de mai*, orris and lily of the valley for the middle note; and sandalwood, vetiver, styrax, vanillin and nitromusk in the base note.

By the time Beaux composed *No. 5*, he used for the top note, ylang-ylang, neroli (spicy sweet floral) and aldehydes with 10–12 carbons, including branched 2-methylalkanals, especially 2-MNA; for the middle, a mixture of jasmine, rose, lily of the valley and iris; and for the base note sandalwood, vetiver, musk ketone (see p535), vanilla, civet and oak moss.

AND 2-METHYLUNDECANAL IS TOTALLY SYNTHETIC?

It has been found in nature in very small amounts in the oil of kumquat, a citrus fruit resembling an orange.

WHAT MAKES CHANEL NO. 5 SO SUCCESSFUL?

It is partly due to marketing, and the beautiful women associated with it – Marilyn Monroe, Catherine Deneuve, Nicole Kidman, Audrey Tautou, …

But the smell is special; it is said that Beaux believed that he had created something that reminded him of snow-covered Russian landscapes. Others have compared its effect on the perfume to 'lemon juice on strawberries'. And that is down to the aldehydes, especially 2-methylundecanal.

Chapter

$$\boxed{39}$$

Monosodium Glutamate
(and the Fifth Flavor)

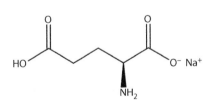

Monosodium glutamate (MSG)

MSG is often associated with
Chinese food, but it's just as
common in Western fast food

REMIND ME OF THE FIRST FOUR FLAVORS …

Sweet, sour, bitter and salty.

AND THE FIFTH?

It's called *umami*, which means 'pleasant savoury taste' in Japanese. It's been described as a meaty taste with a long-lasting, mouth-watering and coating sensation over the tongue. In chemical terms, the *umami* taste comes from special receptors in the tongue responding to the carboxylate anion of glutamates.

GLUTAMATES?

Glutamates are molecules that are naturally occurring in many fruits and vegetables. They are simply the ionized form of glutamic acid, which are created whenever glutamic acid is put into water.

Glutamic acid

The ionized glutamate form, in which it exists in biological systems

The salt, monosodium glutamate (MSG)

Tomatoes are particularly high in glutamates, as are many of the meats we eat. Animal proteins contain between 11% and 22% glutamic acid by weight. Glutamate plays a major role in the biosynthesis of several key amino acids, and it is a major breakdown product of proteins. Most of the non-essential amino acids in our bodies (such as alanine and aspartate) derive their α-amino groups from glutamate. To obtain glutamate as solid crystals, you simply add to it a positive ion of a corresponding metal, such as sodium or potassium. The sodium salt is called monosodium glutamate (MSG).

THAT'S THE FOOD ENHANCER?

Yes, MSG is widely used as a flavor enhancer in a range of foods, and is said to impart the *umami* flavor to foods. It is very peculiar, in that it does not act by adding a specific flavor of its own (as salt and sugar do), but instead stimulates and increases the sensitivity of taste receptors, thus 'multiplying' or enhancing the existing taste of food.

Kikunae Ikeda

A packet of MSG from the
Ajinomoto company

WHO DISCOVERED IT?

Glutamic acid was discovered in 1908 by Professor Kikunae Ikeda from Tokyo Imperial University who studied the brown crystals left behind after evaporating large amounts of *kombu* broth, made from a local seaweed. When he tasted it (something chemists generally don't do in the lab!), he found that the taste reminded him of many different foods, including seaweed, but also meat. He came up with the name *umami* for this savory taste. He then made the glutamate salts of sodium, potassium and ammonium, and found they all tasted of *umami*, but most had a metallic aftertaste. The best-tasting salt was (mono)sodium glutamate, and he patented this as a food additive. He then helped to set up a company in Japan to produce MSG, which was called *Ajinomoto* meaning 'essence of taste'. In many countries of the world, MSG is still known simply as *ajinomoto*, but it also goes by other names such as *Vetsin* and *Ac'cent*.

WHY DOES MSG WORK AS A FLAVOR ENHANCER?

As meat ages, the proteins within it start to decompose to form a number of different breakdown products. Among these are MSG, plus inosine monophosphate (IMP) which is formed when ATP decomposes (see p1). Together, these two compounds produce a very meaty flavor, and they are the main components responsible for the taste of meat. Different meats contain different relative amounts of MSG and IMP, and thus taste differently. For example, beef and pork contain roughly the same amount of IMP, but beef contains about twice as much MSG as pork. Manufacturers

Inosine monophosphate

found that by adding these two chemicals, protein-rich meals could be made to taste even meatier. However, IMP is much more expensive to make than MSG, so MSG is used as the sole additive. It's often used to enhance the flavor of meat dishes, fish and surprisingly even mushrooms. Many mushrooms also contain a large number of proteins composed of glutamic acid constituents. This might account for their slightly meaty flavor and the fact they go so well with meat dishes. For many years, MSG was associated with Chinese cooking as it was used in Chinese restaurants as a substitute for salt. In fact, this is a very healthy option, as studies have shown that in order to make food palatable, much less salt needs to be added to food that has also had a small amount of MSG added, than otherwise. Doctors recommend that reducing salt intake helps lower blood pressure and prevent cardiac problems, so replacing lots of salt with a pinch of MSG might be one way to do this without having to put up with bland food.

BUT WHAT ABOUT 'CHINESE RESTAURANT SYNDROME'?

As with all food additives, MSG has been extensively studied in numerous safety tests for nearly 100 years and no evidence has been found that it is unsafe to eat. However, this hasn't stopped anecdotes and rumors spreading about potential medical problems linked to MSG. These all began in April 1968 when Robert Ho Man Kwok wrote a letter to the *New England Journal of Medicine* complaining of symptoms such as numbness of the neck, back and arms, plus weakness and palpitations, which he said occurred whenever he ate at a local Chinese restaurant. He suggested various reasons, such as the alcohol used when cooking with wine, the sodium content or the MSG added to the food, and coined the phrase 'Chinese Restaurant Syndrome' (CRS). Despite all these possible causes, MSG became linked with these symptoms as the sole reason, and CRS later became known as 'monosodium glutamate symptom complex'. Although the symptoms were vague and unspecific, on hearing of CRS, many other people also claimed to be suffering the same or similar symptoms after visiting Chinese restaurants. These claims received a lot of publicity in the media, who, as usual, sensationalized the story without doing any proper scientific checks. Chinese restaurants, worried about loss of business, started putting signs in their windows or on their menus stating 'No MSG used' or 'MSG-free'. These signs are actually a bit misleading, because if the dishes contained meat, mushrooms, wheat or tomatoes, then they would be packed full of natural glutamates anyway, and so could never be MSG-free.

Sign often found in Chinese restaurants

So Is CRS Real?

Since MSG was first used as an additive there have been dozens of safety tests on it by countries worldwide, and no evidence has ever been found of any toxicity or health risks. After the CRS story, many health bodies, including the US FDA, performed a large number of double-blind experiments to find out if CRS was real. They used as test subjects people who were convinced they were susceptible to MSG as well as normal people, and gave them all food which did and didn't contain MSG, without telling them which was which. Often the symptoms appeared when the person had not eaten MSG – suggesting the person was allergic to some other ingredient, and had simply blamed MSG for lack of better information. Sometimes, a supposedly susceptible person unknowingly ate MSG but showed no effects – suggesting that any symptoms might be psychosomatic. In any case, under these conditions, and in every test performed, and in every country, no association was ever found with the CRS symptoms and MSG.

REALLY?

Yes, but this is not surprising. On average, a person consumes between 10–20 g of bound glutamate and about 1 g of free glutamate every day. The only chemical difference between them is that the bound glutamate comes in the form of a salt, such as MSG, while the other does not. And once ingested, the bound glutamate salts dissolve in the aqueous digestive liquids and becomes free glutamate. So it doesn't make any difference in what form we eat it (meat, mushrooms, or MSG), the glutamate ends up as the same molecule in the end. Furthermore, the human body synthesizes about 50 g of free glutamate every day, much of which is then used to make proteins, and at any one time the body contains about 10 g of free glutamate. So it's hard to see how a few milligrams of MSG added to a meal could be a health risk. The LD_{50} for rats (the amount they'd have to eat orally for 50% of them to die) for MSG is 15–18 g/kg body weight, which is 5 times greater (*i.e.* safer) than that for table salt. If this were to be scaled to human levels, a 70-kg person would need to eat over a kilogram of pure MSG for it to be fatal. So it's not surprising that all food regulatory bodies consider MSG to be completely non-toxic when eaten at normal levels.

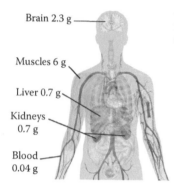

Brain 2.3 g

Muscles 6 g

Liver 0.7 g

Kidneys 0.7 g

Blood 0.04 g

There is approximately 10 g of free glutamate in the human body at any one time, with most being stored in the muscles and brain.

Compare this to the levels found per 100 g in typical foodstuffs:

• Parmesan cheese	1.680 g
• Scallops	0.014 g
• Mushrooms	0.042 g
• Corn	0.106 g
• Chicken	0.022 g
• Beef	0.010 g
• Tomato	0.246 g
• Chicken *chow mein* with added MSG	0.5 g

Source: www.glutamate.org

So Are All the 'MSG-Free' Signs Going to Disappear?

Probably not in the near future. Despite all the scientific evidence, it takes a long time for food scandals to be disbelieved by the suspicious (and litigious) general public. And with the advent of the Internet, rumors and uncorroborated anecdotes spread far more easily than the results of properly conducted scientific studies.

So You're Saying MSG Is 100% Safe?

Nobody can ever say that. It is known that the human body uses glutamates as neurotransmitters in the brain (see p169 and p451), and they are believed to be linked to memory retrieval. In fact, every major organ in the body contains glutamate receptors. The main concern with glutamates is the very fast rate at which they are absorbed into the blood by the alimentary canal. This can lead to glutamate levels in the blood rising very rapidly in a short time period, and some scientists are worried this may lead to overstimulation of the brain and neurotoxicity. Extremely high doses of glutamates have been shown to cause brain damage in rodents, but there is no evidence that the same happens in humans, and especially not at the very low concentrations of MSG present in an average chicken *chow mein*!

WHAT'S THE *TAKE-AWAY* MESSAGE (IF YOU PARDON THE PUN)?

As usual, it's be careful what you read and who you believe. Only properly conducted, large-scale, double-blind, scientifically rigorous tests should be considered as appropriate evidence. Unverified anecdotal stories should not be sufficient to start (or maintain) a food scare, although in today's scandal-hungry world sometimes that's all the popular media require.

Chapter

40

MORPHINE, CODEINE AND HEROIN

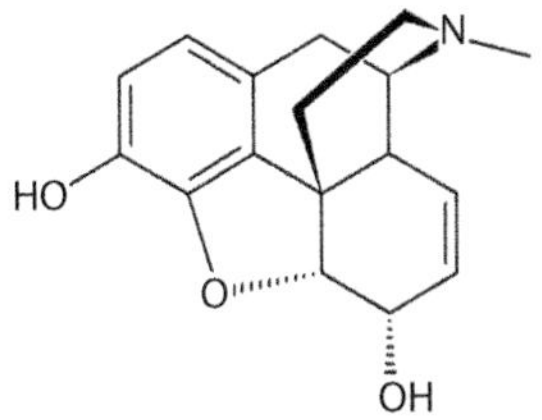

Morphine

Morphine? A Molecule That's Good at Changing Its Shape?

No, morphine, not morphin'.

It's a Painkiller, Right?

Morphine is the best molecule available for dulling acute and chronic severe pain. It takes its name from Morpheus, the Greek god of dreams.

How Long Has It Been Used?

Morphine itself was not isolated until 1804. It is one of a number of alkaloids found in the milky sap obtained when unripe seedpods of the opium poppy (*Papaver somniferum*) are scraped. On drying, this becomes a yellow-brown paste, which is known as opium. This contains over 20 alkaloid molecules – the most abundant being morphine – but the only other one with sedative properties is

A red opium poppy (*Papaver somniferum*) and its seedpod

A seedpod oozing opium

codeine, which only differs from morphine by replacing an OH group with an OCH_3 group.

The opium poppy is native to Asia Minor (the Eastern Mediterranean) but is now cultivated over a much wider area which extends into China and Southeast Asia. The ancient Sumerians in Mesopotamia traded opium as far back as 4000 BC. And opium came before morphine as a painkiller.

HOW DID ADDICTION START?

Early in the sixteenth century, a Swiss-born doctor usually known as Paracelsus popularized an extract of opium in brandy, which was known as laudanum. A century and a half later, the Oxford-educated doctor Thomas Sydenham promoted laudanum as a general cure-all. It became very popular in the eighteenth century, not least because it was half the price of beer, as well as being a treatment for cholera and dysentery. And it was safer than drinking the water in many places, which made it a boon to the poor.

A German bottle of laudanum from the nineteenth century. The label says it could be used to cure a variety of ailments, including headaches and diarrhea, and could even be given to 3-month-old babies!

Its use was taken up by a string of writers – Byron, Keats, Shelley and Southey dabbled, but Samuel Taylor Coleridge and Thomas de Quincy had a more serious drug problem, and Wilkie

Collins wrote *The Moonstone* while under its influence. Sadly, Lizzie Siddal, the original 'Supermodel' (for the Pre-Raphaelites) died of a laudanum overdose. But opium had a worse downside than that.

WHAT WAS THAT?

Opium was addictive, and its addiction caused wars. Tobacco was introduced to China toward the end of the sixteenth century, and its use quickly became a habit. Once tobacco was banned by the Chinese in 1644, opium took its place as a recreational drug among the Chinese merchants, much of this being supplied by the British East India Company. Its export offset the cost of tea imports to Britain. Chinese attempts to stop the opium trade led to the Opium Wars of 1839–1842 and 1856, following which Hong Kong became a British Crown Colony until 1997.

SO WHERE DOES MORPHINE COME IN?

Morphine was isolated at roughly the same time by Jean-François Derosne, Armand Séguin and Friedrich Wilhelm Sertürner, but as Sertürner's work was publicized by Gay-Lussac, he usually gets the credit. People thought that morphine was less addictive than opium and so could replace it. But it was found that morphine was not too effective orally (we now know that it is rapidly metabolized in the liver and intestine).

HOW DID THEY GET ROUND THAT?

Fortunately – or unfortunately – the hypodermic syringe was invented just in time (1853) to be used to inject measured doses of morphine into wounded soldiers in the American Civil War and the Franco-Prussian War. Hundreds of thousands of combatants in the American Civil War were said to have become addicted to it. One unfortunate problem is caused by the fact that it can cross the placental barrier, so babies born to mothers who are on morphine can go through a long period of withdrawal.

How Does Morphine Work?

Morphine and related molecules have exactly the right three-dimensional shape to fit into protein receptors in the brain and spinal cord. These receptors were first located by Solomon Snyder and Candace Pert at Johns Hopkins University, Baltimore, in 1973, as well as by others soon afterwards. For years scientists wondered why the brain should have receptors seemingly designed to fit molecules from a poppy extract, which seemed highly unlikely. They realized that these receptors were not there to bind morphine, but to bind unknown painkilling molecules with a similar structure to the opioids, that must be produced naturally by the brain. The search began for these natural painkillers, which turned out to be small proteins named enkephalins and endorphins.

According to one (simplified) view, the receptor has three main areas that bind morphine and its derivatives: a negatively charged site which binds the protonated amine nitrogen (A); a cavity into which the projecting piperidine ring (B) fits; and a flat surface to which the flat benzene ring (C) gets attached. For the opioid painkiller to be active, *all* these features must be present, as well as a quaternary carbon (D) attached to the piperidine ring and a C_2 spacer (E), part of the piperidine ring, between (A) and (D).

Isn't Codeine a Painkiller Too?

Codeine

Codeine is the second most abundant alkaloid in poppies, and the only other one that is a painkiller. It can be obtained from the Iranian poppy, *Papaver bracteatum*, though most is made from morphine. When taken orally, about 10% of codeine is demethylated by the liver enzyme sparteine oxygenase into morphine, and this is where the sedative effects originate. Codeine is a prodrug in effect; morphine is the effective agent.

How Do You Make Morphine?

The opium poppy (*Papaver somniferum*) makes morphine starting from the amino acid tyrosine. Morphine's structure was not finally solved until 1925 by Sir Robert Robinson. The first total synthesis in a chemistry lab was first achieved by Marshall and Gates in 1952. Making it in the lab is much more complicated and expensive than the natural route, so commercially morphine is obtained from poppies.

What's Morphine Got to Do with Heroin?

Heroin is made by acetylating both the –OH groups of morphine, a synthesis first achieved by Charles Romley Alder Wright at St Mary's Hospital, Paddington in 1874, but it was not until 1898 that Heinrich Dreser of the German Bayer company repeated this. To begin with, it was seen as a non-addictive morphine replacement in cough suppressant medicines, being given the name heroin because of its 'heroic' properties. This optimism was short-lived and sadly misplaced. Like codeine, heroin is a prodrug, being metabolized to morphine in the body. Replacing the OH groups in morphine by ester groups makes heroin less hydrophilic; as it is more lipid-soluble, it crosses the blood–brain barrier more readily, so that heroin supplies morphine to the brain faster.

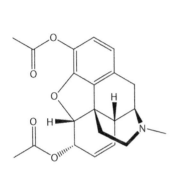

Heroin

Pre-WWI Bayer
heroin bottle, originally containing
5 g of heroin substance

In many parts of the world, heroin is used by doctors as a way of administering morphine, not least to the terminally ill, though they call it by its medical name of diamorphine (short for diacetylmorphine) so as not to worry the patients!

However, an overdose of heroin often proves fatal. In an infamous UK murder case in the 1990s, the serial killer Dr. Harold Shipman used it to kill over 250 elderly patients who were in good health. He would administer lethal overdoses of diamorphine, then sign patients' death certificates, and then forge medical records indicating they had been in poor health.

Heroin, of course, is notorious for its use as a recreational drug, and is known colloquially as smack, skag, horse, brown or H.

Jim Morrison   Janis Joplin   Sid Vicious

It is taken because of the feelings of relaxation and intense euphoria it induces. Most illegal heroin is in its freebase form (as opposed to medical heroin which is the crystalline hydrochloride salt). Freebase heroin can be injected or, because of its lower boiling point, smoked in a pipe. But in any form, heroin is highly addictive, and repeated use makes the user require more and more to obtain the same thrill. This often leads to overdoses which have proved to be fatal for many users, including famous names in the world of entertainment, such as Janis Joplin, Sid Vicious (Sex Pistols), Jim Morrison (The Doors) and Gram Parsons (The Byrds), not to mention Talitha Getty and Paula Yates. And heroin has formed the *leitmotif* of films like *The French Connection* and *Trainspotting*, as well as being a major source of funds for the Taliban.

IS IT TRUE THAT PEOPLE CAN FALSELY TEST POSITIVE FOR HEROIN?

Yes. Many firms test employees to see if they are abusing drugs, and all these tests rely on detecting morphine, irrespective of whether it entered the body as morphine or as heroin. It is now known that people who have eaten products like poppy-seed bagels can give false positives, because modern analytical techniques can detect the very low levels of morphine (or codeine) present in poppy seeds. In some cases people have even lost their jobs or had their babies taken away from them!

CAN'T YOU TELL THE DIFFERENCE BETWEEN SOMEONE WHO'S TAKEN MORPHINE AND SOMEONE WHO'S TAKEN HEROIN?

When heroin is metabolized in the body, there is a short-lived metabolite called 6-monoacetylmorphine (6-MAM), created by removal of one of the acetyl groups. It is believed that 6-MAM may be the source of the intense rush of euphoria associated with heroin taking, before it is ultimately converted to morphine. 6-MAM can only arise from heroin, but it persists in the body for less than a day, so it can only be detected within 24 h of that person taking heroin.

6-MAM
(6-monoacetylmorphine)

Nalbuphine

I'VE HEARD THAT MORPHINE WORKS BETTER ON MEN THAN ON WOMEN

People have noticed that women in childbirth prefer a painkiller called nalbuphine to morphine, while this order of preference is reversed for men in pain. It has been suggested that men and women are 'wired' with different pain circuits. Morphine binds to one type of receptor while nalbuphine binds to another type. It thus appears that nalbuphine and similar drugs have very limited analgesic ability in males and may in some cases enhance the pain.

WHY DO MEN AND WOMEN DIFFER IN THEIR RESPONSE TO PAIN?

Who knows? It may be that it reflects the fact that men and women experience pain in different situations (*e.g.* fighting versus childbirth). Perhaps one day painkillers will be gender-coded.

Chapter

41

NANDROLONE

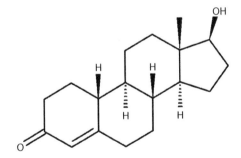

Somebody's been overdoing the
nandrolone!

British sprinter Linford Christie, who controversially failed a test for nandrolone in 2000 (© Press Association Images)

IT'S ONE OF THOSE STEROIDS THAT ATHLETES TAKE, ISN'T IT?

It is a steroid, yes, but it is only taken by just a few athletes and some other sportsmen.

NAME SOME NAMES

At different times, British athletes who have tested positive include Linford Christie (left), Mark Richardson and Dougie Walker, also the US world shot-put champion C.J. Hunter (who was then married to Marion Jones), the Jamaican sprinter Merlene Ottey and the German middle-distance runner Dieter Baumann. But there have also been soccer stars like Jaap Stam, Christophe Dugarry and Edgar Davids; the Czech tennis player Petr Korda and the English player Greg Rusedski; and Pakistani cricket fast bowlers Mohammad Asif and Shoaib Akhtar ('The Rawalpindi Express').

The medalists in the 200 m backstroke at the Moscow Olympics, with Rica Reinisch (center) claiming the gold, with a little chemical help!

SO IT HAS ONLY BEEN USED QUITE RECENTLY?

Sadly, no. In the 1970s and 1980s, nandrolone and its

derivatives were extensively administered by East German coaches and doctors to teen and pre-teen female swimmers. At the 1976 Montreal Olympic games, East German women won 11 of the 13 swimming gold medals. But medal-winning performances were bought at the expense of what it did to the bodies of the swimmers who were too young to know or consent to what was done to them. Brave women like Rica Reinisch and Karen König have testified in court about the long-term detrimental effects of these drugs upon their health.

How Do People Take It?

Most usually they inject a solution (in a suitable edible oil) of an ester of nandrolone, either nandrolone decanoate (also known as *Deca-Durabolin* or *Depot-Turinabol*) or nandrolone phenylpropionate (*Durabolin*). These are more easily assimilated and are then hydrolyzed in the body.

Nandrolone decanoate   Nandrolone phenylpropionate

Why Do Sportsmen Take It?

Nandrolone is very similar in structure to the male hormone testosterone (see p489 and p507), differing by one methyl group, and has many of the same effects in terms of increasing muscle mass, without some of the more unwanted side effects such as increased body hair or aggressive behavior. As such, it has been actively examined in clinical tests as a possible treatment for wasting diseases, and to strengthen and increase body tissue and musculature in HIV-infected men. It helps the body to build muscle by making more protein. It is also believed to enable athletes to recover faster from the effects of a training session.

Testosterone

19-Norandrosterone

HOW DO DRUG TESTERS KNOW THAT ATHLETES HAVE TAKEN IT?

Once in the body, nandrolone is metabolized, and the metabolism product of this molecule, called 19-norandrosterone, is excreted from the body in urine, making it easy to obtain samples, and this is the molecule that testers look for. However, nandrolone does occur naturally in the human body in tiny amounts, so the International Olympic Committee (IOC) set a limit of 2 ng per ml of urine for men and 5 ng per ml for women as the maximum concentration thought possible to occur in human body by 'natural means', and if this is exceeded the drug test is considered positive. Since some samples given by athletes have shown levels up to 100 times higher than this, the conclusion is that the athletes must have been taking extra quantities of the drug to enhance their performance.

Had They?

Some sportsmen came up with good excuses!

Bobsleigh racer Lenny Paul said that his positive test was due to eating a plate of spaghetti bolognese which contained beef from cattle that had been fed steroids.

Track-and-field athlete Daniel Plaza said that he had prolonged sex with his pregnant wife, as pregnant women may produce nandrolone. In fact, the female pregnancy hormone, progesterone, is structurally quite similar to nandrolone.

Progesterone

Tennis player Petr Korda blamed eating veal; it was calculated that he would have had to eat 40 calves a day for 20 years to achieve the test levels.

The German middle-distance runner Dieter Baumann gave test results that varied with the time of day; he claimed that the nandrolone was in his toothpaste.

But some people giving 'positive' tests were only just above the threshold. It was suggested that hormonal levels in some animals used for food may have been responsible for some athletes testing positive, and that some nutritional supplements may have been contaminated with nandrolone – though IOC rules say that what goes into an athlete's body is the responsibility of the athlete.

More strikingly, scientists at Aberdeen University found that a combination of allowed dietary supplements, such as creatine, together with hard cardiovascular exercise could lead to norandrosterone levels over the legal maximum. And a few cases have also been found of norandrosterone being formed by reaction between androsterone and etiocholanolone in stored urine samples.

SO THE TESTING FOR NANDROLONE IS UNRELIABLE?

No, the testing is reliable; it is just there to detect a nandrolone metabolite, without any concern for how it got there. Care needs to be taken over what is eaten in the run up to testing (*e.g.* avoiding pig offal), while a recent review lays down some guidelines to be followed in testing, which includes considerations such as possible pregnancy for women, and whether permitted medicines that give rise to the same metabolites had been taken.

Chapter

42

NICOTINE

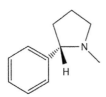

What's the Problem?

Worldwide, smoking tobacco causes over four million smoking-related deaths a year, due to coronary heart disease, emphysema and cancers of the mouth, throat and lung. It is also a contributory cause of many other cancers. Cigarette smoke contains around 4000 different compounds and its carcinogenic effect is due to chemicals such as benzo[a] pyrene, not nicotine. So nicotine does not cause these deaths directly, but *addiction* to nicotine does.

Why Is It Addictive?

When someone smokes tobacco, most of its nicotine content is destroyed. But the small amount that survives is rapidly carried – in less than 10 s – in the blood from the lungs to the brain, where it binds to 'nicotinic' acetylcholine receptors, and stimulates the release of neurotransmitters. One of these, dopamine, is a 'reward chemical' (see p169), and that seems to be behind the addiction, as the small time delay leads smokers to associate the drag on the cigarette with the pleasure sensation. When their nicotine levels drop, the smoker experiences withdrawal symptoms and opens the pack for another cigarette. Mark Twain once said: '*To cease smoking is the easiest thing I ever did. I ought to know, because I've done it a thousand times*'.

Just How Toxic Is Nicotine?

The lethal dose for a human is around 50 mg. Nicotine is rapidly absorbed, not just through the lungs, but also through the skin and mucous passages. A cigarette contains less than 10 mg of nicotine, but most of that is oxidized to other compounds on being smoked, with maybe 1 mg of nicotine being delivered to the smoker's lungs. Cigars contain much more nicotine, up to 100 or 200 mg; if all this was extracted and injected, the recipient would die! However, most of this nicotine is destroyed in combustion, together with the fact that most users smoke cigars slowly and do not inhale, means that a sub-lethal dose is delivered. Quite fortuitously. In fact, since nicotine gum usually comes in 2 or 4 mg doses, someone may actually get more nicotine from these, from snuff, or nicotine patches, than by smoking a cigarette.

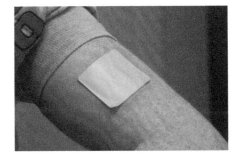

A nicotine patch

Of course, people trying to give up smoking often resort to nicotine patches, because nicotine is absorbed through the skin. A famous example of this was reported in 1933 – an American florist sat on a chair on which some nicotine-containing insecticide had been spilt; it soaked through the trousers he was wearing, then through his skin. He was hospitalized for several days, though the hospital did not know what caused his problem. When he was discharged, his clothes – which had been kept in a paper bag – were returned to him. When he put them on, he succumbed to a second dose of nicotine poisoning, necessitating a second hospitalization.

DID HE SURVIVE?

Yes, he was lucky – some people have died from absorbing nicotine via this route.

What Does the Body Do with That Nicotine?

Cytochrome P450 enzymes in the liver, particularly one called CYP2A6, metabolize it, mainly to cotinine (the name comes from an anagram of nicotine).

Cotinine

Cotinine levels in the blood or urine can be used as a check on whether that individual has been smoking in the last day or two, as this molecule has a lifetime in the body of around a day. Humans have one of three versions of the gene that codes for CYP2A6; each gene affords a slightly different version of the enzyme. The most common form of CYP2A6 is most efficient at destroying nicotine, and these people tend to be the heaviest smokers.

So If You Are Efficient, You Stand the Greatest Chance of Lung Cancer?

Could be. Smoking is such a drag.

Why Do People Put on Weight When They Stop Smoking?

For years people have known that smoking reduces appetite. Smokers are, on average, 2.5 kg lighter than non-smokers. In recent research using mice which were given doses of nicotine for a month, the mice cut their food intake by 50% and lost 15–20% of their body fat. Professor Marina Picciotto of Yale University has found that this is linked with influencing the hypothalamic melanocortin system by binding to the receptor in the brain that is associated with addiction.

Is Nicotine Always a Bad Thing?

Not if you are a tobacco plant (*Nicotiana tabacum*) which can contain 2–4% nicotine in its leaves. It makes nicotine to protect itself against insect predators. When the insects absorb and inhale it, nicotine binds to their acetylcholine receptors, paralyzing the insect. It is good for the plant, but

Benzylacetone

bad for most insects. Even in tiny doses it is fatal to them. People have used nicotine as an insecticide since the early eighteenth century.

A wild tobacco plant from the Mojave Desert in the United States (*Nicotiana attenuata*) uses very low doses of nicotine in its nectar. The plant uses sweet-smelling benzylacetone to attract insect pollinators, but the bitter taste of the very small amounts of nicotine deters visitors such as hawkmoths and hummingbirds from taking too much nectar.

You Said 'Most Insects'?

Along with tomato, tobacco forms a sizeable chunk of the diet of the tobacco hornworm *Manduca sexta*, and escapes harm by binding the nicotine. The cigarette beetle, *Lasioderma serricorne*, appears to excrete nicotine and is unharmed by it.

But Doesn't It Kill Bees Too?

Not nicotine itself, but there are a number of insecticides based on nicotine that have recently been implicated in the recent serious decline in the bee population. These insecticides

A beehive full of dead bees that died of starvation. This could be caused by a harsh winter. But one effect of the neonicotinoids is to disrupt the bees homing sense, so they cannot find their way back to the hive. Without bees bringing food back, the hive starves

are collectively known as neonicitinoids ('new nicotine-like chemicals') which act on the nervous system of insects, but which have a much lower health risk to mammals than older sprays. They are water-soluble and are applied to the soil, then taken up into the plant as it grows. In effect, the whole plant is turned into a poison factory, and the insects die on contact with almost any part of the plant.

A report published by the European Food Safety Agency (EFSA) in January 2013 concluded that these pesticides posed a 'high acute risk' to pollinators, including honeybees. Although the evidence is not conclusive, the EU banned the use of three of these neonicotinoid pesticides, clothianidin, thiamethoxam and imidacloprid (currently the most widely used insecticide in the world), for 2 years from 2014 onwards. Time will tell if the bee population recovers.

Clothianidin   Imidacloprid   Thiamethoxam

ARE THERE ANY BENEFITS OF NICOTINE TO HUMANS?

Possibly. Sufferers from Tourette's syndrome are helped by using a nicotine patch to stimulate the haloperidol medicine they are taking. It is also believed that smokers are less likely to contract Parkinson's disease, but smoking is a rather drastic way to get those benefits.

IS THERE A WAY AROUND THIS?

Cotinine may be the answer. A research team in Augusta, Georgia, led by the late Dr. Jerry Buccafusco, discovered that cotinine may help enhance memory, and protect brain cells. The next step is to see if cotinine can help protect against Alzheimer's and Parkinson's. It could have the benefits of nicotine, without having the addictive (and other) side effects of nicotine.

AND HOW LONG HAVE PEOPLE BEEN SMOKING TOBACCO?

Well, archeologists detected nicotine in a late Mayan flask (*ca.* 700 AD), but it goes back much further than that, perhaps to 5000 BC. It seems that smoking is a long-term addiction for the whole human race!

Chapter

43

Nitrous Oxide, N₂O
(Laughing Gas)

$$N\equiv\overset{\oplus}{N}-\overset{\ominus}{O} \qquad \overset{\ominus}{N}=\overset{\oplus}{N}-O$$

'I am sure the air in heaven must be this wonder-working gas of delight'.

WHO SAID THAT?

It was Robert Southey (1774–1843), who was the Poet Laureate of the United Kingdom for the last 30 years of his life.

AND HE WAS TALKING ABOUT NITROUS OXIDE?

Yes – which was pretty new then. N_2O is a colorless, almost odorless gas, that was discovered in 1772 by the English scientist and clergyman Joseph Priestley (who was also famous for being the first to isolate other important gases such as oxygen, carbon monoxide, carbon dioxide, ammonia and sulfur dioxide). Priestley made N_2O by heating ammonium nitrate in the presence of iron filings, and then passing the gas that came off through water to remove toxic by-products. The reaction he observed was

$$2NO + H_2O + Fe \rightarrow N_2O + Fe(OH)_2$$

Joseph Priestley   Sir Humphry Davy

The French chemist Charles-Louis Berthollet (1748–1821) made N$_2$O by heating ammonium nitrate in 1785. Sir Humphry Davy repeated this in 1800, and it remains the current industrial method for making N$_2$O, by heating ammonium nitrate at 245–270°C, with a risk of explosion if heated above ~290°C.

$$NH_4NO_3(l) \rightarrow N_2O(g) + 2H_2O(g)$$

So What Did They Do with It?

After initial trials, Priestley thought that N$_2$O could be used as a preserving agent, but this proved unsuccessful. Priestley noted that it would support combustion of a candle, but a mouse exposed to it would die.

Following Priestley's discovery, Humphry Davy of the Pneumatic Institute in Bristol, England, experimented with the physiological properties of the gas. On April 16th 1799, Davy noted the effects of inhaling N$_2$O, obtaining what he described as a '*semi-delirious trance*'. Perceptively, he commented, '*As nitrous oxide in its extensive operation appears capable of destroying physical pain, it may probably be used with advantage during surgical operations in which no great effusion of blood takes place*'.

He even administered the gas to visitors to the institute, and after watching the amusing effects on people who inhaled it, coined the term 'laughing gas'! For the next 40 years or so, the primary use of N$_2$O was for recreational enjoyment and public shows. The so-called 'nitrous oxide capers' took place in traveling medicine shows and carnivals, where the public would pay a small price to inhale a minute's worth of the gas. People would laugh and act silly until the effect of the drug came to its abrupt end, when they would stand about in confusion. Many famous people (of their

time) and dignitaries from Clifton and Bristol came to inhale
Davy's purified nitrous oxide for recreational purposes. A poster
for these events proudly proclaimed:

> '*Gallons of gas will be prepared and administered to*
> *all in the audience who desire to inhale it. The effect*
> *of the gas is to make those who inhale it, either laugh,*
> *sing, dance, speak or fight, according to the leading*
> *trait of their character. Men will be invited from the*
> *audience to protect those under the influence of the gas*
> *from injuring themselves or others. Probably no-one will*
> *attempt to fight*'.

And it goes on to say, '*The gas will be administered only to gentlemen*
of the finest respectability. The object is to make the entertainment in every
respect, a genteel affair'.

Davy even demonstrated the effects of N_2O upon members of the
audience at Royal Institution lectures, with effects satirized by
Gillray, the caricaturist.

Satirical cartoon by James Gillray showing a Royal Institution lecture on
pneumatics with Davy holding the bellows

So Nitrous Oxide Became an Anesthetic?

Not immediately. For the next 40 years or so, it was a laboratory curiosity, used mainly for public entertainment both in Europe and the United States. The story goes that at that time, a medical school dropout called Gardner Quincy Colton went around the US putting on nitrous oxide exhibitions. In 1844 he happened to put on a demonstration in Hartford, Connecticut, and in the audience that day was a local dentist named Horace Wells. Dr. Wells watched with interest as one of the volunteers, a man named Samuel Cooley, inhaled the gas, and, while still under the effects of the N_2O, injured his leg when he staggered into some nearby benches. When he went back to his seat next to Dr. Wells, Cooley appeared to be unaware of the injury until the effects of the gas wore off. Dr. Wells immediately realized that N_2O might possess painkilling qualities, and so after the demonstration, Wells approached Colton and invited him to participate in an experiment the next day. Colton agreed and administered nitrous oxide to Dr. Wells while another local dentist extracted one of Wells' molars. Dr. Wells experienced no pain during the procedure, and N_2O as a dental and medical painkiller had arrived.

So That Was It?

Early surgical operations, such as the one shown in the photo, used ether as an anesthetic. This was dangerous because ether is highly flammable. So a new anesthetic was needed that was not so dangerous to use, and nitrous oxide looked like the ideal candidate.

However, there was no happy ending to this story. In January 1845, Dr. Wells demonstrated his discovery of the effects of nitrous oxide at the Harvard Medical School in Boston. A patient was anesthetized and a tooth was extracted, but during the

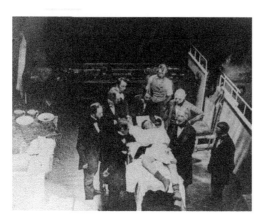

A re-enactment of the first operation to use ether as an anesthetic in Boston, MA in 1846

demonstration the patient complained that he felt some discomfort. Even though the experiment had been successful (in that the patient had only felt slight discomfort and *not* excruciating pain), the suspicious audience were unhappy, and booed Wells from the stage. This public humiliation eventually led to Dr. Wells losing his reputation as a professional dentist, and finally to his suicide three years later. Ironically, 15 years after his premature death, his discovery would be adopted by dental practices worldwide, and Wells would be given the accolade – the 'Discoverer of Anaesthesia'.

WHY IS IT SUCH A GOOD ANESTHETIC?

Nitrous oxide is a very safe and popular agent still utilized by dentists today. It is much less toxic than alternatives, such as chloroform, with far less risk of explosion than ether. Its effects are felt quickly and people get over it fast, too. The main use for N_2O is usually as a mild sedative and analgesic. It helps to allay anxiety that many patients may have toward dental treatment, and it offers some degree of painkilling ability. This use has passed into popular culture.

FOR EXAMPLE?

The Mack Sennett silent comedy *Laughing Gas* (1914) starred Charlie Chaplin as a dentist's assistant, pretending to be a dentist, while the plot of P.G. Wodehouse's novel *Laughing Gas* (1936) revolves round a switch of personalities between two bodies while they are simultaneously under anesthetic at the dentist's.

How Does It Work as an Anesthetic, Anyway?

A lot of research has gone into this, and no one is quite sure. N$_2$O has been suggested to affect various ligand-gated ion channels, and causes the release of opioids (molecules similar to morphine), and it is believed that it these opioids that send you to sleep.

So That Is the Main Use of N$_2$O?

Actually, its main use is as a propellant; its unreactivity means it can take the place in aerosol cans of the chlorofluorocarbon gases (CFCs) which can damage the ozone layer (see p125). It is used as an aerating agent for 'whipped cream' as it is unreactive, tasteless and soluble in vegetable fats under pressure. It 'undissolves' when pressure is released, causing the aerated 'whipped' effect. It is non-toxic in the quantities used and even has a European food additive E number (E942).

Why Is It Used in Racing Cars?

N$_2$O is normally pretty unreactive, but when heated to above 600°C it decomposes exothermically to N$_2$ and O$_2$:

$$2N_2O(g) \rightarrow 2N_2(g) + O_2(g)$$

If this reaction occurs in the combustion chamber

of an automobile, 3 moles of gas would be produced from 2 moles, providing an extra boost to the piston, as well as liberating more heat. It also has a number of other benefits. The increased oxygen provides more efficient combustion of fuel, the nitrogen buffers the increased cylinder pressure controlling the combustion, and the latent heat of vaporization of the N_2O reduces the intake temperature. Therefore, liquid N_2O is occasionally injected into the fuel line of racing cars to give more power to the engine and to give the car exceptional acceleration.

So Nitrous Oxide Is All a Big Joke?

No, there are two particular problems with nitrous oxide, and neither is anything to laugh at. One is its effect on the ozone layer and also on global warming.

Meaning?

N_2O is formed by the action of bacteria in soils and the oceans, so it is present naturally in the atmosphere, but sewage plants, livestock and the breakdown of fertilizers are responsible for more N_2O released. In terms of its greenhouse effect, N_2O is in much the same league as CO_2 (see p61) and methane (see p323). It's also been suggested that N_2O may be the dominant man-made ozone depleter, as when stratospheric N_2O is decomposed, it can form NO, which reacts with ozone (see p387).

And the Other Problem?

People are concerned that clubbers are abusing it, in a throwback to the days of nitrous oxide capers. It has been used at events like the Notting Hill Carnival and the Glastonbury Festival, as well as nightclubs. People pay a small sum of money for a 'hit' lasting a couple of minutes – it is very short-lived. When a 23-year-old

Englishman named Daniel Watts was found dead at his home in September 2006, he had a bag over his head and there was a large cylinder of nitrous oxide was beside him. It is thought that he placed a bag over his head to increase and lengthen the effects of the N$_2$O. The bag stopped his inhalation of oxygen, and death from asphyxia was the result. Maybe N$_2$O's not such a laughing matter after all …

Chapter

44

1-Octen-3-ol

or 'Mushroom Alcohol'

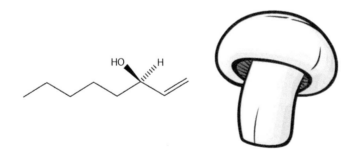

What's Special about This Substance?

Well, some of its molecules smell of mushrooms.

What Do You Mean, Some? They Either Do or Don't

Because this molecule has a chiral carbon, with four different groups attached, it can exist as two optical isomers, denoted *R* and *S*. One smells of mushrooms, and the other smells rather mouldy. (The 'oct' in its name means it has eight carbons in the chain, the number 1 refers to the double bond being between the first and second carbons in the chain, and the '3-ol' tells you that the –OH group is attached to the third carbon.)

(*S*)-(+)-1-octen-3-ol
Mouldy, grassy smell

(*R*)-(–)-1-octen-3-ol
Mushroom smell

Why Do the Two Isomers Smell Differently?

Although the molecules have the same atoms joined up in the same sequence, the atoms are oriented differently in space, so they 'plug into' and activate a different combination of olfactory receptors in the nose. (*R*)-(–)-1-Octen-3-ol is the main component of mushroom flavor, and is indeed known as 'mushroom alcohol' – mushrooms only make this isomer. It is largely synthesized in the cap and gills of the mushroom.

Why Do Mushrooms Smell of This?

Linoleic acid is a dominant fatty acid in mushrooms (see p287). Enzymes in the mushroom break it down and 1-octen-3-ol is one of the products. The process involves two enzymes, generating 1-octen-3-ol and 10-oxo-*trans*-8-decenoic acid (ODA). In plants, different enzymes break down linoleic acid in other ways, leading to molecules like hexenal (see p237). 1-Octen-3-ol and ODA can also be produced with a crude enzyme homogenate from mushrooms in a bioreactor, and this process is used industrially to make 1-octen-3-ol.

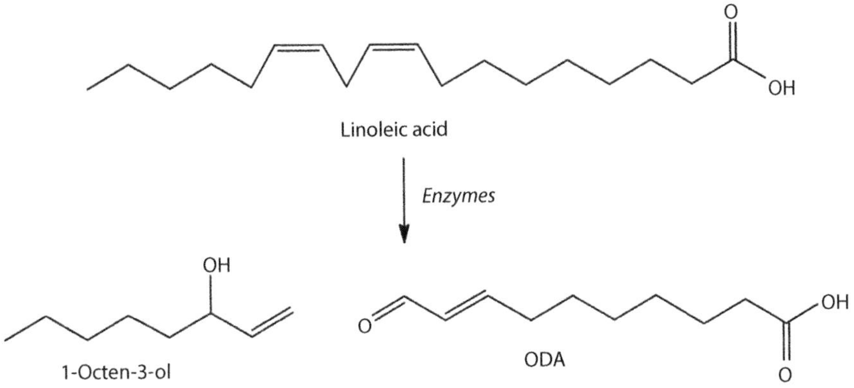

Linoleic acid

Enzymes

1-Octen-3-ol

ODA

A future commercial source could be a member of the mint family, *Melittis melissophyllum*. The plant itself contains little of the compound, but on hydrodistillation of the flowering aerial parts, the essential oil obtained was found to contain a large amount (43–54%) of 1-octen-3-ol.

Melittis melissophyllum

So You Only Meet 1-Octen-3-ol in Mushrooms?

No, apart from mushrooms (and truffles), it crops up in many other foodstuffs, notably giving the odor of Camembert cheese a mushroom note. It has also been found in blue cheeses, and in some fruit sources, such as raspberries and orange juice oil, in elder flowers and in Australian prawns and sand-lobsters.

Mushrooms...   ...Camembert cheese...   ...and prawns...
...all contain 1-octen-3-ol

But Is It a Nice Smell?

Some mosquitoes seem to love it! 1-Octen-3-ol is in the breath and sweat of mammals, most notably cows. 1-Octen-3-ol attracts mosquitoes, especially in conjunction with CO_2, the two components having a synergistic effect, and has been used in mosquito traps. There is an active study of molecules like 1-octen-3-ol which may either attract or repel insect pests, including tsetse flies and the Scottish biting midge. The neuron for the octenol receptor in the Yellow Fever mosquito (*Aedes aegypti*) has recently been examined and found to be over 100 times more sensitive to the (*R*)-isomer of 1-octen-3-ol than to the (*S*)-isomer.

Many insects seem to like 1-octen-3-ol, and one millipede (*Niponia nodulosa*) emits 1-octen-3-ol, along with geosmin, possibly as an alarm pheromone. On the other hand, *Clitopilus prunulus* mushrooms are rejected by the coastal Pacific Northwest banana slug, *Ariolimax columbianus*, possibly because they don't like the smell of 1-octen-3-ol.

"Eeeew, mushrooms....how disgusting!"

Is 1-Octen-3-ol the Only Molecule to Smell Like This?

1-Octen-3-ol has a very low odor threshold (~1 ppb), but its oxidation product, 1-octen-3-one, has an even lower threshold and a very strong smell of mushrooms. It's been suggested that this molecule also contributes to mushroom smells and possibly to the smells of other foodstuffs.

1-Octen-3-one

When a sweaty hand touches an iron object, a 'musty' metallic smell can often be detected. 1-Octen-3-one is the key odorant causing this, along with various aldehydes and ketones. Back in the 1990s, people living near a Canadian car-painting factory were surprised by a smell of mushrooms pervading the neighborhood. This was traced to 1-octen-3-one formed by reaction of methanal in paint resins with methylamylketone (heptan-2-one) used as a solvent.

Chapter

45

OXYGEN

(AND OZONE)

WE CAN'T DO WITHOUT IT, CAN WE?

Life on Earth depends on it. It's not just that it is found in most of
the compounds associated with living things, but most organisms
need it to live. It is essential to aerobic respiration.

WHY IS THAT?

In aerobic respiration, this reaction occurs:

$$C_6H_{12}O_6 + 6O_2 \rightarrow 6CO_2 + 6H_2O \text{ (+ lots of energy)}$$

It requires oxygen molecules (as well as glucose, see p193), and
produces energy, which is used to power muscle contractions
essential to movement, and also to maintain body temperature (if
you are a mammal or a bird). The energy is also used to drive other
chemical reactions in the body, making molecules like amino acids
and proteins. The more energy required, the more energy is used,
which is why mammals need hearts to pump oxygenated blood,
and require hemoglobin as an oxygen carrier (see p227).

WHERE DOES ALL THIS OXYGEN COME FROM?

In a word, photosynthesis. Green plants (that includes green algae)
as well as marine cyanobacteria do it. They absorb light, mainly
thanks to chlorophyll, which drives the conversion of carbon
dioxide and water into glucose (and thence other carbohydrates)
and oxygen (see p81).

$$\text{(Energy from sunlight +) } 6CO_2 + 6H_2O \rightarrow C_6H_{12}O_6 + 6O_2$$

This is almost the exact reverse of respiration, so you can see that
the net effect of the two reactions is to take energy in the form

of sunlight and convert it into chemical energy that our bodies require to survive.

Before life evolved, the Earth's atmosphere was composed of mostly those gases released from volcanoes, such as CO_2, SO_2, N_2 and methane. But about 2.6 billion years ago photosynthetic cyanobacteria evolved that began to create lots of oxygen as a waste product. For many millions of years, most of this oxygen was chemically captured by iron in the ground (forming rust, iron oxide) or other carbonaceous matter to form CO_2. But

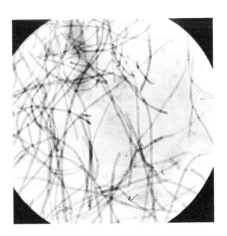

Cyanobacteria such as these were responsible for creating most of the free oxygen in the atmosphere

eventually these oxygen sinks were filled up, and free O_2 molecules started flooding into the atmosphere. This is known as the Great Oxygenation Event, and was the trigger which allowed all of the rest of life on Earth to evolve, using O_2 as the oxidizing agent to power their metabolisms.

I SEE, AND NOW THERE'S LOTS OF OXYGEN IN THE ATMOSPHERE?

O_2 now makes up 21% of the air, most of the rest being nitrogen (78%) and the noble gases (mainly argon), plus a tiny fraction of CO_2. It's interesting to think that if the O_2 content of the atmosphere were only a few percent lower, respiration would be far less efficient and many higher life forms could not survive. And if it were only a few percent higher, then combustible material would burn much easier – forest fires would become unstoppable, and much of the planet would become a burnt wasteland. Luckily for us, 21% is just about the right balance between the two extremes.

BUT CAN WE BREATHE 100% OXYGEN?

Yes, but only for a short period of time. Oxygen is actually considered toxic at concentrations above about 50% at normal air pressure. The reason for this is that although oxygen gas is composed almost entirely of O_2 molecules, it also contains a very small number of oxygen atoms and ozone molecules, which biologists call 'oxygen free radicals'. These are very reactive, and can destroy proteins and membranes in the delicate linings of cells in the lungs, causing them to leak fluid. Eventually the person 'drowns' in their own bodily fluids. This is why hospital patients on ventilators or scuba divers have to be very carefully monitored to ensure the elevated oxygen levels they breathe are not above the safe limits. However, if the air pressure is lowered, then the toxic limit for oxygen increases. For example, the astronauts in the Gemini and Apollo programs breathed 100% oxygen for up to 2 weeks with no problems, but that was because

of the low total pressures used, which meant that the actual concentration of dissolved O_2 in their blood was not much different to that of a person at sea level.

Scuba divers can breathe normal air mixtures, but often use *nitrox*, which is air with 32%–36% O_2 and less N_2, which helps prevent decompression sickness ('the bends'). To remove N_2 without running the risks associated with increased oxygen concentration, helium is added. The resultant mixture of $O_2/N_2/He$ is called *trimix*, and when the nitrogen is fully substituted by helium, it's called *heliox*, which is often used for very deep dives.

WHY IS OXYGEN GAS SO REACTIVE?

That depends on the temperature. Fortunately, at room temperature, many fuels do not react (*e.g.* wood, petrol) in the absence of a spark, but when the activation energy is exceeded, they burn well.

Take the reaction between methane (the main ingredient of 'natural gas', see p323) and oxygen. For many of us, we use this reaction every time that we cook something or use our central heating. Fuels like methane give out a lot of heat when they burn, because a lot of energy is released when very strong H–O bonds (bond energy 463 kJ mol^{-1}) and C·O bonds (bond energy 740 kJ mol^{-1}) are formed, even allowing for the energy input needed to break strong O·O bonds (bond energy 498 kJ mol^{-1}).

Combustion of methane requires breaking of 4 × C-H bonds and 2 × O = O bonds (+2640 kJ mol^{-1}), but makes 2 × C = O bonds plus 4 × O-H bonds (−3332 kJ mol^{-1}), giving a net energy release of −692 kJ mol^{-1} (This is an approximation to the accurate value of 890 kJ mol^{-1} because we used average bond energies rather than the specific values for the molecules involved.)

MAGNESIUM DOESN'T FORM COVALENT BONDS, YET Mg BURNS VERY BRIGHTLY INDEED

That's true. Magnesium forms magnesium oxide when it burns in oxygen which is made up of small Mg^{2+} and O^{2-} ions. Because the ions are so small yet strongly charged, they have a very strong attraction for each other. This means that they are tightly bonded together in magnesium oxide. So MgO has a very high lattice energy (around -3900 kJ mol^{-1}) which provides a driving force for this very exothermic reaction. It also has a very high melting point (around 2850°C) for the same reason.

HOW DO YOU MAKE OXYGEN?

Because air is a mixture, not a compound, the components can be separated by making use of the differences between their boiling points: nitrogen (77 K, −196°C) and oxygen (90 K, −183°C) in the fractional distillation of liquid air. Liquid nitrogen and liquid oxygen also differ in their colors, colorless and pale blue, respectively. Oxygen has a small but significant solubility in water, essential to marine and aquatic life. Its solubility decreases with increasing temperature, so a (warm) tropical fish tank requires an aerator.

A tube of liquid oxygen, with bubbles of O_2 gas

WHY IS LIQUID OXYGEN BLUE?

Because it feels cold? No, only joking. It is because some frequencies of visible light, mostly in the red and orange, can be absorbed by liquid oxygen. The reflected or transmitted light is depleted in reds and oranges and so looks bluish. The exact mechanism is quite complicated and involves simultaneous absorption of two photons by two different oxygen molecules, but we won't go into that now.

What Use Is Oxygen, Apart from Our Breathing It?

Over 100 million tonnes of oxygen is made each year in industry. Most of that is used to make steel, but other uses include supporting combustion and generation of high temperature (oxyacetylene cutters) and in life support, the latter in health care (oxygen tents and masks, emphysema treatment) and by divers, in aircraft and in spaceflight.

You can also make oxygen by electrolysis of water into its component parts, O_2 and H_2. These gases can be stored and then recombined again in a controlled way to produce power, with water as the only waste product. This forms the basis of hydrogen-powered engines and fuel cells, and is one of the big hopes for clean energy generation in the next few decades.

What about Ozone?

Ozone, O_3, is an allotrope of oxygen, which is less stable than O_2. It is a strong oxidizing agent and that makes it harmful to life. You can sometimes smell its chlorine-like odor near the seaside, after lightning storms, or by electrical equipment like photocopiers (another good reason for not standing round the office photocopier), because it is formed by the action of electrical discharges upon O_2 molecules. At ground level, that is bad, but when it is formed in the upper atmosphere, then it's good.

How Does That Happen?

Ultraviolet light generates O atoms by splitting O_2, and they react with another molecule of O_2 forming a molecule of O_3. These ozone molecules absorb other wavelengths of UV light which decomposes them back to $O + O_2$, continuing the cycle. That is why the 'ozone layer' around 30 km above the Earth's surface filters out (harmful) UV light, which could really damage our skin.

And That Is Where CFCs Come In?

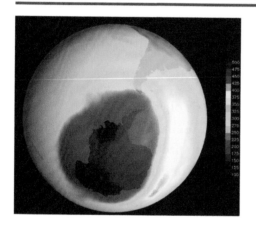

NASA satellite image of the 'ozone hole' which forms over the polar regions each year as a result of ozone depletion due to reactions with CFCs

Yes, back in the mid-1970s, it was found that the chlorofluorocarbons (see p125) used in refrigerants, solvents and aerosols survived long enough to reach the upper atmosphere, whereby the UV light them broke up and formed very reactive chlorinated radicals. These reacted with the protective ozone molecules and destroyed them. This led in particular to the 'holes' in the ozone layer over the Earth's polar regions, allowing increasing amounts of UV light to reach much of the Earth's surface. Frank Sherwood Rowland and Mario Molina, who made this discovery, were awarded the 1995 Nobel Prize for Chemistry, along with Paul J. Crutzen. Rapid international action led to the 1987 Montreal protocol phasing out CFCs so that the ozone layer could recover.

You Say That Ozone at Ground Level Is Bad

Yes, it is very bad for the respiratory system. It is generated on hot sunny summer days when the sunlight acts on mixtures of hydrocarbons and nitrogen oxides emitted from car exhausts to form peroxyacylnitrates (or PANs, with general formula $RCOOONO_2$ with R = alkyl). These are the compounds which typically make your eyes water on a hot summer's day in city streets. Ozone is formed from them when they decompose.

Is Ozone Any Use?

Apart from its irreplaceable role as a filter in the atmosphere, commercially, it can be used as an alternative to chlorine to disinfect water supplies and also to kill bacteria in hospital operating theaters. Organic chemists find it a useful reagent in chemical synthesis.

Why Is Ozone Bent: After All, CO_2 Is Linear?

The central oxygen atom in O_3 has two more electrons than the carbon in CO_2. These have to be accommodated as a lone (non-bonding) pair. The repulsions between the three sets of electron pairs around that central atom are minimized by keeping them as far apart as possible. It is not a perfect triangle as there are 'on average' three electrons between each pair of oxygen atoms, so it can be described as having the average of two 'resonance' structures.

FINALLY, DOES OXYGEN HAVE ANYTHING TO DO WITH THE PHONE COMPANY, O_2?

Indirectly... O_2 is the trading name of a group of telecommunication companies that come under the umbrella of Telefónica Europe, plc. that sell broadband and cellphone services worldwide. The company was re-branded as 'O_2' in 2002 suggesting the idea that the company supplied services that were essential, much the same as oxygen is essential for life.

Chapter

46

OXYTOCIN

Tyr

OH

H₂N

Cys

S

S

Pro

Cys

Leu

Asn

Gln

Ile

HN

HN

NH

NH

NH₂

NH₂

H₂N

Gly

Oxytocin

TABLOID HEADLINES?

No, the world's number 1 science journal, *Nature*.

WHAT HAS THIS GOT TO DO WITH OXYTOCIN? I THOUGHT THAT WAS ALL TO DO WITH PREGNANCY AND FEEDING BABIES?

Breastfeeding releases oxytocin which helps bonding between baby and mother

It is, but there is a lot more to it than that. In humans, oxytocin is produced in the hypothalamus and released from the posterior pituitary gland. There are receptors in many parts of the brain, as well as elsewhere in the body, such as the reproductive system. In lactating mothers who are breastfeeding, oxytocin causes milk to be released in the breasts, so that the infant can feed at the mother's nipple. Oxytocin is also responsible for the dilation of the cervix during birth, and for contractions during labor. Its name actually derives from 'quick birth' in Greek.

It has been used in maternity wards as the drug *Pitocin*, where it has been given to heavily pregnant women to induce labor, although it is not as effective as the natural release of oxytocin. It has now been found to have a role in the male reproductive system, too, being synthesized in the male testis, epididymis and prostate.

WHO DISCOVERED IT?

Oxytocin was discovered by a British pharmacologist named Henry Dale in 1909. It was first synthesized in 1953 by Vincent de Vigneaud, an American chemist who was awarded the 1955 Nobel Prize for Chemistry for making oxytocin, the first polypeptide hormone to be synthesized. He also synthesized the related hormone vasopressin, which controls the body's retention of water.

IS THAT ALL THAT OXYTOCIN DOES?

Human studies suggest that oxytocin levels are elevated by massage and by close intimacy, whether cuddling or by intercourse. Oxytocin is implicated in the mother–baby relationships, and in the very action of falling in love. You might say that it adds a whole new meaning to the phrase 'chemical bonding'. For this reason it's often referred to as 'the love hormone'.

Cuddling and intimacy increase oxytocin levels and strengthen relationships

SO WHERE DOES THE LOVE-RAT ENTER THE STORY?

Scientists have been investigating the prairie vole (*Microtus ochrogaster*) for well over 20 years. Like humans, it is one of the 3% of mammals that form lifelong monogamous partnerships, and share tasks like raising their offspring. They also have exciting sex lives. Researchers were struck by the differences

between the prairie vole and its close relatives, the montane
vole (*Microtus montanus*) and the meadow vole (*Microtus pennsylvanicus*), with the leading research being carried out by
C. Sue Carter, now at the Research Triangle Institute in North
Carolina. Though they are genetically extremely similar, the
other voles do not form monogamous relationships, and the male
does not take on a parenting role. In short, scientists asked 'Why
do voles fall in love?'.

The prairie vole (*Microtus ochrogaster*) has a love life that closely mimics
that of humans (With permission of Ryan Rehmeier)

It turns out that female prairie voles have many oxytocin
receptors in their brains, while males have similarly many
receptors for both oxytocin and vasopressin. However, there are a
lot fewer in the montane vole. Blocking these receptors in prairie
voles prevents them from forming their usual pair bonds, while
introducing the vasopressin receptor gene into male meadow
voles made them monogamous (though the role of a single gene
in this is disputed).

It began to look as if oxytocin was implicit in sexual fidelity,
and that prairie voles were a model for human love. But

Oxytocin is important in fidelity ... and infidelity

paternity tests on prairie voles then found that up to a
quarter of the litters had *not* been fathered by the life
partner. So they are socially monogamous but not genetically
monogamous. As Professor Ophir and his co-researchers at
the University of Florida commented: '*Somewhat ironically, this
distinction between prairie voles and other monogamous rodents – the
dissociation of social and sexual fidelity – leads us to suggest that
prairie voles are even better models of human attachment than has
been appreciated*'.

THAT'S IT FOR OXYTOCIN, THEN?

Because oxytocin reduces fear and promotes trust, scientists have
studied people playing 'investment games', in which subjects were
asked to give money to a trustee to invest on their behalf. Subjects
who had inhaled oxytocin from a nasal spray were found to show
greater levels of trust than control groups. Indeed, subjects whose
trust had been abused in the game were able to 'forgive and
forget' if they had received oxytocin. Scientists suggested that
oxytocin might be able to help people with autism and sufferers
from shyness. Others found that oxytocin kept 'partnered' men at

a greater distance than bachelors from an attractive woman, more support for the 'fidelity' theme.

Oxytocin is implicated in feelings of trust within one's own group, but also in feelings of prejudice toward outsiders

But things don't seem to be quite that simple. Evidence is mounting that the effects of oxytocin depend upon the person with whom the subject interacts. Another 'game' study found that people exposed to oxytocin were more favorable to their own team but more likely to punish competitors, and, indeed, others have found that oxytocin promotes favoritism to peoples' own ethnic group and prejudice against others from different groups. When oxytocin sprays were administered to someone with borderline personality disorder, they became less trusting. So if people are oversensitive, then oxytocin may make them overreact to social cues. It's quite ironic that the so-called love hormone could also encourage prejudice and bigotry!

It seems that oxytocin isn't a universal panacea, and we are only part of the way to understanding this fascinating molecule.

Chapter

47

PARACETAMOL/ ACETAMINOPHEN

(THE PAINKILLER KNOWN BY THE BRAND NAMES TYLENOL OR PANADOL)

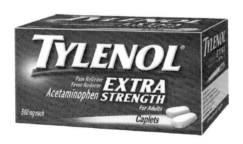

Paracetamol in the form of Tylenol® pills… (McNeil Consumer Healthcare Division of McNeil-PPC, Inc., Maker of Tylenol®. With permission)

… and paracetamol the molecule

WHY THE TWO NAMES FOR IT?

There are actually several names for it, which are mostly contrived from contractions of the full chemical name (see 'Name Game'). In the United States, Canada, Japan, South Korea, Hong Kong and Iran it's known as 'acetaminophen', while in the rest of the world it's called 'paracetamol'. The two best-known brand names are called 'Tylenol' and 'Panadol', both of which were introduced in the 1950s.

The paracetamol 'Name Game'

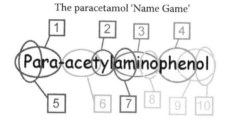

Many drugs and medicines get their common name or trade name from contractions of the full chemical name. See if you can work out how some of the various names for paracetamol have been derived. The first one has been done for you. Answers are in the Bibliography.

Paracetamol	1 + 6 + 7 + 10
Acetaminophen	? + ? + ?
Tylenol	? + ?
Panadol	? + ? + d + ?
A.P.A.P	? + ? + ? + ? + ? + ?

IS PARACETAMOL DANGEROUS?

Why do you ask?

WELL, YOU READ OF PEOPLE USING IT IN SELF-POISONING?

All chemicals are dangerous, in the wrong hands and in the wrong amount. After all, Paracelsus is supposed to have said '*The dose makes the poison*. Paracetamol is actually safer than most other drugs so long as it's taken in the correct dosage.

How Long Has It Been Around?

In the late nineteenth century, the emergence of an industry using aniline-based dyes prompted chemists to look for medicines based on aniline (phenylamine). First on the scene was acetanilide, but its toxic side effects saw it being rapidly replaced by phenacetin. Phenacetin remained a widely used painkiller until after World War II, but concern about its safety saw it being replaced on the market by paracetamol during the 1960s. As it happens, both acetanilide and phenacetin owe their effects as drugs to the fact that they are metabolized into paracetamol in the body.

        Acetanilide   Phenacetin

Phenacetin was very popularly used in medicines like 'headache cures' for many years. It was commonly combined with codeine (see p347) in medications. Following severe injuries in a plane crash in 1946, the American business magnate and aviator Howard Hughes was heavily dependent upon painkillers and it has been suggested that their phenacetin content may have caused kidney failure.

How Does Paracetamol Work?

There are several theories. One is that when it is metabolized and loses an acetyl group, it forms 4-aminophenol, which conjugates with a fatty acid called arachidonic acid (see p423) to form N-arachidonoylphenolamine (AM404). This is a naturally occurring cannabinoid (a group of molecules which activates specific receptors in the brain and spinal cord, and whose most famous member is tetrahydrocannabinol (THC), the primary psychoactive compound found in cannabis (see p501). AM404 binds to these receptors (specifically the CB1 receptors) and blocks them, preventing other molecules binding there and triggering pain signals.

AM404

Over 90% of a dose of paracetamol is safely metabolized in the body, but about 5% is turned into N-acetyl-p-benzoquinone imine (NAPQI). It has been suggested that NAPQI acts at other receptors in the spinal cord and prevents the COX-2 enzyme (see p37) from forming inflammatory compounds. So paracetamol both helps to relieve pain and reduce inflammation.

WHY IS PARACETAMOL TOXIC?

Paracetamol itself isn't, but its metabolite, NAPQI, is very toxic. Happily, the body contains a substance called glutathione (GSH) which mops up small amounts of NAPQI. But if an overdose of paracetamol is taken, then the body's supply of GSH becomes depleted, and the excess NAPQI can attack the liver, often with fatal consequences. In the United Kingdom, paracetamol is responsible for some 30,000–40,000 visits to casualty departments a year, with around 200 fatalities.

N-Acetyl-*p*-
benzoquinone imine
(NAPQI)

Glutathione (GSH)

Is There a Treatment for Paracetamol Overdose?

L-Methionine and *N*-acetyl-L-cysteine are both effective in replenishing the body's GSH. It has been suggested that if methionine was added to every paracetamol tablet, this would prevent many overdoses (both accidental and deliberate), but this is currently only done with a small percentage of the paracetamol tablets on sale mainly due to the higher added cost. Antidotes are only successful if administered within 8–10 h of taking the paracetamol (ideally within 2 h). Part of the problem with paracetamol overdoses is that there are few symptoms in the early stages, and someone having a change of heart after taking an overdose may think that the lack of any apparent symptoms means there is no reason to seek treatment – fatally – because liver failure symptoms can take 16–36 h to appear.

What's This about Paracetamol Being Used to Kill Snakes?

Brown snakes (*Boiga irregularis*) arrived on the Pacific island of Guam in the 1940s, and have wiped out much of the island's wildlife. In a bid to conserve

The mice are very tasssty...

...but I can't stomach their parachutessss...

what's left, the plan is to use paracetamol-fortified dead mice, dropped by parachute. About 80 mg of paracetamol is a lethal dose to these snakes, and dropping the mice so they land in the trees will stop other wildlife from being poisoned. So far the plan is only in the trial stage.

Chapter

48

PENICILLINS

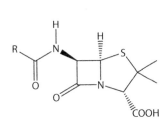

"We doctors call penicillin a 'wonder drug' because whenever we wonder what illness a patient has, we simply give them penicillin!"

WHY NOT JUST 'PENICILLIN'?

There are many different penicillins – there's penicillin G and penicillin V, for a start. Then there's amoxicillin and ampicillin, there's methicillin …

OK, POINT TAKEN, BUT WHY MORE THAN ONE?

Penicillins share a core structure, where the R group varies.

The penicillin 'core' (also known as a 'penam')

SO WHY ARE THEY SO SPECIAL?

They are probably one of the most important medicines ever discovered. In fact many of you reading this book may have had your lives saved, or the lives or your parents or other relatives saved, by penicillin. It's hard to imagine now, in our modern antiseptic world, what life was like before antibiotics such as penicillin. Before about 1950 there were no real cures for infections, and people regularly died from infections they had contracted from trivial routes, such as insect bites or being scratched by the thorns of a rose bush. Every hospital in the developed world had a 'septic ward' where the infected patients languished, either until their natural body's immune system successfully fought off the infection, or they died. The doctors could do very little to help these patients, except ease their suffering. Charles Fletcher, a doctor associate of Howard Florey, described this terrible situation as

'Every hospital had a septic ward, filled with patients with chronic discharging abscesses, sinuses, septic joints, and sometimes meningitis ... Chambers of horrors, seems the best way to describe those old septic wards. Carbuncles oozing pus – abscesses on the body the size of a cup – were not uncommon. There was little treatment. About half the people who entered the wards exited on gurneys to the morgue'.

All this changed when penicillin was discovered.

Sir Alexander Fleming

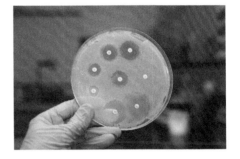

Mould (white spots) growing on a
plate alongside bacterial colonies.
The bacteria are repelled by
the mould and die off forming
a dead-zone ring around each
mould spot

How?

Alexander Fleming's original observation in 1928 was of a fungal mould (*Penicillium notatum*) that exuded a substance that seemed to kill bacteria that were growing on nutrient agar. Fleming called the substance penicillin. This lucky observation was the first antibiotic, and the start of our ongoing war with infectious bacteria. However, Fleming didn't realize the significance of his discovery, and simply wrote up his findings in a paper and then forgot about it. Eleven years later, the paper was 'rediscovered', and the observation was taken up by Howard Florey, Ernst Chain and a team of co-workers in Oxford, who showed that penicillin also worked when given to patients as a medicine, and cured a range of infections that hitherto might have killed the patient. In 1945, Florey, Chain and Fleming shared the Nobel Prize for Medicine for their 'joint' discovery of penicillin.

So, Problem Solved

In 1943, a laboratory worker checks on one of the 4000 flasks containing corn steep liquor and spores of *Penicillium* mould

Not really. The problem was that extracting the tiny amounts of penicillin from moulds grown on Petri dishes in the lab was time-consuming and difficult. World War II was raging and thousands of soldiers were dying from infections that could potentially be cured by a simple dose of the new wonder drug – but there simply wasn't enough of it. In June 1942, there was enough penicillin in the whole of the United States to treat only 10 patients! The US and UK governments made it a priority to find a way to mass produce penicillin, and a worldwide search was found for a source of mould that would yield the best-quality penicillin. It was eventually found on a mouldy cantaloupe melon in a market in Illinois in 1943. Research on fermentation of corn steep liquor, together with the new process of deep-tank fermentation, allowed the cantaloupe penicillin to go into large-scale production. By June 1945, nearly 650 billion units per year were being produced, and these made a huge difference to the number of amputations and deaths caused by infected wounds among Allied soldiers. It is often reported that many of these doses were ring-fenced and used to cure Allied soldiers of sexually transmitted diseases, such as syphilis and gonorrhea, which although not life-threatening, compromised the fighting efficiency of otherwise healthy soldiers.

SO, NOW WAS THE PROBLEM SOLVED?

Not quite. The early forms of penicillin were excreted very rapidly, with about 80% of the dose being cleared from the body within a few hours of it being taken. Indeed, penicillin was so rare and valuable during WWII that the urine from patients who were being treated with penicillin was often collected and the penicillin isolated from it and reused! To solve this desperate shortage, researchers needed something that would prevent the penicillin being excreted so quickly. They eventually found that the molecule probenecid did the job; when this was taken together with penicillin, the penicillin was retained for much longer, while the probenecid was preferentially excreted. Nowadays, the advent of more effective types of penicillin has made the use of probenecid unnecessary.

Probenecid

DIFFERENT TYPES OF PENICILLIN?

Yes. While in the United States, Heatley found that the addition of different chemicals to the fermenting liquor created different penicillins. Most significantly, adding phenylacetic acid gave rise to penicillin G, later called benzylpenicillin.

Penicillin G   Penicillin V

SO DOES IT MATTER WHICH ONE YOU TAKE?

It certainly does. For one thing the earliest penicillins like benzylpenicillin (penicillin G) are decomposed by acid. The contents of our stomachs are strongly acidic, so that these penicillins could not be taken orally but had to be injected. It was found that when phenoxyacetic acid was added to the fermentation broth, the resulting phenoxymethylpenicillin, commonly known as penicillin V, was fairly acid-stable and could be taken by mouth. Subsequently many more oral penicillins like this have been developed.

HOW DOES PENICILLIN WORK?

Until 1945, no one had a clue. That year, the Oxford crystallographer Dorothy Hodgkin solved the structure of penicillin F (the version made by the Oxford group), finding that it contained a four-membered β-lactam ring, which was the clue to how penicillin worked. As any high-school chemist can tell you, a four-membered ring is normally unstable and very reactive. The β-lactam ring binds to the enzyme peptidoglycan transpeptidase, which the bacterium needs to build its cell walls, so the bacteria grow bigger but cannot divide, and eventually burst.

β-Lactam

Penicillin F

If It Kills Bacteria, Why Doesn't Penicillin Kill Us?

Simple, animal cells do not have cell walls, so we do not have peptidoglycan transpeptidase and therefore penicillin is not toxic to humans.

So the Answer to Infections Is Simple?

No, ever since antibiotics like penicillin were first used, bacteria began to appear that were resistant to them. Florey and Abraham had even noticed bacterial resistance to penicillin in 1940. In his 1945 Nobel lecture, Fleming commented:

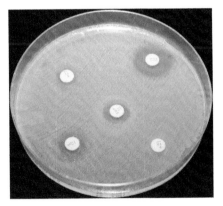

Some of the bacterial colonies growing in this petri dish are no longer being killed by the mould spots – they have developed resistance to the antibiotic

'*Then there is the danger that the ignorant man may easily underdose himself and by exposing his microbes to non-lethal quantities of the drug make them resistant*'.

Bacteria adapt rapidly so that mutations can arise that alter the receptor sites, making it impossible for penicillin to bind to them. There are other bacteria that have the capacity to make the enzyme penicillinase. This is a β-lactamase enzyme that catalyzes the hydrolysis of the β-lactam ring, thereby destroying the activity of the penicillin. So, for example, penicillin G is now ineffective against *Staphylococcus aureus*.

How Can We Get Around Antibiotic-Resistant Bacteria?

One way is by making bulkier penicillins that are too big to reach the penicillinase active site and so do not bind to it. Methicillin is an example, where the *ortho*-methoxy groups protect it from β-lactamase enzymes. Methicillin isn't ideal, though, as it is acid-sensitive and therefore has to be injected, and it has rather low activity. You still hear it mentioned in the context of methicillin-resistant *Staphylococcus aureus* (MRSA), the name given to infections that are resistant to penicillin and other β-lactam antibiotics.

Methicillin

It is a balancing act, though, as if the groups are *too* bulky, the penicillin cannot attack the enzyme controlling the cell wall synthesis.

Another way is to use a combination of clavulanic acid with the penicillin. Clavulanic acid is structurally similar to penicillins, but does not have any activity against bacteria. What it does is to attach itself to a serine residue in the β-lactamase enzyme, thus permanently inhibiting it.

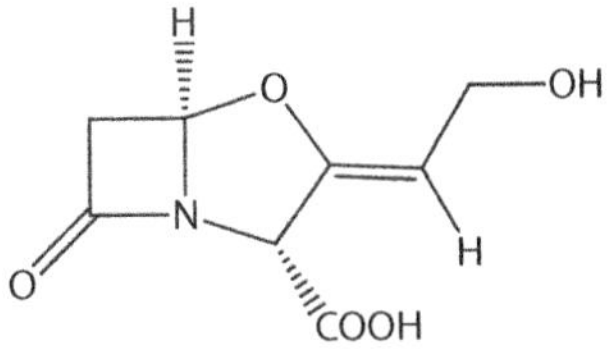

Clavulanic acid

It is often used in combination with some semi-synthetic penicillins like amoxicillin.

SEMI-SYNTHETIC PENICILLINS?

In 1957, an American chemist named John Clark Sheehan devised the first synthesis of penicillin, making penicillin V. One of the compounds he made *en route* was 6-aminopenicillanic acid. This contains the essential core of a penicillin molecule, and can be used as the starting material for making other penicillins by 'semi-synthesis', using acylation to insert the side chain, as shown in the reaction below.

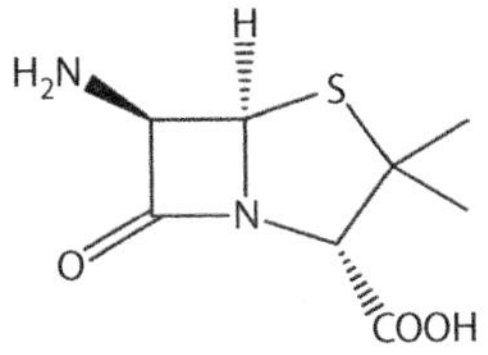

6-Aminopenicillanic acid

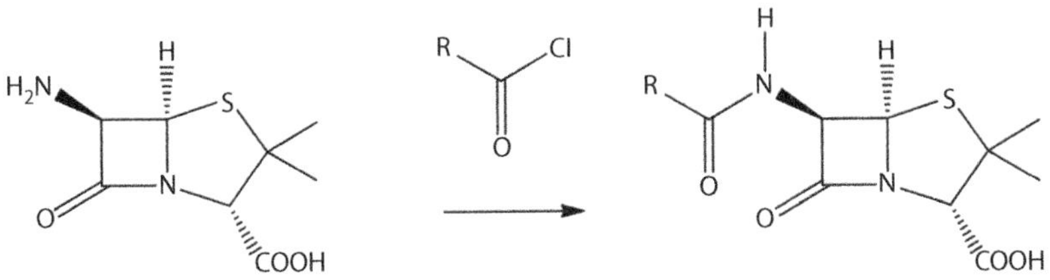

This approach has led to molecules like ampicillin and amoxicillin, which can be used orally. Unlike the earlier penicillins, which are only available against Gram-positive bacteria, the −NH$_2$ side chains in the R group gives the molecules the ability to penetrate the outer membrane of Gram-negative bacteria, so they have a wider spectrum of availability.

Amoxycillin   Ampicillin

Another trick is to use penicillin pro-drugs, like pivampicillin, which masks the polarity of the $-CO_2H$ group. This makes the molecule easier to absorb through the wall of the gut, whereupon the protecting group is hydrolyzed on metabolism in the body, forming ampicillin, the active antibiotic molecule.

Pivampicillin

But the battle against antibiotic-resistant bacteria is still being fought, with new drugs such as vancomycin (see p543) one of the latest to join the fray. In fact scientists are becoming increasingly worried that the bacteria are winning – they are evolving faster than we can develop new antibiotics. The worst-case scenario is that in a few years' time we will no longer have any effective antibiotics, and we may have to bring back the septic wards once more. A terrifying thought.

Chapter

49

PROSTANOIC ACID AND PROSTAGLANDINS

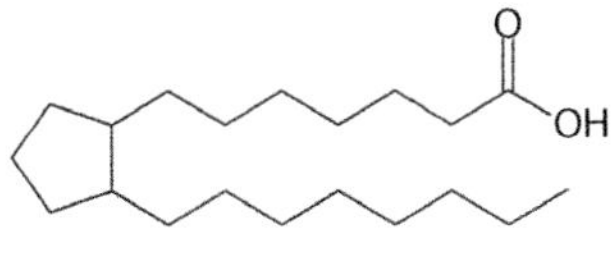

Prostanoic acid

Anything to Do with the Prostate Gland?

'Nothing to worry about, Mr. Smith. I'm just going to take a sample of your prostaglandins ...'

Actually, yes. Prostaglandins were named because they were first detected in human seminal fluid obtained from the prostate gland. They are important biologically active molecules that resemble hormones in their effects, but are chemically quite different. They were not detected for many years because they occur in such low concentrations and have such short half-lives. Prostaglandins are present in almost all cells and tissues, and they are one of the most potent of all biological agents. As little as a nanogram can be biologically active!

Prostanoic acid – the basic building block of the prostaglandins

What's the Difference between Prostanoic Acid and Prostaglandins?

Prostaglandins are a group of molecules based upon the 20-carbon structure of prostanoic acid, which is a fatty acid containing a five-membered cyclopentane ring (carbons 8–12). They are named as PG (for prostaglandin), followed by another letter from A–H depending upon the oxidation state and side groups of the cyclopentane ring. A subscript usually indicates the number of double bonds in the structure. For example, PGE_2 means a carbonyl group at carbon-9 and an –OH group attached to carbon-11, with two double bonds in the tail groups. A very slight change in their structure or side groups can completely alter their biological function.

What Do They Do?

In general, the importance of the prostaglandins is due to them being able to create specific effects in the body. They can regulate the activity of certain smooth muscles, causing them to contract or relax. They can also induce or prevent secretions from glands, especially endocrine glands. Finally, they can affect blood clotting and flow. Through these actions they are capable of affecting many aspects of human physiology, and this accounts for their remarkable versatility.

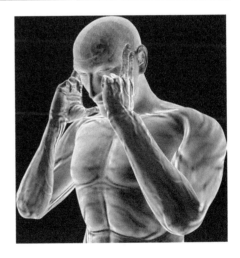

Prostaglandins are one cause of headaches and migraines

Prostaglandins can raise or lower blood pressure, and regulate gastric secretions, helping digestion. They are also responsible for fevers and inflammations, and if these inflammations occur in the blood vessels around the brain, they result in headaches, or the visual effects and pain associated with migraine. Other prostaglandins cause uterine contractions, and so can be deliberately administered to induce labor. Prostaglandins are also used to inhibit the secretion of stomach acids in people with peptic ulcers, and treat diabetes and atherosclerosis. Prostaglandins relax certain muscles and are used to relieve asthma and treat high blood pressure.

THEY SOUND IMPORTANT! BUT HOW DOES THE BODY MAKE SUCH A COMPLEX MOLECULE?

Well, it doesn't make it from scratch. It starts with a fatty acid, such as linoleic acid, which can be found in vegetable oils (see p287). This is then converted to arachidonic acid using enzymes in the liver. Other enzymes then set to work on arachidonic acid, attaching an oxygen molecule between carbon-9 and 11 to form an endo-peroxide, as well as linking together carbon-8 and 12 to form the cyclopentane ring. Various hydrolysis reactions then turn the peroxide into carbonyl or hydroxyl groups, producing one of a number of primary prostaglandins (*e.g.* PGE_2). This primary compound is the precursor for a whole range of other prostaglandins, which are made by subtly changing the structure or by adding or removing side groups.

HOW DO THEY WORK?

When body cells are injured, an enzyme makes prostaglandins, such as PGE_2, which creates inflammation, fever and pain. The

exact effect depends upon where the damaged cells reside –
damage or irritation to skin cells can lead to eczema, to cells in the
lungs can initiate asthma, and to cells in the blood vessels in the
brain can cause headaches. Although these seem like bad effects,
they are designed to help the body recover from the trauma.
Inflammation allows white blood cells to enter the damage area
and kill bacteria, so preventing infection; fever raises the body
temperature above the optimal value for bacterial growth, slowing
their reproduction rate and giving the body time to fight any
infection; and pain encourages the person to leave the injured area
alone and give it time to heal.

CAN THESE EFFECTS BE BLOCKED?

Yes, a good example is the drug aspirin (see p37) which acts
by blocking the active site on the enzyme, and so prevents the
synthesis of the prostaglandin PGE_2. Thus some headaches
and other pains or inflammations can be relieved by aspirin.
However PGE_2 has a second function: it also protects the cells of
the stomach wall by inducing the cells to make a protective layer
of mucus. Thus, aspirin also inhibits this function, making the
cells of the stomach wall more likely to be damaged and less able
to repair themselves. Taking too much aspirin, therefore, can
lead to stomach problems. The elucidation of the mechanisms by
which aspirin works and by which the main prostaglandins are
biosynthesized won the 1982 Nobel Prize for Medicine for John
Vane (UK), Sune Bergstrom and Bengt Samuelsson (Sweden).

ARE THERE ANY OTHER USEFUL PROSTAGLANDINS?

$PGF_{2\alpha}$ causes contractions in the muscle of the uterus and is used
both for abortions and to induce labor. A variation of this, used for
the same purposes, goes under the brand name *Carboprost*.

The 1982 Nobel Prize for medicine winners: biochemist Sune Bergstrom
of Sweden, left; pharmacologist John Vane of Great Britain, center;
and biochemist Bengt Samuelsson of Sweden, right, await the award
presentation at the Concert Hall in Stockholm, Sweden (© Press
Association Images. With permission)

$PGF_{2\alpha}$ if R = H. *Carboprost* if R = CH_3

PGE_1 is a powerful vasodilator, that is, it dilates the blood vessels
allowing blood to flow more freely. It is used to prevent gangrene,
and in keeping blood vessels near the heart open in cases of angina
and stroke. The only difference from PGE_2 is the change of the
double bond to a single bond between carbon-5 and 6.

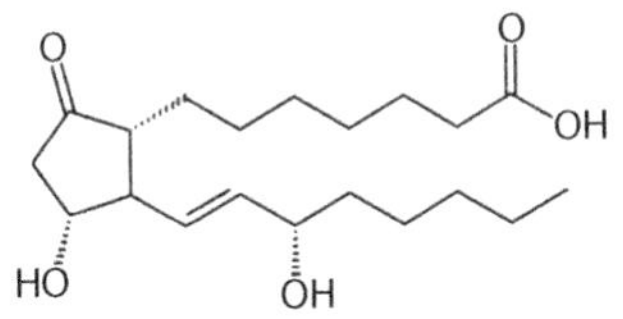

PGE$_1$

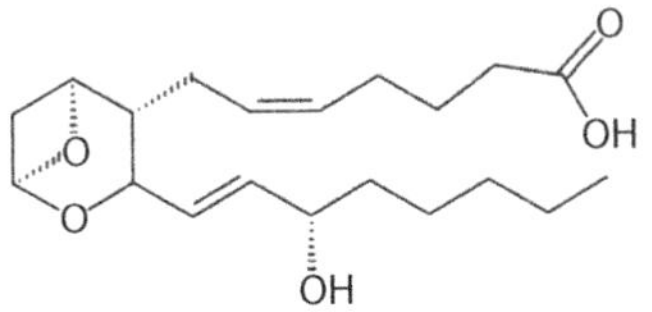

TXA$_2$

The action of blood platelets on arachidonic acid gives a number of products, including some fatty acids, plus two new compounds that are called thromboxanes (given the shorthand TX). An example is TXA$_2$ which is a potent muscle contractor and also causes rapid blood clotting. It also appears to be important in maintaining a healthy system of blood circulation in the body. The main structural change is an increase in the size of the ring from five to six members by the insertion of an oxygen atom, and the addition of a peroxide bridge across the ring (although this is rapidly oxidized to two –OH groups).

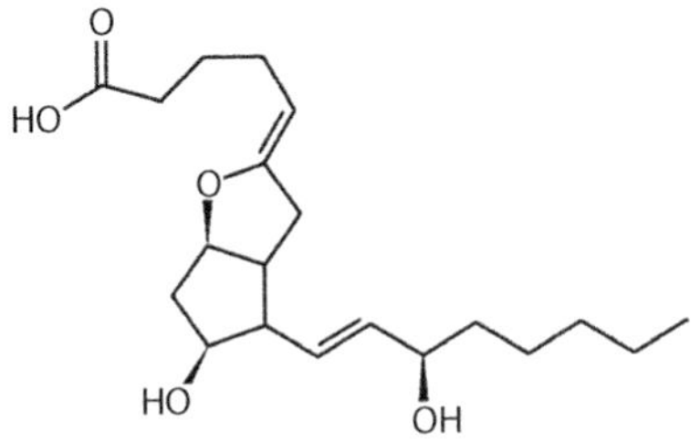

PGI$_2$

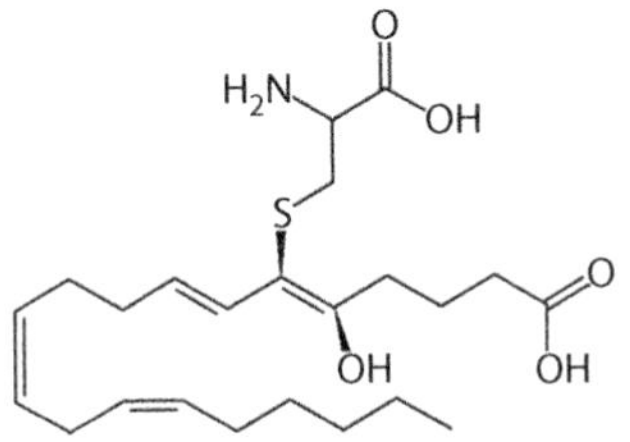

LTE$_4$

Another compound, named prostacyclin (PGI$_2$), has the opposite properties to TXA$_2$ – it relaxes muscles and prevents blood clotting. It is amazing that two similar molecules, made from an identical precursor, should have such potent, yet entirely opposite properties! In PGI$_2$, the uppermost tail has been bent around and joined to the OH group on the ring via an ether bond.

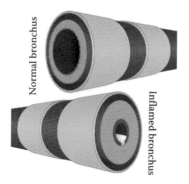

Normal bronchus

Inflamed bronchus

Irritation of cells in the lungs by pollen, smoke particles or dust; SRS-A is produced which causes the bronchioles to constrict and start an asthma attack·

The effect of white blood cells on arachidonic acid produces leukotrienes (named after leukocytes and given the shorthand LT). There are a number of these, an example of which is SRS-A (which is actually a mixture of three leukotrienes, one of which is called LTE_4), and they are implicated in the occurrence of asthma. SRS-A is made, along with histamines, when cells in the lungs are damaged or irritated, triggering an allergic response and the contraction of the bronchioles leading to an asthma attack.

Various drugs (such as *Montelukast* and *Zafirlukast*) are now being developed which block the cycle of synthesis of these leukotrienes and so prevent the asthma attack. The main difference from the prostaglandins is that the five-membered ring hasn't formed, and so the compound is still a straight-chained fatty acid. However, there is one branching side group, attached via a sulfur link, consisting of an amino acid derivative.

Chapter

50

PSILOCYBIN AND MESCALINE
THE MAGIC MUSHROOM MOLECULE
(ALONG WITH MESCALINE
FROM CACTI)

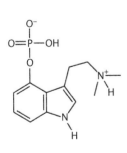

Psilocybin

A type of magic mushroom
(*Psilocybe mexicana*)

ARE ALL MUSHROOMS MAGIC MUSHROOMS?

No, the ones that matter are varieties that include *Psilocybe mexicana* and *Stropharia cubensis*. Many varieties of these are associated with temperate regions of the world, which include the United States, Mexico, Europe and the United Kingdom, notably *Psilocybe semilanceatao*, the 'liberty cap' mushroom. 'Magic mushrooms' became fashionable in the United Kingdom in the 1970s as a 'natural' and legal alternative to LSD; in most countries their possession or consumption is not illegal, but their preparation or deliberate cultivation is – in the United Kingdom that makes them a Class-A drug.

WHY ARE THEY CALLED MAGIC?

That's because psilocybin, the main active ingredient in the so-called 'magic mushrooms', has hallucinogenic properties.

Psilocybin

This molecule, and the related molecule, mescaline, have been known for centuries by the Aztecs in Mexico, who used them in tribal rites, believing the vivid, colorful hallucinations had religious significance. Indeed, the mushroom was so important to the Aztecs that they named it *teonanácatl*, meaning 'God's flesh' or 'God's meat'. This mushroom was said to have been distributed to the guests at the coronation of the emperor Montezuma in 1502 to make the ceremony seem even more spectacular. The Aztecs even had professional mystics and prophets who achieved their inspiration by eating other hallucinogenic plants, such as the mescaline-containing peyote cactus (*Lophophora williamsii*). During the nineteenth century, the use of peyote cacti in tribal rituals spread north to the natives of North America, such as the *Comanches, Kiowas* and the *Mescalero Apaches*, from where mescaline obtained its name. Over the years, the use of these drugs in religious rites became fused with Christianity, and even today some tribes believe that God put some of His power into peyote, and Jesus was the man who gave the plant to the Indians in a time of need. There is still a living cult in the hinterland of Mexico in the province of Oaxaca.

How Long Have We Known of These Drugs in the West?

Their effects were first described in Europe by Francisco Hernandez, the court physician to King Philip II of Spain, who traveled to Mexico with Cortez. He remarked that '*those who eat or chew it* [the peyote cactus] *see visions either frightful; or laughable ... terrifying sights like the Devil*'. Mescaline was identified in 1896 but was first made famous by Aldous Huxley in his 1954 essay 'The Doors of Perception', when he referred to '*a slow dance of golden lights. A little later there were sumptuous red surfaces swelling and expanding from bright nodes of energy that vibrated with a continuously hanging, patterned light*'. It alters senses of vision, hearing, time and space. On the other hand, details of the mushroom ceremony remained secret until the 1950s; in 1956

The peyote cactus (*Lophophora williamsii*) from which mescaline is derived

psilocybin was identified chemically by Albert Hofmann, the discoverer of LSD (see p293).

WHY DO PEOPLE TAKE THEM?

Mescaline is especially noted for producing distortions of reality as well as states of meditation. They produce intensification of visual experience, described as 'flooding with vivid colored images'; mescaline is said to produce a more visual effect than LSD.

HOW DO THEY WORK?

Psilocybin and mescaline are psycho-active because they closely resemble the structures of neurotransmitters, such as serotonin (see p455) and norepinephrine (see p9), that convey impulses from one nerve to another, especially in the brain. The hallucinogenic molecules fit into the same receptors as the neurotransmitter leading to false signals being created.

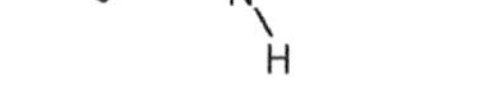

Psilocybin

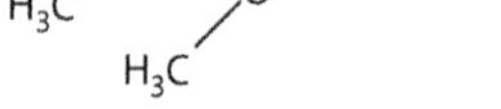

Mescaline

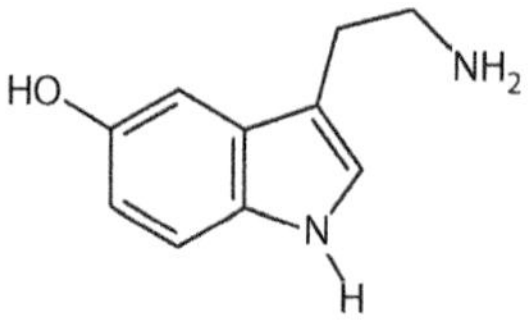

Serotonin

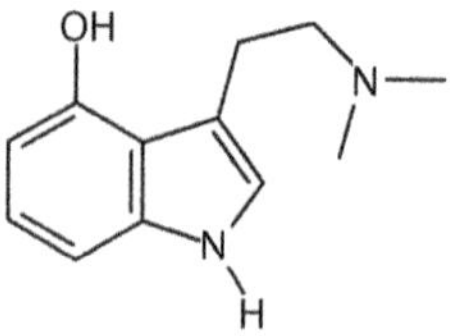

Psilocin

Many hallucinogens are variations of important biological substances called indole-amines, like serotonin. These contain an indole ring structure which is simply a six-membered benzene ring fused to a five-membered ring containing nitrogen. Hallucinogenic drugs like these activate various receptors in the central nervous system, and specifically cortical neurons.

Mescaline does not have the indole ring, but as shown in the diagram on p430, its structure can be represented so as to suggest its relation to the ring. In fact, mescaline has a structure which is closely related to phenylethylamine, the parent of the amphetamine family (see p309). It is therefore not too surprising that it acts on the brain. However, the presence of the three methoxy groups attached to the benzene rings appears to be needed for hallucinogenic activity on mescaline; these may make the molecule more lipophilic (soluble in fats), and so better able to penetrate the fatty membranes that protect nerves and nerve endings. This allows the molecules to more readily penetrate the central nervous system, and hence makes them more potent. It needs 2–3 h for the effects to come on and these can last for 12 h. The effects depend upon individuals, and maybe on their mood at the time of ingestion.

Psilocybin is a pro-drug, as it becomes active when the PO_4 group is removed in the body, converting it into psilocin, which is the true psycho-active drug.

Do These Drugs Produce 'Religious Experiences'?

Aldous Huxley certainly seemed to think so. In his book *Heaven and Hell,* he says,

> '*Reading these accounts* [of drug-induced visionary experiences], *we are immediately struck by the close similarity*

The Ecstasy of St. Teresa in Rome. This statue is meant to represent the moment of religious ecstasy when encountering an angel. Or was it a hallucination due to an imbalance in brain chemistry?

between induced or spontaneous visionary experience and the heavens and fairylands of folklore and religion'.

Recent double-blind studies support the view that psilocybin can induce experiences similar to mystical experiences and enhance autobiographical recollection. It has potential in psychotherapy either to facilitate the recall of salient memories or to reverse negative cognitive biases. There is evidence that a strong dose of psilocybin can affect the personality for a year, possibly for longer. And if psilocybin can produce religious experiences, then it's quite possible that naturally occurring psychoactive molecules like dopamine or serotonin could do the same, if their concentrations in the brain become unusually low or high for some reason. It could be argued that many of the ancient prophets and mystics revered by different religious all over the world just experienced an imbalance in their brain chemistry, rather than a true supernatural intervention. Either way, psilocybin and other psychoactive molecules are definitely food for thought (literally, one might say!).

Chapter

51

QUININE
(AND SYNTHETIC ANTIMALARIAL DRUGS)

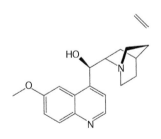

Quinine

What's the Title All about?

It's a reference to a Bob Dylan song, *The Mighty Quinn*, about Quinn the Eskimo.

> *But when Quinn the Eskimo gets here*
> *Ev'rybody's gonna jump for joy*
> *Come all without, come all within*
> *You'll not see nothing like the mighty Quinn.*

Why Choose a Title That Only Appeals to Over-50s Bob Dylan Fans?

Well, quinine is a mighty molecule, and it's an important drug.

Why?

For many years, quinine was the drug of choice in treating malaria. Malaria is reckoned to have killed more people in World War II than either bullets or bombs. Even today, it causes around a million deaths a year.

How?

Malaria is a disease caused by protozoa that are carried by the female *Anopheles* mosquito. When the mosquito feeds, it inserts a sharp tube, like a hypodermic needle, through the victim's skin to suck their blood. But in doing so, the protozoa carried by the mosquito enter the victim's bloodstream and infect them with malaria.

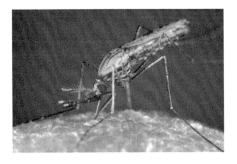

Anopheles gambiae mosquito feeding

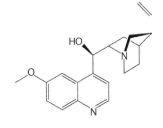

Quinine

WHY THE NAME?

It was once thought to be due to 'bad air', hence the name malaria was derived from an Italian phrase (*mal aria*).

HOW LONG HAS QUININE BEEN KNOWN?

Sixteenth-century Spanish settlers in South America found that natives had used the powdered bark of the Cinchona tree to treat fevers for hundreds of years. Jesuits then brought the bark back to Europe around 1630.

Harvesting the bark of the Cinchona tree in Java

Quinine was isolated from the bark in 1820.

AND ITS USE SPREAD THROUGH EUROPE?

Patchily. Malaria was widespread in Northern Europe from the fifteenth to the nineteenth centuries. Known as the 'Ague', it was endemic in parts of England like the Cambridgeshire Fens and the Kent and Essex marshes. Its disappearance is probably due to factors like improved marsh drainage and better housing and sanitation.

It is said that Oliver Cromwell's final illness in 1658 was fatal because he refused to take 'Jesuit's bark' because of Protestant

Oliver Cromwell

suspicion of something associated with Roman Catholics. Such suspicion was encouraged by Robert Talbor, an apothecary who in 1672 published a book entitled *Pyretologia: A Rational Account of the Cause and Cures of Agues* that said *'Beware of all palliative cures, and especially of that known as Jesuits' Powder'*. Ironically, Talbor had successfully cured King Charles II of England and the Dauphin, son of Louis XIV of France, with a potion, which turned out to be 'Jesuits' Powder' dissolved in wine.

As European colonization spread throughout the world, demand for quinine grew. At first, the cultivation of Cinchona was monopolized by the Spaniards, but in 1865, Cinchona seeds were smuggled out of Peru to start its cultivation in the Dutch settlement of Java, and the use of quinine spread. Undertakings like building the Panama Canal could not have happened without it.

It was only at the end of the twentieth century that the cause of malaria was understood; Sir Ronald Ross won the Nobel Prize for Medicine in 1902 for his work on malaria.

Is It True That There Is Quinine in Gin and Tonics?

Gin was invented by a Dutchman, Franciscus Sylvius, in the mid-seventeenth century. He flavored spirits with oil of juniper (which in French is *genièvre*) calling it *genever*, which in time was corrupted to gin. Juniper oil contains quinine, which is almost insoluble in water (about 0.05 g per 100 ml of water) but it is readily soluble in ethanol. It is alleged that the British Empire-builders took their daily dose of quinine dissolved in ethanol in the form of gin, using lemon or

Gin and tonic contains small amounts of quinine – but not enough to prevent malaria

lime to help mask the bitter taste of the quinine. Tonic water came on the scene later (Schweppes introduced it around 1870). There is about 20 mg of quinine per 6 fluid ounces of one leading brand of gin.

So You Can Take G&T for Malaria?

The recommended daily dose of quinine for malaria is 600 mg, three times daily, for up to a week. At that rate you would be drinking up to 100 gin and tonics a day. This would certainly cure you of malaria, but could cure you of life first. Nice try, though.

How Does Quinine Work?

The flat ring fits between the bases in the DNA of the protozoa and interferes with transcription, blocking cell replication. Thus, they can't reproduce and eventually die.

But Better Synthetic Drugs Took Its Place?

Drugs like quinacrine and chloroquine were developed to counter quinine shortages. When the Japanese invaded Java in 1941, quinine supplies to most of the world stopped, so these alternatives were welcome.

Quinacrine (marketed as *Mepacrine* or *Atabrine*)

Chloroquine (marketed as *Aralen* or *Resochin*)

Universally?

'Tokyo Rose', an American-born and educated radio broadcaster (her real name was Ikuko Toguri; she was a UCLA graduate) who broadcast Japanese propaganda to American troops, told them that quinacrine would also make them sterile. The result was that many American troops did not take their antimalarial tablets, and caught malaria. A long and expensive campaign featuring 'Annie the mosquito' was required to persuade the soldiers to take antimalarial precautions. There was a similar problem in the campaign in Burma. When the use of quinacrine was enforced, malaria rates dropped below 5%.

Ikuko Toguri 'Tokyo Rose'

But they had real disadvantages as well. Quinacrine turned the skin yellow, and sometimes caused psychotic behavior changes or depression. A famous example is that of Major-General Orde Wingate, later to command the Chindits in Burma in WWII, who contracted malaria and then became depressed while taking higher than recommended doses of quinacrine, and attempted suicide by stabbing himself in the neck. He survived, only to die in a plane crash some years later. Despite these side effects, quinacrine was a very successful antimalarial drug for that time. A related molecule, chloroquine

Still frame taken from a 1944 US Army video about malaria prevention ('Private Snafu and Anopheles Annie' [1944])

was better, but as time went on, the malarial parasite developed a resistance to both these drugs.

Mefloquine (also known as *Lariam*) came on the scene in the 1960s, and was widely used by American troops during the Vietnam War to treat chloroquine-resistant strains of *P. falciparum* malaria. Resistance to mefloquine, too, has developed over the years, and it remains controversial, with claims for rare but unpleasant side effects, including depression, mood disorders and permanent personality changes.

Mefloquine (*Lariam*)   Atovaquone

Atovaquone is a relatively recent discovery and is part of a popular antimalarial drug called *Malarone* when it is taken in conjunction with *Proguanil*. But at present its high cost means it is not an option for the poor.

So Quinine Is Back in Favor?

Yes, it is still used to treat severe *falciparum* malaria. The big development recently has been a new batch of Chinese drugs. Chinese herb extracts (in particular, *qinghao*) have long been used to treat fevers, and it was found that an extract from this called Artemisinin (*qinghaosu*) works against chloroquine-resistant malaria. Molecules developed from it, like arte-ether, could be even better. But that is another story (see p29). And the other plan of attack is to develop molecules such as DDT which repel mosquitoes (see p135).

AND QUINN THE ESKIMO?

Despite global warming on the agenda and fears of malaria spreading north, Quinn will still probably just need quinine for his gin and tonic.

Chapter

52

SODIUM HYPOCHLORITE
(BETTER KNOWN AS BLEACH)

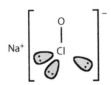

Is Bleach the Chemical That's Used to, Er, Bleach Hair?

No, that would be hydrogen peroxide, as in a peroxide blonde (see p245).

What about Bleaching Teeth?

No, that's also hydrogen peroxide, or a diluted version of it in the form of carbamide peroxide.

Ok, So What Actually Is Bleach?

Household bleach is actually a mixture of chemicals. Its main constituent is a solution of ~3–6% sodium hypochlorite (NaOCl), which is mixed with small amounts of sodium hydroxide, hydrogen peroxide and calcium hypochlorite. Its main use is to remove color, whiten or disinfect clothing or surfaces, and is invaluable in most modern kitchens and bathrooms.

When you swim in a public swimming pool, you're actually swimming through a dilute bleach solution!

Sodium hypochlorite is used on a huge scale in agriculture, and areas such as chemical, paint, lime, food, glass, paper, pharmaceuticals, synthetics and waste disposal industries. It is often added to industrial waste water to reduce odors, since NaOCl neutralizes H_2S and ammonia. It is also used to detoxify the cyanide baths used in metal plating processes, and to prevent algae and shellfish growth in cooling towers. It is also used to purify water supplies and swimming pools.

WHO INVENTED IT?

Liquid bleaching agents based on sodium hypochlorite were developed in 1785 by the Frenchman Claude Louis Berthollet while working in the village of Javel near Paris. He later founded the Javel company and introduced his new wonder cleaning agent to the population under the name '*liqueur de Javel*'. At first, it was used to bleach cotton, but soon became a popular compound for bleaching other clothing materials since it was quickly found that the sodium hypochlorite could remove stains from clothes at room temperature. In France and Canada, sodium hypochlorite solution is still known as '*eau de Javel*', while in other countries it has various trademarked names such as '*Clorox*' or '*Domestos*'.

Claude Louis Berthollet
(1748–1822)

Bleach is still known as *eau de Javel* in many French-speaking countries

WHAT DOES IT LOOK LIKE?

Sodium hypochlorite is a white powder which dissolves in water
to give a slightly yellowish solution with a characteristic odor.
Different concentrations of sodium hypochlorite have different
potencies in terms of their bleaching effect. For domestic use,
bleach usually contains 5% sodium hypochlorite, giving it a pH of
around 11 and making it mildly irritating to the skin. Concentrated
bleach (10–15% sodium hypochlorite) is highly alkaline (pH ~13)
and now is so corrosive that it can burn skin on contact.

HOW IS BLEACH MADE?

Barthollet's original production method involved passing chlorine
gas through a sodium carbonate solution, but the resulting
solution of sodium hypochlorite was quite weak. In fact, addition
of chlorine gas to water gives both hydrochloric acid and
hypochlorous acid:

$$Cl_2 + H_2O \rightleftharpoons HOCl + HCl(aq)$$

Addition of salt to this mixture allows formation of the aqueous
sodium hypochlorite solution. From the equilibrium, you can see
that addition of acids to this solution will drive the reaction to the
left, with chlorine gas being evolved. Therefore, to form stable
hypochlorite bleaches, the equilibrium must be driven to the right,
and this can be accomplished by adding an alkali, such as sodium
hydroxide.

A more effective production method was invented in the 1890s
by E.S. Smith which involved the electrolysis of salt solution to
produce NaOH and Cl_2 gas, which was then mixed together to
form NaOCl. Nowadays, the only large-scale industrial method for
production of NaOCl is called the Hooker process, and is just an

improved version of Smith's electrolysis process. In this, Cl_2 gas is passed into cold dilute NaOH solution, forming NaOCl, with NaCl as the main by-product. The disproportionation reaction (the Cl_2 is simultaneously oxidized and reduced) is driven to completion by electrolysis, and the mixture must be kept below $40°C$ to prevent the undesired formation of sodium chlorate.

$$Cl_2 + 2NaOH \rightarrow NaCl + NaOCl + H_2O$$

How Does Bleach Work?

Sodium hypochlorite is very reactive, and actually unstable. Left exposed to the atmosphere, chlorine gas evaporates from the solution at a considerable rate, and if it is heated, the sodium hypochlorite falls apart into salt and oxygen. This also happens when it comes into contact with acids, sunlight, certain metals, and many gases, and is one of the reasons why bleach can be used on a large scale – after use it decomposes to benign products (salt and water) which can be flushed into the drainage system without problem.

Bleach works by several methods. The hypochlorous acid (HOCl) component is a very strong oxidizing agent (even stronger than Cl_2 gas), and can react with and destroy many types of molecules, including dyes. Also, the hypochlorite ion decomposes into chloride and a highly reactive form of oxygen, which is why household bleach gradually goes off if left for long periods:

$$2ClO^- \rightarrow 2Cl^- + O_2$$

The HOCl (and to lesser extents the Cl_2 and active oxygen) can then attack the chemical bonds in a colored compound, either completely destroying the chromophore (the part of the molecule that gives it its color), or converting the double bonds in the chromophore into single bonds, thereby preventing the molecule from absorbing visible light (see p73).

When it reacts with microbes, sodium hypochlorite causes proteins in the cells to aggregate and the microbes to clump together and die. It can also cause cell membranes to burst. This broad-spectrum attack makes bleach effective against a wide range of bacteria.

"Mrs Smith - Bleach may kill 99% of all known germs, but it won't touch computer viruses!"

Sodium hypochlorite is alkaline, and household bleach also contains NaOH to make the solution even more alkaline. Two substances are formed when sodium hypochlorite dissolves in water. These are hypochlorous acid (HOCl) and the hypochlorite ion (OCl$^-$), with the ratio of the two being determined by the pH of the water.

SOUNDS DANGEROUS?

Actually, bleach is generally very safe if handled with respect. In 2002, the Royal Society for the Prevention of Accidents estimated that there are about 3300 accidents needing hospital treatment caused by sodium hypochlorite solutions each year in British homes. Most of these were due to drinking bleach by mistake (often children drinking it from an unlabeled bottle), but many were also due to handling errors. Sodium hypochlorite reacts with many reagents, even sunlight, to produce chlorine gas, which in enclosed environments can be a severe lung irritant. Because household bleach also contains NaOH (caustic soda), contact with the skin will cause burns due to the NaOH destroying the fatty tissue and oils. This process is known as saponification, and is the method to manufacture soap. The slippery feel of bleach on skin is due to saponification of the skin oils and destruction of tissue!

Mixing bleach with some other household chemicals can be hazardous due to unwanted side reactions. Adding acid (even vinegar) to bleach generates chlorine gas, as mentioned above, whereas mixing bleach with ammonia solutions (even urine!) can produce chloroamines, which are toxic. So remember, if you put bleach down your toilet, flush it before you pee!

Reaction of bleach with some household products, such as surfactants and fragrances produces chlorinated volatile organic compounds (VOCs), such as carbon tetrachloride (CCl_4) and chloroform ($CHCl_3$), which can also be harmful to health. Nevertheless, the benefit gained from cleaning and disinfecting household areas probably outweighs any potential harmful effect from these VOCs.

Chapter

53

SEROTONIN

Serotonin

'No physiological substance known possesses such diverse actions in the body'.

WHO SAID THAT?

Irvine Page. He was talking about the molecule serotonin, which he and some colleagues had just isolated a few years before in 1948.

Serotonin

IS SEROTONIN REALLY THAT USEFUL?

Well, it has a remarkable number of actions within the body. In the intestines, it makes the muscles contract, forcing food along the digestive tract. When released into the bloodstream, it acts as a vasoconstrictor, contracting blood vessels to prevent bleeding and start blood clotting. It can also encourage certain cells to grow, which helps with wound healing.

But perhaps its major function is as a neurotransmitter involved in the transmission of electrical signals from one nerve to another across the synaptic gap (see p169). It is found throughout the brain and central nervous system, where it helps to regulate appetite, sleep cycles, and even mood. It helps to maintain a 'happy feeling' and prevents anxiety and depression. It also helps with memory and learning.

SO IT'S A MOLECULE OF HAPPINESS?

You could say that. It provides 'reward' of feelings of happiness and contentedness to advanced animals (including humans) when they find themselves in situations that are beneficial to their survival or to the survival of their genes. For example, when food resources are plentiful and the animal is well fed, serotonin levels increase and the animal is content. Conversely, if it is starved, serotonin levels fall and the animal becomes anxious and distressed, and is triggered to look for food.

Having a stable food supply, and eating food, increases serotonin levels and makes you feel happy

Rutting bull elks fighting for social dominance. The reward for the winner is more food, and more mates, and increased serotonin levels!

Because most advanced social animals have to compete against their peers for food, for shelter, and for a mate, those with a higher social rank get more of these resources than lower-ranked animals. The serotonin level in a dominant alpha-male, for example, is higher than in subordinate animals. The 'top dog' feels happy about this, while the others are discontent, and continually challenge the leader. Thus, serotonin might be partly responsible for our competitive nature and constant urge to outdo our neighbors. It could even be argued that serotonin (along with dopamine, see p169) were the driving forces for the evolution of mankind, and even for all of our scientific, artistic and cultural progress.

Is It Only Found in Animals?

No, many insects and plants use serotonin as well. For example, serotonin is the trigger for swarm behavior in locusts. Many stinging insects, such as wasps and scorpions, have serotonin as part of their venom which they inject into their prey. The sudden large increase in serotonin levels in the region of injection site is interpreted by the prey's body as an indicator of bleeding, and the body experiences these signals as pain.

A wasp stinger, showing a droplet of venom, containing various poisonous compounds, including serotonin

A 'deathstalker' scorpion (*Leiurus quinquestriatus*), whose venom also contains serotonin

So That's Why Wasp Stings Hurt?

Yes – which is exactly what the wasp wanted! Also, for smaller prey, the serotonin in a scorpion sting can make the local blood vessels constrict so much that its prey dies – which may be exactly what the scorpion wanted!

What about in Plants?

It's found in mushrooms, walnuts and various fruits and vegetables, such as bananas, pineapples, kiwifruit and tomatoes, which have internal edible seeds. It is believed that when animals eat them, the serotonin stimulates the animal's gut and acts as a sort of laxative. The animal rapidly excretes the seeds, so distributing them around the local neighborhood, ready to grow new plants.

Serotonin is also found in some species of coral and also stinging nettles, and causes pain when injected locally by the barbs on the leaves or spines in a similar manner to that of wasp stings.

If Plants Contain Serotonin, Why Don't We Get Hallucinations When We Eat, Say, a Banana?

Eating foodstuff containing large amounts of serotonin will increase its levels in the bloodstream, but serotonin cannot cross the blood–brain barrier. This means that ingesting serotonin directly will not affect the serotonin levels in the brain, although it may work as a laxative!

But I Heard That Your Diet Can Affect Your Behavior?

Serotonin is biosynthesized from an amino acid called tryptophan, which is readily available in many foodstuffs. Unlike serotonin, tryptophan *can* cross into the brain, and be converted into serotonin there. Research is showing that eating foods that are high in tryptophan, such as bananas, milk, eggs, fish, red meat, poultry and peanuts, increase the

Tryptophan

serotonin levels in the brain, and therefore may help with sleep, calm anxiety and even relieve depression. Conversely, poor diets, such as ones that involve constant snacking on junk foods that are low in tryptophan, may even contribute to depression.

So, Low Levels of Serotonin May Cause Depression?

Yes. Recent research has linked low levels of serotonin (and a

related protein that transports serotonin around the brain) to depression in humans and laboratory mice. It can also lead to symptoms such as apathy, fear, feelings of worthlessness, insomnia and fatigue.

If depression is mild, it can sometimes be managed without prescribed medications. A very effective method of raising serotonin levels is with vigorous exercise. Studies have shown that serotonin levels increase with physical activity, and that serotonin production remains elevated for some days after the activity.

For more serious cases of depression, scientists have developed a range of antidepressant drugs which alter levels of serotonin in the brain. The earliest of these were called tricyclic antidepressants (TCAs), but because they had a number of severe side effects they have gradually been replaced by reuptake inhibitor (RI) drugs. RI drugs act on the synapse, and prevent the neurotransmitter molecules from being reabsorbed after they have performed their function. The neurotransmitters, therefore, remain in place, and the nerve stays 'on' for longer. By making the neurotransmitter molecule work for a longer period, it tricks the brain into thinking there are more molecules than there really are. RI drugs can be selective to serotonin (SSRI), to another neurotransmitter called norepinephrine (see p9) (NRI), or to both serotonin and norepinephrine (SNRI).

Amitriptyline – an example of a TCA drug

Fluoxetine (*Prozac*) – an SSRI drug

Venlafaxine (*Effexor*) – an SNRI drug

Reboxetine (*Edronax*) – an NRI drug

Another class of antidepressant drug targets the enzyme monoamine oxidase which breaks down the neurotransmitter molecules (serotonin, dopamine, norepinephrine, *etc.*) after use. By preventing this enzyme from working, the neurotransmitters survive for longer, increasing their effectiveness. The so-called monoamine oxidase inhibitors (MAOIs) have been used to treat

Phenelzine (*Nardil*) – an MAOI drug

depression since the 1950s, but they are now generally regarded as a 'last resort' option, *i.e.* if the patient doesn't respond to the safer RI drugs. The main problem is that MAOIs are notorious for their numerous interactions with other drugs, many of which can be dangerous or even fatal. A number of TV detective shows have exploited this as a plot device: a victim is murdered by a clever killer giving them an apparently innocuous drug, such as a painkiller, which then reacts badly with the MAOI they were taking for depression. The victim apparently dies from natural causes (*e.g.* a heart attack) – unless the clever TV detective can spot how it was caused.

The five classes of drugs described above are prescribed widely to control not just depression, but a whole range of mood and mental disorders, including ADHD, anxiety, post-traumatic stress disorder, anorexia and bulimia, Tourette's syndrome, narcolepsy, insomnia, smoking cessation ... and even chronic hiccups!

CAN THE BRAIN BE TRICKED IN OTHER WAYS?

LSD

Yes – there are various molecules which have structures very similar to that of serotonin, but which *can* pass through the blood–brain barrier. They then mimic the effects of serotonin with a variety of consequences. The most well known of these are the psychedelic drugs, such as psilocybin, mescaline (see p431) and LSD (see p293). These all bind to the serotonin receptors and stimulate the nerves, but because they have a higher affinity for the receptors than serotonin (*i.e.* they stick more strongly), the presence of the molecule blocks the receptor and prevents any more neural messages from traveling along that nerve cell. Instead, later impulses are redirected to other parts of the brain for which they weren't intended, confusing the brain with false signals. This results in feelings of euphoria, altered states and hallucinations. Amphetamines and ecstasy (MDMA) (see p309) also bind to and activate the serotonin receptors (as well as the dopamine receptors, see p169), in these cases leading to changes in alertness and feelings of well-being.

Chapter

54

SKATOLE

Skatole

WHAT'S THAT?

The smell of human excrement.

YUCK, HOW DISGUSTING!

Actually, skatole is an important molecule, and crops up in some surprising places.

Ice cream contains skatole

SUCH AS?

Ice cream, would you believe?

YOU'VE JUST PUT ME OFF ICE CREAM FOR LIFE

It is used in very small amounts as a flavoring material in ice cream. The skatole put in ice cream is man-made. No, let me rephrase that … The skatole put in ice cream is synthetic; it has no human connections.

HOW CAN IT BE USED IN ICE CREAM?

The odor of skatole depends on its concentration. At low concentrations, skatole actually has a rather nice, sweet smell; it crops up in orange blossom and jasmine in small amounts. It's also used by *Zantedeschia aethiopica*.

What's That?

It's an Arum lily which makes skatole to attract pollinating insects. It has flowers on the central part (the *spadix*) in male and female zones. The faint scent attracts crawling insects and bees which pollinate the flowers. Other Arum plants use it, including *Arum creticum*

Arum lily (*Zantedeschia aethiopica*)

and *Arum nigrum*. It also gives the mushroom *Coprinus picacens* its characteristic odor.

All the Same, though ...

It's used in perfume too, as a fixative.

What Does That Mean?

Perfumes are mixtures of molecules, generally with molecular mass below 300, of varying volatilities. Their 'top note' is made of volatile substances with instant impact. Once this disappears, the 'middle note' is longer lived, while the 'base note' which may linger for days or weeks, and may make up 50% of the perfume, are deep odors that may not be smelt for some time. The fixative is a molecule with a deep odor that has a low volatility and which helps to reduce evaporation of the more volatile components (like musk). This means that the perfume keeps longer.

How Did That Use Come About?

Civet cat

Civet cats, mainly found in Africa, India and South-East Asia, produce an oily substance called 'civet' from the perineal glands in their abdomen which they use to mark their territory. Civet has a strong smell, due partly to skatole, and also due to civetone (9(Z)-cycloheptadecenone). When diluted, its smell becomes more musky (and less objectionable). Demand from the perfume industry for civet has been met by a synthetic version. Chinese civet cats (wrongly) attracted publicity in 2004–2006 as a cause of SARS outbreaks.

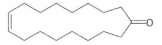

Civetone

Just How Smelly Is Skatole?

It is supposed to be more unpleasant than mercaptans, with a threshold for detection of 1.7×10^7 molecules per cm^3. This sounds a lot, but it is over two orders of magnitude less than pyridine, which itself is pretty objectionable.

WHAT DOES SKATOLE LOOK LIKE?

Pure skatole is a white crystalline material, believe it or not.

AND THE WORD SKATOLE?

It is derived from the Greek σκωτ or σκατ (skat), meaning dung. It gives meaning to the adjective scatological, meaning filthy or obscene.

SO LOTS OF ANIMALS PRODUCE SKATOLE, NOT JUST HUMANS?

Skatole produced by male pigs can cause 'boar taint', which makes pork from these sources unattractive as food. Skatole, along with indole itself, is responsible for the objectionable smell of giraffes (there is a plus side for the giraffe, as it repels ticks).

Indole   Skatole

The Southern house mosquito (*Culex quinquefasciatus*) is highly sensitive to it, and uses it to guide its egg-laying.

HOW IS SKATOLE FORMED?

It is produced in the digestive tract by bacterial breakdown of the amino acid tryptophan.

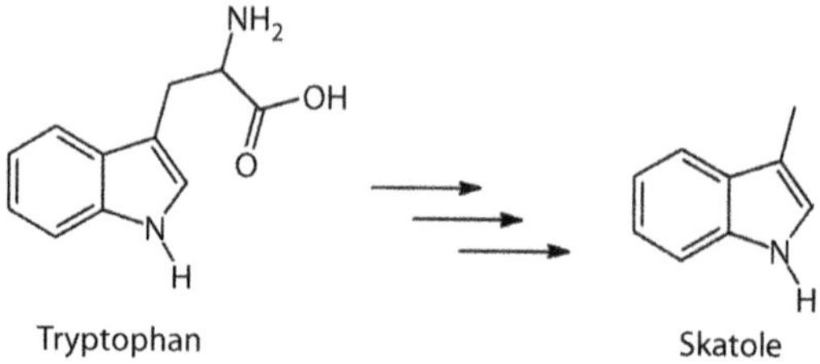

Tryptophan   Skatole

So Do We Know How It Ends Up in Ice Cream?

The presence of skatole can be detected in milk and in other dairy products, doubtless from breakdown of some of the tryptophan in protein. So next time you're eating ice cream (especially chocolate ice cream!), please don't dwell over what it might contain …

Chapter

55

SUCROSE

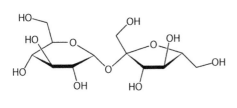

Sucrose

You got the peaches, I got the cream
Sweet to taste, saccharine …
Do you take sugar? One lump or two?

Def Leppard
Pour Some Sugar on Me

SO I GATHER FROM THOSE RATHER DODGY LYRICS THAT SUCROSE IS SUGAR?

Yes, sucrose is what's commonly referred to as table sugar, and its name comes from the Greek word for sugar (*sucrum*). It's strange to think now, but until about 200 years ago sugar was very rare in the Western world, and a very valuable luxury.

SO HOW DID PEOPLE SWEETEN FOOD?

Usually with honey. But this doesn't work well for baking, and it's messy to add to drinks like coffee or tea. Once the people in the newly developing Western economies tried sugar, and found out how useful, convenient and tasty it was, the demand grew hugely. Within a few years in the late 1700s, it had gone from being a luxury to an essential food item. In the United States, frontier families relied on jams, jellies and preserves to keep them alive in the period from summer harvest through the harsh winters to the 'starving time' before the first harvest. And sugar is an essential ingredient in the manufacturing of alcohol, and in the nineteenth century some workmen were partly paid in alcohol allowance rather than cash. It could be argued that sugar was so influential that it changed the course of world history.

HOW SO?

Sugarcane grows best in hot, humid climates, such as the islands in the Caribbean, and from around the beginning of the eighteenth century, a race began among European countries to colonize and exploit these areas as fast as possible. But the work in the sugar plantations was harsh, inhumane and labor-intensive. A supply of cheap labor was needed, and this fueled the slave trade from Africa to work on sugar plantations and in sugar refineries. In the West, millions of slaves were shipped from Africa to the West Indies and Southern States of the United States, while in the East, indentured laborers, mostly from India, were taken to East Africa and Southeast Asia. Sugar was

A 1787 engraving from Voltaire's *Candide* depicting the scene where Candide and Cacambo meet a maimed slave of a sugar mill near Surinam. The slave has had his hand chopped off for getting his finger caught in a millstone and his leg removed for attempting to escape. The caption is the slave saying, 'It is at this price that you eat sugar in Europe'

so valuable that wars were fought over access to Caribbean islands and piracy was rife. The standard drink of the Royal Navy (and of pirates) used to be rum, made from distillation of sugarcane juice or its by-products, such as molasses. It's been said that much of the wealth of the early British Empire came from its sugar plantations, and that this money helped bankroll the Napoleonic Wars against the French, as well as pay for many of the scientific and engineering advances that led to the Industrial Revolution. The slave trade changed the population demographics of many parts of the world, and even nowadays, it can be said that the varied mix of ethnicities and nationalities found in most modern countries can be traced back to the upheavals brought about by global trade in sugar (with help from tobacco and cotton, of course).

So What Exactly Is Sucrose?

To a scientist, the term 'sugar' refers to a class of molecules called saccharides. Monosaccharides, such as glucose contain only one sugar unit per molecule; disaccharides contain two sugar units bonded together, such as in sucrose; and polysaccharides are polymers containing many sugar units, as in cellulose or starch (see p193).

Sucrose is made from joining glucose, a six-membered ring sugar (left), to fructose, a five-membered ring sugar (right)

So Sucrose Is a Disaccharide?

Yes, it's composed of two smaller sugars bonded together, in this case glucose and fructose, linked via an ether bond.

Why Do We Use a Disaccharide When a Monosaccharide Is Simpler?

Well, we use both glucose and fructose as sweeteners in food too. But the economic reason is because the sugar-rich crops (sugarbeet and sugarcane) contain 10–20 times more sucrose than either glucose or fructose, so it's easier to extract and purify. It also turns out that the absorption of both of these sugars from the intestine into the bloodstream is significantly enhanced if they are ingested in equal amounts. And in sucrose they are in exactly the correct 1:1 ratio.

WHY DO WE NEED AN EFFICIENT ABSORPTION OF SUGAR?

In the past, efficiently and quickly metabolizing food into energy might make the difference between successfully running away from danger and being eaten ourselves! But nowadays, of course, it's becoming a large (no pun intended) problem. Sugar is now so commonplace, and eaten so often, that obesity levels are rising all over the modern world, bringing with it associated problems of heart disease and diabetes (see p255).

The two main sources of table sugar are sugarcane ...

... and sugarbeet, after harvesting (left) and cleaning (right)

CAN ANYTHING BE DONE ABOUT IT?

Lots of exercise and a carefully controlled diet help, such as cutting down on sugar intake. One of the best ways to do this is to use artificial sugar substitutes, rather than sugar.

SUCH AS?

There are a number of sugar substitutes, or artificial sweeteners on the market; the most popular are saccharin, aspartame, sucralose, and cyclamate. Many of these were discovered by accident, when a chemist did something you're never supposed to do in a chemistry lab – lick your fingers. In the case of sodium cyclamate, the graduate student had put his cigarette on the side of the lab bench (yes, you were allowed to smoke in chemistry labs in 1937!), and when he put it back in his mouth it tasted sweet. The advantage of these artificial sweeteners comes from the fact that they are often many times as sweet as sucrose, which means you don't need to use very much of them, and also that they are not metabolized in the same way as sugar, so you don't get fat. They also don't cause tooth decay.

BUT WHAT ABOUT THE HEALTH RISKS?

That's a long and complicated story. In order to be allowed into foodstuffs, chemicals like sweeteners need to undergo a rigorous set of safety tests. Almost all of the sweeteners have been controversial, with claims being made about them being unsafe or that the testing wasn't done properly, and counter claims that they are safe. Much of the criticism came from vested interests, such as the sugar manufacturers, worried about the diminishing sugar market.

REALLY? TELL ME MORE …

One of the first artificial sweeteners was saccharin, first produced in 1878. When it started to become popular, the US Department of Agriculture (USDA) investigated its safety. However, Harvey Wiley, director of the bureau of chemistry for the USDA thought it might

ruin the lucrative US sugar industry, and
did all he could to ban it. In a meeting
with President Theodore Roosevelt, Wiley
said, '*Everyone who ate that sweet corn was
deceived. He thought he was eating sugar, when
in point of fact he was eating a coal tar product
totally devoid of food value and extremely
injurious to health*'.

Saccharin

But Roosevelt, who had become increasingly overweight due to his
sedentary lifestyle and sugar-rich diet, had himself been prescribed
saccharin for many years by his doctor. Roosevelt replied angrily,
'*Anybody who says saccharin is injurious to health is an idiot*'.

And with that, saccharin was allowed as a food additive … for a
while. The USDA attempted to ban saccharin several times in the
1950s and the 1960s, but the lack of medical evidence showing any
harmful effects prevented this. Then, in the early 1970s studies in
laboratory rats linked saccharin with bladder cancer in rats, and all
food containing saccharin was immediately labeled with a very off-
putting warning. It took 30 years for these findings to be checked,
and then it was found that the link with bladder cancer was specific

USE OF THIS PRODUCT MAY BE
HAZARDOUS TO YOUR HEALTH.
THIS PRODUCT CONTAINS SACCHARIN
WHICH HAS BEEN DETERMINED TO CAUSE
CANCER IN LABORATORY ANIMALS.

The infamous warning label which used to be put on all foodstuffs
that contained saccharin. These were finally removed after new tests
showed that saccharin did *not* cause cancer in humans

to rats, and did not apply to humans. In December 2000, the *Sweetness Act* finally removed the requirements to put a warning label on any foodstuffs containing saccharin, and it was removed from the list of possible carcinogens.

WHAT ABOUT THE OTHER SWEETENERS?

Aspartame is an artificial sweetener that's 200 times as sweet as sucrose, and was discovered in 1965 and was first sold under the brand name *Nutrasweet*. It's often used in soft drinks or cereals and chewing gum, but because it is a peptide, it can hydrolyze back to its constituent amino acids at high temperatures. As such, it's not generally used in cooking or baking.

Aspartame

Aspartame, too, has not been free of safety controversies. It was originally granted approval for use in foodstuffs by the US Food & Drug Administration (FDA) in 1974, but then banned after problems with the safety tests were uncovered. The tests were checked and found to be ok, so aspartame was unbanned in 1981. Since December 1998, a hoax email spread on the Internet which cited aspartame as the cause of various diseases, including cancer and multiple sclerosis. Although it was fictitious, and all the scientific claims were wrong, many unsuspecting people read it and believed it. Nowadays, aspartame has been deemed to be safe for human consumption by more than 90 countries worldwide. Indeed, FDA officials have described aspartame as 'one of the most thoroughly tested and studied food additives the agency has ever approved'.

ANY OTHERS?

Sucralose was discovered in 1976 by researchers Leslie Hough and Shashikant Phadnis working for the Tate & Lyle sugar company. Hough asked Phadnis to *test* a new sucrose derivative, but Phadnis misheard him and thought he said 'taste' it. On doing so, he found the compound was very sweet, and sucralose was born, and marketed worldwide as *Splenda*. It was first approved for human consumption in 1991 in Canada, and is now approved in over 80 countries. Unlike aspartame, sucralose is not heat sensitive and can be used in cooking, frying and baking.

Sucralose (*Splenda*)

Chemically, sucralose is unusual in that it is a chlorinated version of sucrose, and unlike many of the other artificial sweeteners the controversies for sucralose are mainly about its marketing rather than its safety. *Splenda* was marketed using the slogan '*Splenda is made from sugar, so it tastes like sugar*'. Although it is, indeed, made from sugar (by chlorinating either sucrose or raffinose), the US Sugar Association objected to this slogan and filed five separate false-advertising claims against the manufacturers Merisant and McNeil Nutritionals. The case was settled out of court in the United States, but the slogan remains banned in France. Many independent studies have shown that sucralose is safe, but this didn't prevent the US Sugar Foundation funding a project at Duke University in North Carolina, which (unsurprisingly) found evidence that high doses of sucralose given to rats contributed to obesity, destroyed 'good' intestinal bacteria and prevented prescription drugs

"Good news, sir. We've finally found a way to discredit that new sweetener... if we force-feed a rat with twice its body weight of the stuff it drops dead! That's a health risk!"

from being absorbed. However, other studies did not find the same effect in humans, and the report was criticized by an independent scientific panel as 'not scientifically rigorous and is deficient in several critical areas that preclude reliable interpretation of the study results'. Nevertheless, this hasn't prevented the Sugar Foundation continuing to use the Duke paper in their anti-sucralose web pages and literature.

AND THE TAKE-HOME MESSAGE?

If the story of sugar substitutes tells you anything, it's to be careful what you read and who you believe. Even supposedly rigorous scientific studies in respected journals or official reports from government bodies may be suspicious when vested interests are involved. The aspartame Internet hoax was read by millions more people than the proper scientific report which refuted it, and while most people will remember the saccharin warning labels, very few will have heard that they've been revoked. With the rapid growth of the Internet, it may become increasingly difficult to distinguish between information and misinformation.

Chapter

56

SWEATY' ACID, (*E*)-3-METHYL-2-HEXENOIC ACID

(*E*)-3-Methyl-2-hexenoic
acid

I Can't Find That in Google?

You won't, and it isn't worth looking under 'armpit odor' either.

So Which Molecule Is That?

This is (*E*)-3-methyl-2-hexenoic acid (a.k.a. *trans*-3-methyl-2-hexenoic acid). It is one of 30 or more smelly molecules in body odor, but this is reckoned to be 'simply the best'.

(*E*)-3-Methyl-2-hexenoic
acid

Where Does It Come From?

The apocrine glands, found especially in the armpit and groin region, produce an oil that contains a range of organic molecules. To begin with, this oil has no odor, but bacteria soon get to work on it, releasing a blend of saturated and unsaturated carboxylic acids.

Why Is It Odorless to Start With?

The acids are not present in the oil as free carboxylic acid molecules; instead they are bound to a glutamine residue as a 'conjugate'. *Corynebacteria*, which are present on the skin, break the conjugates down, releasing the carboxylic acids, so we start to smell.

WHAT ARE THE OTHER COMPOUNDS THAT CONTRIBUTE TO THE SMELL OF SWEAT?

Two more compounds that have been identified as contributors to it are 3-hydroxy-3-methylhexanoic acid ('Spicy') and 3-methyl-3-sulfanylhexan-1-ol, whose relative amounts seem to be gender-specific.

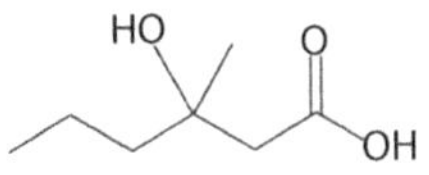

3-Hydroxy-3-methylhexanoic acid

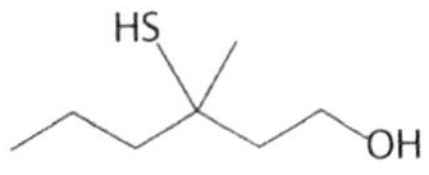

3-Methyl-3-sulfanylhexan-1-ol

And then there are the normal carboxylic acids, both straight chain and branched chain, which are found in sweat ...

SUCH AS?

Butanoic acid is a good candidate. Over 20 years ago, one of the authors of this book gave a class of keen 15-year-olds a little butanoic acid to test its reactions. The lab smelled of it for days, unpleasantly – rather sweaty and vomity. A French teacher who later used that room to register her classes wouldn't speak to him for months!

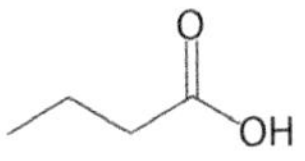

Butanoic acid

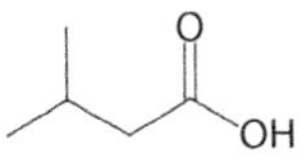

3-Methylbutanoic acid

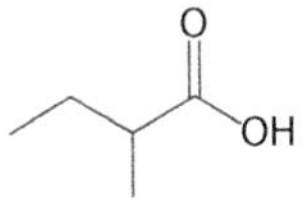

2-Methylbutanoic acid

Another example, 3-methylbutanoic acid, traditionally called isovaleric acid, occurs widely in natural esters, which themselves have quite a pleasant smell. The free acid, though, is another

Limburger cheese

matter; its pungent smell is described as cheesy or sweaty feet! In fact, several carboxylic acids are found in small amounts in cheeses, like Limburger. The acids in Limburger, such as 3-methylbutanoic acid, attract mosquitoes as much as sweaty feet do.

Then there is the isomer, 2-methylbutanoic acid …

THAT SMELLS SWEATY, TOO?

Yes and no.

IT'S GOT TO BE ONE OR THE OTHER, SURELY?

2-Methylbutanoic acid has a chiral carbon atom (marked *). That means it has two optical isomers. The *R*-isomer has a smell described as cheesy and/or sweaty. In contrast, the *S*-isomer apparently smells fruity and sweet.

(*R*)-(−)-2-Methylbutanoic acid, cheesy, sweaty smelling (left) and (*S*)-(−)-2-methylbutanoic acid, fruity, sweet smelling (right)

Chapter

57

TAXOL (PACLITAXEL)

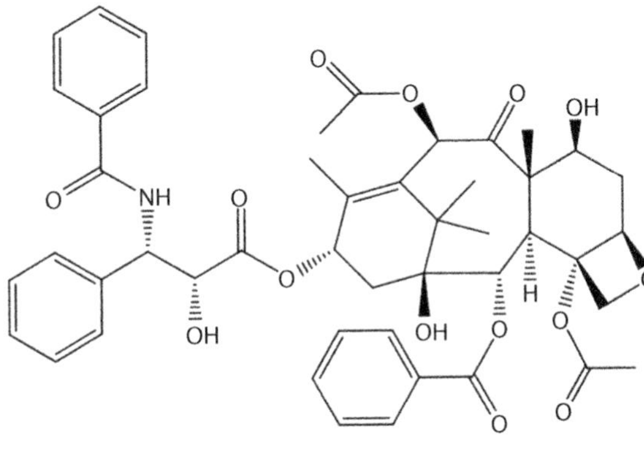

Taxol

SOUNDS LIKE A MOLECULE MADE BY A GOVERNMENT?

Pacific yew tree (*Taxus brevifolia*)

Not quite, although the controversies surrounding it could easily have come from any governmental department. The molecule was originally found in the Pacific Yew tree, *Taxus brevifolia* (*taxus* is the Latin word for 'yew'). Similarly, the European yew tree is *Taxus baccata*, the Himalayan yew *Taxus yunnanensis* and the Chinese yew *Taxus chinensis*. It was given the name taxol by its discoverers, but is now known by the generic name *paclitaxel*, because in 1992 the pharmaceutical company Bristol-Myers Squibb, who had (controversially) acquired the exclusive right to develop the drug commercially, managed to (also controversially) copyright the name taxol.

SURELY THE YEW IS POISONOUS?

Indeed, nearly 2000 years ago, Julius Caesar reported the suicide of Catuvolcus, king of the Eburones, who had taken it. Sadly, people still use it as a means of committing suicide.

WHAT IS IT USED FOR?

It is a drug used on cancer patients, especially for treating ovarian, breast and lung cancers and Kaposi's sarcoma.

How Was It Discovered?

Back in 1958, the American National Cancer Institute began a programme to look for new useful drugs, both synthetic and natural ones. This included screening natural sources like plants and marine life. On August 21st 1962, an American botanist named Arthur Barclay collected bark samples from Pacific yew trees in Washington State. Its cytotoxic properties were discovered in 1964, so the hunt began to find the molecule responsible. Taxol was identified – and named – by Munroe E. Wall and his colleague, Mansukh Wani, at the Research Triangle Institute in North Carolina in the autumn of 1966. Full details of its structure were published in 1971.

So It Soon Came into Use?

Animal testing was carried out first. In 1978, results of tests of taxol against human tumor cell lines (breast, colon, lung) transplanted to mice were reported and showed considerable promise. Phase-1 clinical trials in humans started in 1981 and Phase-2 began in the mid-1980s. Results against breast cancer and small-cell lung cancer were very encouraging, and trials against ovarian cancer showed several long-term remissions. Taxol was approved for treatment of drug-resistant ovarian cancer in 1992 and for use against breast cancer in 1994.

How Does Taxol Work?

Anti-cancer drugs face a tricky conundrum – how to kill cancer cells without killing all the normal body cells nearby, and with them the patient. The answer is to find something that distinguishes cancer cells from normal ones, and target that attribute. Cancer cells, almost by definition, are cells which multiply uncontrollably fast – much faster than ordinary cells. Therefore, many anti-cancer drugs target the mechanism of cell division and prevent this from happening. This slows down or even kills cancer cells while (hopefully) having little or no effect upon the much slower-dividing normal cells. In the case of taxol, it binds to a molecule within the cell nucleus responsible for initiating the cell division process, β-tubulin. In doing so, it inhibits cell division and cancer cell growth – in effect the cancer cells die of old age before they can multiply.

IF IT IS SUCH A GOOD DRUG, WHY THE DELAY?

Well, the testing took some time, and also there were problems. Taxol is insoluble in many solvents, though in the end, a mixture of ethanol and a surfactant (Cremophor EL) was found to work. There were also environmental problems, and yet more controversies.

Cutting off the bark of a Pacific yew tree to extract the taxol

Northern spotted owl (*Strix occidentalis*) (US Fish and Wildlife Service, National Digital Library)

WHY?

When you take the bark off the Pacific yew to extract the taxol, it usually kills the tree! Besides which, there is not much taxol in it – the bark from a whole tree contains about 350 mg of taxol, which is enough for one dose, but you need around 3 g for a course of treatment, so you'd need to kill half a dozen mature trees per patient. There was also the problem that the Pacific yew tree takes up to 200 years to grow – and is also home to the endangered Northern spotted owl (*Strix occidentalis*). So there were cancer patients demanding the trees be cut down, and environmentalists opposing this on the basis of conservation.

SO HOW DID THEY GET AROUND THAT?

In Europe, Pierre Potier, of the French National Center of Scientific Research (CNRS) at Gif-sur-Yvette, was famous for looking for medicines in nature (which he termed *le magasin du bon dieu* – the Good Lord's shop). In 1981 he discovered the answer was right under his nose. The taxane core at the heart of taxol also existed in a compound called 10-deacetylbaccatin III, which was found in the needles of the fast-growing European yew (*Taxus baccata*), growing freely in the CNRS park. These needles were a renewable resource that could be harvested, and which permitted a much shorter synthetic route.

Close-up photo of the European yew, showing its needle-like leaves and cones

10-Deacetylbaccatin III could be turned into taxol in only four steps (a process called 'semi-synthesis'). Molecules like this are responsible for the resistance of yew to woodworm and insects in general – yew isn't just poisonous to humans!

The semi-synthetic drug *taxotere* was developed as a result of Potier's work, and this also is now being used to treat cancer.

10-Deacetylbaccatin III   Docetaxel (*Taxotere*)

How Else Can You Make Taxol?

It has got a really complicated structure, with fused four-, six- and eight-membered rings, and this posed a big challenge to organic chemists. They took it up, though, and two total syntheses were reported almost in the same breath in early 1994, from the Holton group at the University of Florida and the Nicolaou group at Scripps Research Institute and the University of California, San Diego. Subsequently, others have reported total syntheses, with the shortest being that of Wendler's group (37 stages, 0.4% overall yield). Total synthesis is, of course, very expensive, and not a commercial option. Commercially, it is manufactured at present by biological methods, especially plant cell culture fermentation processes.

How Does the Yew Tree Make It?

It uses a series of enzyme-controlled reactions. The step responsible for the creation of the unusual core is a cyclization reaction which changes geranylgeranyl diphosphate into taxa-4,11-diene. This then undergoes a variety of processes involving oxidation, acylation, and addition of a side chain to form taxol. The structure of the enzyme responsible for this, taxadiene synthase, was reported in January 2011.

Geranylgeranyl
diphosphate

Taxa-4(5),11(12)-diene

In 2010, scientists at MIT and Tufts University reported that they had engineered *Escherichia coli* bacteria to produce taxadiene, as well as the next molecule in the biosynthetic pathway, taxadien-5α-ol. An obvious next step would be to find ways of making baccatin III, the useful starting material for taxol synthesis.

How Expensive Is It?

A course of treatment for breast cancer is reported to cost around $6000 (£4000).

THAT'S QUITE A LOT OF MONEY

Yes, the increasing costs of medical drugs is becoming a major issue for governments all over the world, whose tax-payers demand treatments that might keep them alive, but which cost more than the patient, or sometimes even the government, can afford.

Note to Readers: Most chemists know this compound as taxol, but it now has the generic name paclitaxel and the registered trade name *Taxol*® (Bristol-Myers Squibb Company, New York, NY). '*Taxotere*' here refers to the drug now generically referred to as docetaxel with the registered trade name *Taxotere*® (Rhone-Poulenc-Rorer Pharmaceuticals, Inc., Collegeville, PA).

Chapter

58

TESTOSTERONE

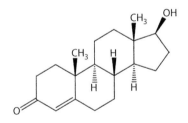

Testosterone

It's a Male Molecule, Yes?

Caster Semenya during the World Athletics Championships 2009

It is actually found in both men and women. Usually there is more of it in men, though it is possible for a woman to have a higher level of testosterone than a particular man. After the South African athlete Caster Semenya won the 800-meters race at the 2009 World Championships in Berlin, she was barred from competing for a year and had to undergo tests. It was said that her testosterone levels were three times those of an average woman, though still below male levels.

Why Is Testosterone Associated with 'Male-Ness'?

It's associated with noticeable changes in males at puberty, such as deepening of the voice, increase in muscles, changes in sexual development and growth of hair.

From Where Does the Body's Testosterone Come?

The body makes it in the male testes from cholesterol (see p91) in a multi-stage, enzyme-catalyzed reaction. The most obvious change is the removal of the long eight-carbon side-chain of the parent cholesterol molecule. Six carbons are removed by an enzyme, leading to progesterone, and another enzyme removes two more.

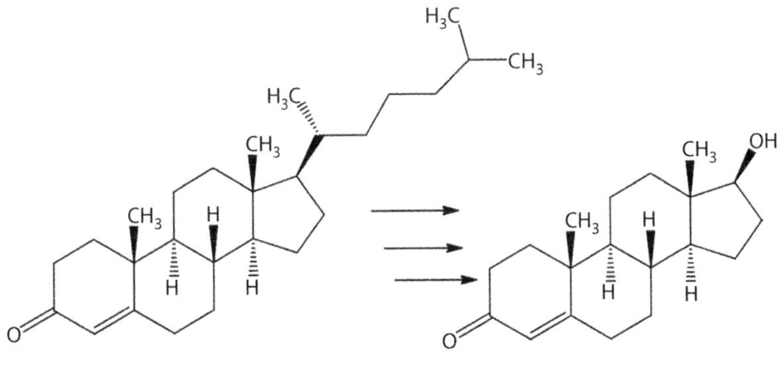

Cholesterol → Testosterone

Outside the body, testosterone was actually first isolated in 1927 from bulls' testes. The first chemists to synthesize it, Adolf Butenandt and Leopold Ruzicka, shared the Nobel Prize in chemistry in 1939.

Compare the structure of testosterone with the two most important female hormones, estradiol and progesterone (see p185), and you'll see how small changes in structure can have a big effect upon bodily function.

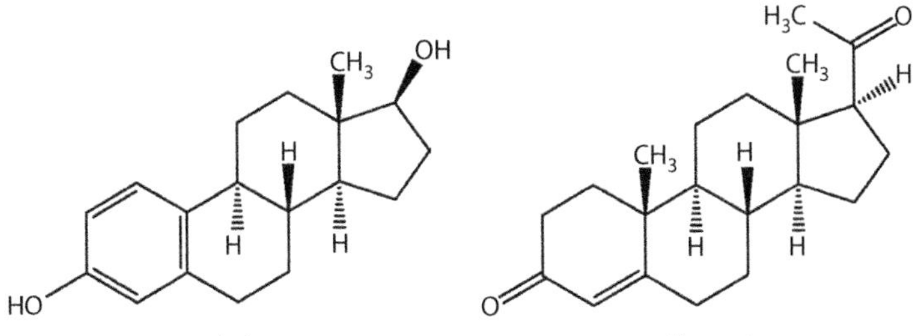

Estradiol   Progesterone

Originally, testosterone and its ester derivatives were used medicinally. By restoring a sick patient's testosterone levels, it speeded up the recovery of their muscles. Esters were first made in the 1930s and are widely used and abused by bodybuilders and sportsmen. One of the most widely used is testosterone enanthate.

Testosterone enanthate

DO OTHER ANIMALS USE TESTOSTERONE?

Most spectacularly in Asian bull elephants. Their testosterone levels can increase by a factor of 60-fold in the periodic condition called *musth*. This is hormonally induced, but no one understands why. During this time the male elephants become very aggressive, and can go on a mating frenzy with any female elephants in season. They have also been known to kill female elephants that aren't in season! Rather than simply a feature of mating, another suggestion is that *musth* is connected with the reorganization of dominance among males in elephant herds.

An elephant in *musth* charging a giraffe

How Does Testosterone Work?

Testosterone binds to the androgen receptor in the cytoplasm of cells. This produces structural changes that enable this complex to enter the nucleus of the cell. There it binds to chromosomal DNA and promotes genes carrying out transcription and translation. This increases protein synthesis in cells and leads to the build-up of muscular tissue that is characteristic of anabolic effects.

Why Do Athletes Take Testosterone and Similar Steroids?

Testosterone was the first anabolic steroid used by athletes to improve performance, and continues to be so used. In the early 1950s, American weightlifters suspected that their Russian opponents were using testosterone and started using it, usually as esters, followed by synthetic anabolic steroids. Athletes have taken testosterone by injection, by oral routes or by absorption through the skin (*e.g.* transdermal patches). By building up muscles which repair faster, the athletes improve the level of their performance and also their

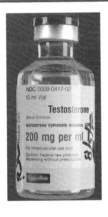

Synthetic testosterone.
Note the warning
on the label which says that
Federal Law prohibits dispensing
without prescription

recovery time. It was estimated that in 1968 one-third of the US track-and-field team for the Mexico City Olympics were using steroids at their pre-games training camp. Steroid testing at the Olympics was not introduced until the 1976 Montreal Games, which was a bit too late as athletes had been using synthetic derivatives of testosterone, such as *Dianabol* since the late 1950s (see p507).

AND TAKING TESTOSTERONE IS SAFE?

Sadly, no. The most obvious symptom of testosterone abuse is that the male testes shrink, as they aren't needed to make testosterone any more. Health problems associated with high levels of steroid abuse include liver and kidney tumors, psychiatric disorder and heart problems.

HOW IS TESTOSTERONE ABUSE DETECTED?

Testosterone has an inactive isomer called epitestosterone, which differs in only one stereogenic center – the OH group. Testosterone is the *S*-isomer and epitestosterone is the *R*-isomer. In most people, they occur in an approximate 1:1 ratio; however, epitestosterone has no anabolic effects.

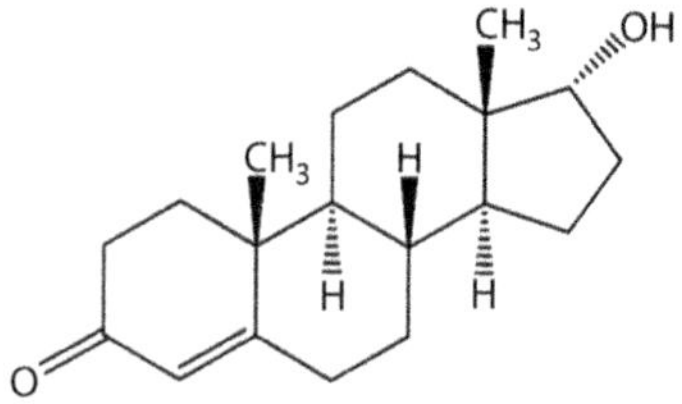

Epitestosterone

Testers in sport look at the relative amounts of testosterone and epitestosterone. In 1982, the International Olympic Commission (IOC) decided that, from 1982, a 6:1 ratio would result in immediate disqualification. The rule now says that if the ratio is above 4:1, additional samples must be tested.

Of course, drug cheats can try to get around this by using synthetic epitestosterone as a 'masking agent', by taking that as a mixture with testosterone, so that the T:E ratio is low enough not to be suspect. But drug testers have a further weapon in their armory. 'Temperate-zone' plants operate by a C_4 biosynthetic pathway, while tropical plants use a C_3 photosynthetic route. Synthetic testosterone is usually made from C_3 plants, like wild yams or soy, while human testosterone is derived from a diet that mixes C_3 and C_4 plants. So testers therefore can carry out a carbon isotope ratio (CIR) test, measuring the ratio of ^{13}C to ^{12}C isotopes in the testosterone sample, to distinguish between natural and synthetic testosterone.

AND SPORTSMEN HAVE BEEN CAUGHT?

One of the best-known people to fall victim to the testing was the American cyclist Floyd Landis. Coming to Stage 16 of the 2006 Tour de France, Landis led the field, but that day he dropped right back, losing 8 minutes. However, the next day Landis destroyed the field and as a result of the comeback went on to finish first in Paris. When a sample taken after

Floyd Landis in the Tour of California 2006

Stage 17 was tested, apparently a very high T:E ratio (reportedly 11 to 1) was detected and it is also reported that he had failed the CIR test. Landis has admitted using performance-enhancing drugs for much of his career, but denies using testosterone or other drugs on the 2006 tour.

The 'best' excuse is that of Dennis Mitchell, a member of the gold-medal-winning US team in the 4 × 100-m relay at the 1992 Barcelona Olympic Games. In a test six years later, his sample proved positive for testosterone. He said that he'd had 'five bottles of beer, and sex with his wife at least four times … it was her birthday, the lady deserved a treat'. This impressed the public, but not the authorities, who banned him for two years.

Does 'Roid Rage Exist?

The jury is out on this. Some studies – but not all – have indicated that steroid abuse does have a link with aggressiveness. Remember the elephants!

Chapter

59

TETRAHYDROCANNABINOL (THC)

Leaf from a *Cannabis sativa* plant

Delta-9-THC

WHAT'S THE DIFFERENCE BETWEEN CANNABIS AND MARIJUANA?

Cannabis is strictly the name given to a genus of plants, though when people say 'cannabis' they usually mean *Cannabis sativa*, the stuff that hemp and also the drug is obtained from. Marijuana is the name of the drug, but most people use cannabis and marijuana interchangeably. There are several types of marijuana, like *bhang* (from the leaves, seeds and stem), *ganja* (from the flowering tops), and *hashish* (pure resin from the flowers).

| Dried cannabis flowers | Hashish | A slow-burning 'joint' filled with cannabis |

HOW LONG HAS IT BEEN AROUND?

The Chinese were very likely the first people to cultivate the plants, around 6000 years ago, in order to get hemp for making rope. Subsequently, other applications like clothing have been found. It seems that medicinal uses were discovered around 2000 years ago. Like many other people of the time in eighteenth-century North America, George Washington grew it to make hemp for rope and sacks, but not, it seems, for consumption.

So How Did It Get into Western Culture?

It was widely used in the Middle East during the first millennium AD (it is not prohibited by the Koran, as it is not 'intoxicating') and from there found its way into Europe. It seems that Indian workers took marijuana to the West Indies, from where it was only a short step to the American continent. It certainly reached France in the early nineteenth century, with the great poet Charles Baudelaire among the users. The medicinal uses of 'tincture of hemp' for a whole range of complaints were known, so it became a component of Dr. J. Collis Browne's *chlorodyne*, introduced into the United Kingdom around 1850. This medicine was claimed to treat dysentery, diarrhea, cholera, insomnia, neuralgia, bronchitis and asthma, among others. Because chlorodyne contained alcohol, morphine and cannabinoids, its powers were irresistible to some users, and it was subsequently reformulated (there are no cannabinoids in the current version).

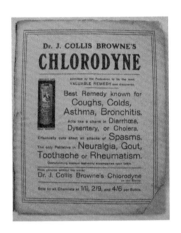

Photograph of an advert for Chlorodyne in a book from 1907

And Pop Culture?

In the 1930s, marijuana was associated with jazz musicians – Louis Armstrong was busted for it in 1931 – while in the 1940s and 1950s cheap paperback fiction ('yellow fiction') regularly featured reefers (marijuana-based cigarettes) leading innocents into lives of shame and depravity. With the expansion in higher education, it reached universities in the United States in the 1950s and 1960s. It is said that when Bob Dylan made his *Blonde on Blonde* album in 1966 he was high on cannabis, and the opening song *'Rainy Day Women # 12 & 35'* includes the words 'everyone must get stoned'. Other Dylan songs like *'Mr Tambourine Man'* were also influenced by it. It was Dylan who introduced the Beatles to marijuana; Paul McCartney's tribute to marijuana, *'Got to Get You into My Life'*, was written for the 1966 *Revolver* album. In February 1967, Keith Richards, the Rolling Stones' guitarist, was arrested and sentenced to a year's imprisonment for allowing cannabis to be smoked at his home – though he was later acquitted on appeal over the trivial detail that there had been no cannabis on the premises. As a Rastafarian, Bob Marley used *ganja* (the Jamaican word for marijuana) and this influenced the development of reggae (with his albums *'Kaya'*, *'Redder than Red'* and *'African Herbsman'*).

Bob Dylan in 1978

In the 1970s, Rastafarians popularized *ganja* as part of their religion and reggae music

A combination of the Vietnam War and the hippy trail to Kathmandu helped spread the pot habit, while ex-President Bill Clinton smoked it when in college but famously did not inhale.

Even the clean-cut housewives' favorite singer, John Denver, didn't escape controversy over the drug, when his song *Rocky Mountain High* was banned by the US Federal Communications Committee in 1973, who thought it was promoting drug use. Eventually, John Denver managed to persuade the FCC that in innocently using the word 'High', he was simply describing the sense of peace and awe he found in the Rockies, and the song was unbanned.

More recently, cannabis use has been the focus of a popular US TV series called 'Weeds', which concerns the story of a widowed mother who begins selling marijuana to support her family after her husband suddenly dies. The show explores the controversial issues and 'problem' of the illegality of the drug.

WHAT PROBLEM?

The use of cannabis for recreational purposes has been illegal in most countries since the start of the twentieth century. This is despite it not being addictive and arguably being far less harmful than many legal drugs, such as alcohol or tobacco. Even so, according to the United Nations World Drug Report 2012, cannabis is the most widely used illicit substance globally, with nearly 1 in 20 of the adult population in the world using it at least once a year! As well as the problem of maintaining the ban on such a widely used substance, the recent advent of 'medical marijuana' has made the issue even more bizarre. In some US states, it is currently illegal to buy marijuana for recreational use, but if you have certain medical conditions, such as multiple sclerosis or chronic pain, it is possible to buy the same marijuana perfectly legally!

WHY DO PEOPLE SMOKE IT?

Smoking it makes people feel happy and self-confident. Its effects include making time seem longer and the world seem more mellow. It affects short-term memory, peripheral vision, reaction time and coordination.

How Is It Made?

Marijuana contains nearly 500 different chemicals. The most important one as far as psychoactive properties are concerned is Δ-9-tetrahydrocannabinol, also known as delta-9-THC ($Δ^9$-THC). Nature, being a brilliant synthetic chemist, just makes the single isomer (–)-$Δ^9$-THC. It is possible to make the other isomer in the laboratory, but (+)-$Δ^9$-THC has no psychoactive properties.

THC is made in the leaves of the cannabis plant, the first step being the biosynthesis of Δ-9-tetrahydrocannabinolic acid (THCA) from cannabigerolic acid (CBGA). A living plant contains very little $Δ^9$-THC, but THCA readily eliminates carbon dioxide to form $Δ^9$-THC, even by just drying the plant.

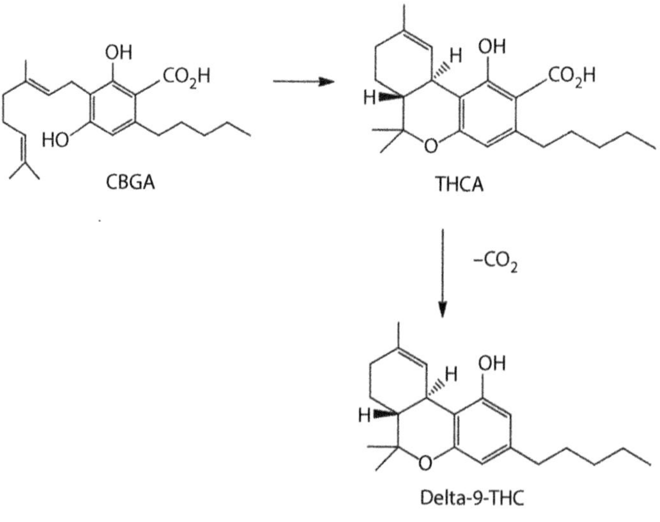

CBGA

THCA

$-CO_2$

Delta-9-THC

How Does It Work?

If cannabis is smoked, Δ^9-THC is transported by the blood from lung to brain within seconds, in the same way as is nicotine (see p363). Δ^9-THC acts on two 'cannabinoid receptors' in the human body. One of them (called CB2) is found in areas like the immune system, but CB1 is prominent in the brain (and is also found in the lungs, liver and kidneys). It is when Δ^9-THC binds to CB1 that hallucinogenic effects are produced. It sets off several cell signalling pathways, and also affects the release of noradrenaline (see p9), a neurotransmitter. There are natural molecules – anandamide is the best known – that also bind to these so-called 'cannabinoid receptors'. In fact, these receptors should really be called 'anandamide receptors', since they are intended for anandamide, not THC. Anandamide, discovered in 1992, is thought to be involved in the regulation of feeding, pain control and the pleasure response. THC just happens to be the right shape to fit the anandamide receptors, thereby fooling the brain by giving it false feelings of pleasure.

Anandamide

Cannabidiol

Delta-9-THC is not the only important chemical in cannabis. By itself, it can produce feelings of panic, but these are alleviated in cannabis by another chemical present called cannabidiol, which is not psychoactive (and may have other useful therapeutic properties).

So It Is Just a Drug That Some People Find Pleasant?

Although most people use marijuana for the 'high', it does have medicinal uses, which were widely studied in the second half of the nineteenth century. Sir John Russell Reynolds, Queen Victoria's personal physician, promoted the medicinal use of cannabis for *dysmenorrhoea* (period pains) in the nineteenth century, but he is unlikely to have recommended it to the Queen. Would she have been amused?

Medical marijuana – you might be ill, but you'll be too stoned to care!

Patients suffering from a variety of illnesses claim that smoking marijuana helps alleviate many of their symptoms. It is reported to reduce nausea and vomiting in cancer patients. When the famous biologist Stephen Jay Gould was being treated for abdominal *mesothelioma* cancer, he smoked marijuana to reduce the effects of the nausea he suffered from his chemotherapy and said 'marijuana worked like a charm'. It acts as an appetite enhancer in AIDS patients, while multiple sclerosis patients say that it reduces muscle cramps, and paraplegics and quadriplegics report it to help control involuntary muscle spasms. An extract from cannabis is reported to lower the pressure in the eye fluid, and this could be useful in glaucoma patients.

As far as commercial sources
go, *Sativex®*, which contains
THC and cannabidiol, is
approved for use in the
United Kingdom as well
as New Zealand, Canada,
Germany and several other
European countries. It
is an oromucosal spray

Nabilone

administered into the mouth for the treatment of spasticity
in multiple sclerosis patients, for neuropathic pain and other
symptoms, and as a painkiller for cancer patients. The pure isomer
(−)-Δ^9-THC is known as dronabinol (trade name *Marinol®*) and is
marketed in the United States as a Schedule III drug, available on
prescription for the treatment of anorexia, nausea and vomiting in
AIDS patients. A synthetic cannabinoid called nabilone (marketed
as *Cesamet* in Canada) finds similar applications.

SHOULD PEOPLE SMOKE IT?

The smoke from a reefer contains over 150 different compounds,
including toxic molecules such as CO, HCN and the carcinogen
benzopyrene. Some of these occur at higher levels than in tobacco
smoke.

HOW DANGEROUS IS IT?

Research suggests that around 9% of cannabis users become
dependent. It is widely believed that there is no link between
cannabis use and moving on to cocaine and heroin; this may be
true of some users, but there are sad stories where it does seem
that it primes the brain to seek other, 'harder', narcotic substances.
Some of the molecules in cannabis, like Δ^9-THC, look to have

useful medical properties, but smoking is a very unsafe way of delivering the molecules – not least because of the risk of illnesses associated with smoking, such as lung cancer and bronchitis. Although no one has yet made any long-term study of marijuana smokers, it is unlikely to be a healthy option.

Chapter

60

TETRAHYDROGESTRINONE (THG)

(AND 'ILLEGAL' STEROIDS)

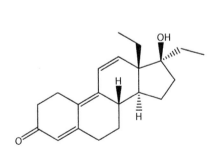

THG

After the big race the tortoise and hare were forced to provide urine samples!

TETRAHYDROGESTRINONE, THAT'S A MOUTHFUL!

Just think 'illegal steroid'.

WHY ILLEGAL?

It was taken for some years by sportsmen, notably some athletes, to get an unfair advantage.

HOW DID THAT HAPPEN?

Well, you have to go back to what some steroids are responsible for, notably testosterone (see p489). It is responsible for the development of male sexual characteristics and also has anabolic effects, building tissue and muscle. After World War II, some athletes started to take it to get an unfair competitive 'edge'. The practice seems to have started with Russian weightlifters and was rapidly taken up by US athletes.

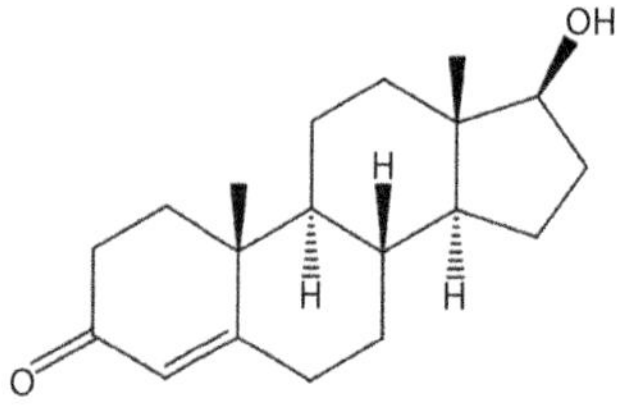

Testosterone

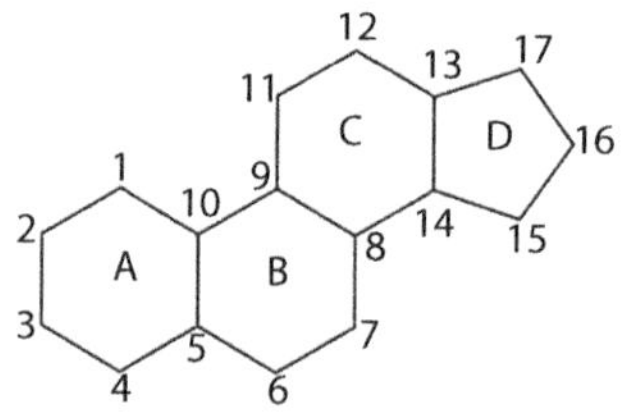

How to label rings and carbon atoms in steroids

SO WHY USE ANYTHING ELSE?

Drug testing of athletes was introduced, so new drugs were needed. These were testosterone-like molecules which had alkyl groups at position 17 in ring D. This made them harder to break down in the body, as these new molecules were tertiary alcohols. It was also harder for enzymes to hydroxylate them, especially at carbons 10 and 16, which meant they didn't get metabolized in the liver as quickly. As a result they had stronger muscle-building effects – but were also more likely to cause liver damage.

Dianabol (methandrostenalone) was released in 1958 as an anabolic steroid that did not interact with the androgen receptor, though it still produced androgenic (masculinizing) effects. The next step was *Turinabol*, which was licensed for clinical use in 1965. *Turinabol* could be taken orally and had the advantage of being metabolized fairly quickly by the body (reducing the risk of detection).

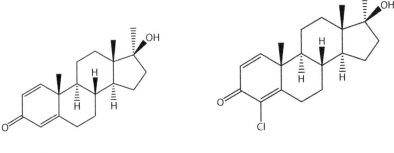

Dianabol
(methandrostenalone)

Turinabol
(4-chlorodehydromethyltestosterone)

Turinabol as well as *Nandrolone* was given to many East German athletes and swimmers. They were told that the blue tablets they were given were vitamins. The women swimmers were spectacularly successful, winning 11 out of the 14 gold medals available at the 1976 Montreal Olympics, but androgenic changes – broad shoulders and deep voices – were also noticed and commented upon (famously an East German coach remarked, 'We came here to swim, not sing').

Later steroids of abuse included *Stanozolol* (found in urine samples taken from the Canadian athlete Ben Johnson at the 1988 Seoul Olympics) and *Nandrolone* (see p357).

Stanozolol   *Nandrolone*

But testing methods got better, and pressure increased to develop an undetectable molecule. This resulted in tetrahydrogestrinone (THG) which had never been made before, so was not on the radar of any of the drug testers. Gestrinone, the 'starting material', was already known. It was developed originally in 1974 as a possible oral contraceptive agent. Catalytic hydrogenation of gestrinone is the obvious route to THG. Careful control of conditions is required to ensure that only the alkyne group is reduced.

Synthesis of THG (right) from gestrinone (left)

WHO MADE IT?

Patrick Arnold was a chemistry graduate and also a bodybuilder, a powerful combination. He dug deep into the literature of steroids in his search.

How Did the Athletes Get It?

Known as 'The Clear', it was supplied by BALCO (the Californian Bay Area Laboratory Co-Operative) to chosen athletes. The leading drug tester, Don Catlin, is reported to have examined some 20 urine samples that tested positive for THG. Athletes known to have taken it include the British sprinter Dwain Chambers and a number of Americans, including the sprinter Kelli White and middle-distance runner Regina Jacobs. Marion Jones, who won three gold medals and two bronze medals at the Sydney Olympics in 2000, admitted THG use in 2007 and was stripped of these medals by the International Olympic Committee (IOC) and the International Association of Athletics Federations (IAAF).

Marion Jones in 2006

Dwaine Chambers in 2012
(© Press Association Images. With permission)

If THG Was Undetectable, How Was It Discovered?

In early June 2003, an athletics coach – now known to be Trevor Graham – contacted an official of the US Anti-Doping Agency, saying that Victor Conte, head of BALCO, had been supplying leading athletes with drugs. The coach promised to send USADA a 'used' syringe that had been thrown away at an athletics meeting. He did just that, and the contents of the syringe were passed to Don Catlin, who at that time was the director of the Olympic Analytical Laboratory at UCLA on June 13th 2003.

Victor Conte standing in front of the BALCO Logo (© Press Association Images. With permission)

Catlin's team tried the normal procedure – they made it into the Me_3Si derivative which made it more volatile in the mass spectrum, but the resulting spectrum was thoroughly confusing. They had one clue, a small amount of the steroid norboletone was present, so they guessed that the unknown had a similar backbone to norboletone.

WERE THEY RIGHT?

In fact, comparison of the mass spectrum of the unknown steroid with known molecules showed considerable similarity with the spectra of trenbolone and gestrinone, indicating common structural features. Detailed analysis led them to deduce the structure of THG, which they then synthesized (from gestrinone)

Norboletone   *Trenbolone*

to check its ID, so to speak. They then tested it on a baboon, showing it could be excreted and then detected in urine, so they were able to devise a drug test to use at competitions.

PRESUMABLY THAT TEST WORKED?

Like a charm as they say. They detected THG in samples taken from four athletes (Kevin Toth, Regina Jacobs, John McEwen and Melissa Price) at the US National Championships in June that year. Dwain Chambers, at that time the European 100 m champion, gave a positive test on a sample given on August 1st 2003. Like the American sprinter Kelli White, he blamed Conte for supplying him with THG ('Victor explained it was a new product on the market that would aid me nutritionally').

WHAT DOES THG DO TO PEOPLES' BODIES?

One look at some of the athletes found to have taken it makes you think that that THG has very strong anabolic effects. More scientifically, a comparison has been made of the binding of four potent androgens, the natural testosterone, together with synthetic dihydrotestosterone (DHT), methyltrienolone and THG, showed that THG had the highest affinity for the human androgen receptor. When the crystal structures were determined of human androgenic receptor with three of these (testosterone, synthetic

DHT and THG), they showed that THG makes more van der Waals' contacts with the receptor than do the other steroids, due to the presence of the 17-ethyl and 18-methyl groups. This explained why THG binds approximately twice as strongly to the human androgen receptor than does DHT.

AND THE OUTCOME?

We know about the banned athletes, but the suppliers ended up in court. In October 2005, Victor Conte was sentenced to four months in prison and another four on house arrest, while in August 2006, Arnold was sentenced to three months in prison.

Chapter

61

TETRODOTOXIN

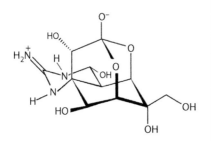

Tetrodotoxin (TTX)

A puffed up puffer fish

Sounds Poisonous!

It most certainly is! Tetrodotoxin (often abbreviated to TTX) is best known for being the poisonous molecule in puffer fish. These belong to the order *Tetraodontiformes*, most famously the genus *Fugu*, which gives rise to the Japanese delicacy of that name. However, it is also found in many other creatures, including the blue-ringed octopus (famous from the James Bond film *Octopussy*), some of the 'poison-dart' frogs from Central and South America (see p177), the rough-skinned newt, certain flatworms and crabs and many other fish. There have been recent poisonings due to people eating some TTX-containing marine snails.

How Does the Puffer Fish Make It?

It doesn't; it probably gets it from certain symbiotic bacteria in its diet (most usually *Vibrio* bacteria), as do the other species which use it. This explains why TTX is so widespread, as animal toxins are usually very specific. Captive-raised puffer fish have much less TTX, but can subsequently accumulate it from bacteria or from wild puffer fish.

Just How Poisonous Is It?

According to poison expert John Timbrell, the lethal dose for an adult human is less than a milligram.

How Does It Work?

TTX blocks sodium channels in nerve membranes by binding to a peptide group at the mouth of the channel. This stops Na^+ ions from entering the cell, which means that the nerves do not fire and signals are not transmitted.

How Do You Know If You've Been Poisoned by TTX?

Within a few minutes of ingesting it, there is a tingling or numbness of the lips and tongue. This spreads to the rest of the face, then to other parts of the body, accompanied by a whole range of other distressing symptoms (vomiting, diarrhea, convulsions, *etc.*). As the paralysis spreads, the victim may become unable to move, but remains conscious and lucid, though speech is affected. Cardiac arrhythmias can occur and the victim eventually dies of asphyxiation. Death usually occurs within 6 h. If the victim is still alive after 24 h, they generally recover.

I Heard That TTX Is Linked to Zombies and Witchcraft?

That's true. In Haiti, witchdoctors were reported to give their victims a 'zombie powder' containing TTX. The idea that a person with TTX-induced paralysis could look dead and therefore be buried, yet could survive, wake up and 'rise from the grave' gave rise to stories of zombies, in particular the case of a man named Clairvius Narcisse. A man of this name was buried in 1962 and someone claiming to be him emerged 18 years later, saying that he had been disinterred alive and spent the intervening time as a slave on a sugar plantation.

Did the myth of zombies arise from victims poisoned with TTX in Haiti?

Do TTX-Fueled Zombies Exist?

The ethnobotanist Wade Davis suggested that TTX was used in voodoo in Haiti, and described this in a popular book '*The Serpent and the Rainbow*'. However, his theory was criticized on the grounds that there was not enough TTX in samples of 'zombie powder' to produce trance-like symptoms, and at present the theory is not generally accepted. The facts have never stopped Hollywood from making a good story. Plotlines involving TTX poisoning, with and without zombification, have been used in a number of TV shows, such as episodes of *Miami Vice*, *Columbo*, the *X-files*, *CSI-NY*, *Nip/Tuck*, and even *The Simpsons*.

Can You Get an Antidote?

There is no known antidote for TTX poisoning, though the novelist Ian Fleming thought that there was. In the closing scenes of the James Bond movie *From Russia with Love*, Rosa Klebb retracts the cap of her shoe to reveal a hidden steel needle, and kicks James Bond on the calf with it. As its TTX coating takes effect, he feels numb and crashes to the floor. In the next Bond book, *Dr. No*, we are told, incorrectly, that an antidote for *curare* poisoning saved him.

The great explorer Captain James Cook was another victim of TTX. In September 1774, he and two of his crew ate some puffer fish and fed the remains to the pigs on board ship. The humans survived the 'extraordinary weakness and numbness' but the pigs did not, probably because the pigs had eaten the more toxic parts of the fish.

But I Thought That Puffer Fish Is Eaten?

So it is. Cooked puffer fish meat is known as *Fugu*, and is an important Japanese delicacy. The problem for the chefs is that cooking does not destroy the tetrodotoxin molecule. Therefore, those parts of the fish that contain TTX, such as the skin, liver, ovaries and intestines, have to be removed by the chef, very, very carefully. In Japanese restaurants, this can only be undertaken by special chefs who have spent three years in training. Each prefecture sets its exams, both written and practical, involving dissecting puffer fish and preparing them

Raw *fugu* (*sashimi*) as served in Japan

for cooking. The pass rate is low, around 30%. Tiny traces of TTX remain, and this can give a delicate tingle in the mouth and tongue, a sensation that is highly valued. Despite all this training up to 100 Japanese still die each year through TTX poisoning, mainly through cooking it at home. *Fugu* is a very democratic food, though the Emperor of Japan is barred from eating it.

WHY DOESN'T TTX KILL THE PUFFER FISH?

It is believed that the puffer fish is immune because a single amino acid mutation has resulted in a modified sodium channel that cannot bind to TTX. It is thought that this is because an aromatic amino acid – either phenylalanine or tyrosine – has been replaced by asparagine, so enabling the otherwise helpless little puffer fish to incorporate TTX into its defense. This immunity also enables the puffer fish to feed off TTX-containing organisms, which are left alone by other species.

WHAT DOES THE FISH USE IT FOR?

The puffer fish swallows sea water to make itself look more impressive by inflating itself. It does have four teeth, which help it feed off reefs or even breaking open molluscs, but are otherwise not much use. After the puffer fish 'puffs up', it floats to the surface, and is helpless until it deflates. So it would be pretty defenseless without TTX. On the other hand, it appears that the blue-ringed octopus uses TTX for both defense and predation.

Common garter snake   Rough-skinned newt
(*Thamnophis sirtalis*)   (*Taricha granulosa*)

Perhaps the most remarkable TTX story concerns the common garter snake, which is engaged in an 'arms race' with the rough-skinned newt. The skin of the newt contains TTX, and so confers protection on it. However, to combat this defense, the snakes have developed some resistance to TTX, probably through mutations in the gene coding for the sodium ion channels, which allows the snakes to eat the toxic newts. You might say that natural selection has favored more poisonous newts, which, in turn, has meant more resistant snakes.

Is TTX of Any Use to Humans?

Intramuscular tetrodotoxin (*Tectin*) was trialled, unsuccessfully, as a treatment for cancer patients in severe pain, but a tetrodotoxin–bupivacaine–epinephrine mixture is being investigated as a nerve block for prolonged local anesthesia.

Chapter

62

THUJONE

THE 'SUSPECT' IN POISONING
BY ABSINTHE

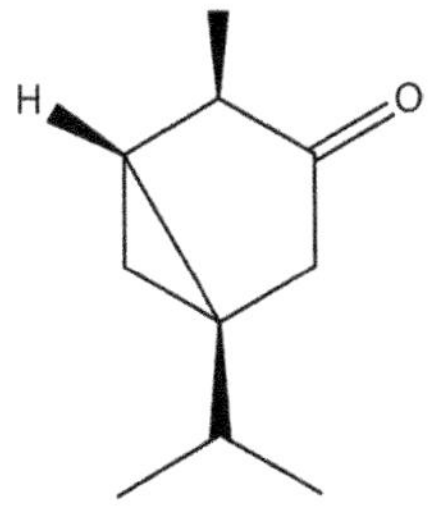

α-Thujone 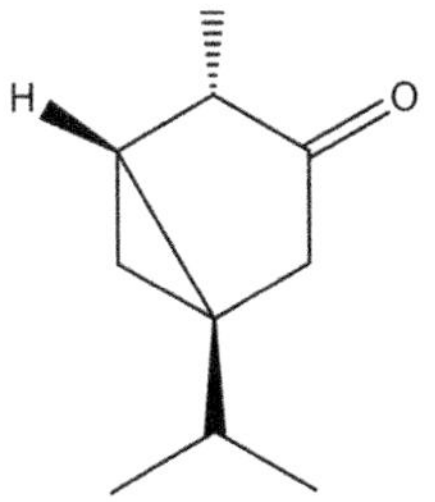   β-Thujone

SUSPECT, WHAT DO YOU MEAN?

Vincent van Gogh was only one – but the most famous one – of a number of people whose deaths were linked with the liqueur absinthe, and thujone was the molecule held to blame.

YOU'VE DRAWN TWO THUJONES

(−)-α-thujone and (+)-β-thujone both occur in nature, with α-thujone being said to be more toxic. They are diastereoisomers, and two more forms are possible, (+)-α-thujone and (−)-β-thujone.

WHAT'S ABSINTHE?

Preparing a glass of absinthe using the traditional method

It is a green alcoholic drink smelling of anise, made by distilling a mixture of alcohol, herbs (notably wormwood) and water. It was also fashionable with the artistic community – names that come to mind include Degas, Manet, Toulouse Lautrec, Verlaine, Baudelaire and of course Vincent van Gogh. Absinthe became the cheapest drink available in France in the second half of the nineteenth century, cheaper than wine. Several artists illustrated absinthe drinkers in cafés. People went off to cafés for their evening tipple. Because it was green, absinthe came to be known as *la fée verte* (the green fairy), and 5 pm was the time to drink it, what they called to *l'heure verte* (the green hour). Nowadays, we'd call it the 'Happy Hour', but there was more to it than that. People drank their absinthe in a very particular way.

WHICH WAS?

A measure of the absinthe was put in a glass (see above figure, 1) and a special slotted spoon placed across the top of the glass, with a sugar cube on it (2). The absinthe tasted bitter (because of the wormwood in the absinthe) and the sugar improved the taste no end. Ice water is then slowly dripped through the sugar cube (3) until it has completely dissolved. As the sugar solution falls into the absinthe, the liquid becomes more polar, and the neutral terpenoid molecules no longer dissolve. These terpenoids came from the herbs used to make the absinthe, including thujone from wormwood, fenchone from fennel, pinocamphone and camphor from hyssop, and anethole from anise. They gave the absinthe its flavor. As they came out of solution, the liquid went milky (4). They called it the *louche effect* (*louche* = cloudy/milky liquid in French).

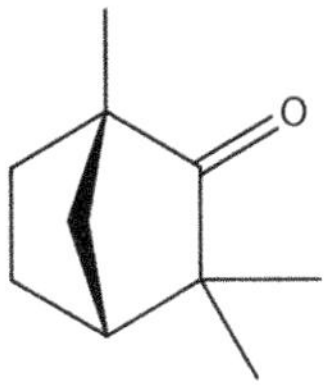

Fenchone 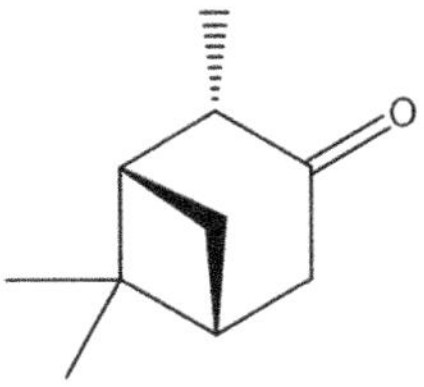   Pinocamphone 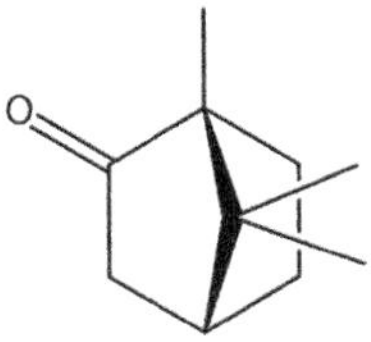   Camphor 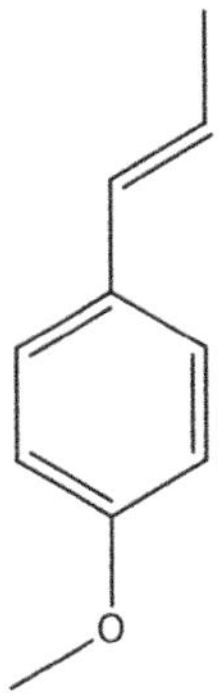   Anethole

WHY WAS ABSINTHE SO POPULAR?

It was the right drink in the right place at the right time, as the French vineyards were decimated by the outbreak of *phylloxera* in the 1860s, because of certain insects feeding on the roots and leaves of grapevines. Unlike wine, absinthe could be made from industrial ethanol, making absinthe cheaper than wine; what's more, it contained 50–70% ethanol.

WHAT WAS THE PROBLEM WITH ABSINTHE, THEN?

The Absinthe Drinker, a famous painting by the Czech artist Viktor Oliva, which shows the drinker being tempted by the 'green fairy'

By the late nineteenth century, health problems associated with absinthe spread. A physician named Dr. Valentin Magnan postulated a complaint called 'absinthism', associated with symptoms like addiction, fits, hallucinations (both auditory and visual) and delirium. And thujone got the blame.

SO WHAT HAPPENED?

The deciding event seems to have been the murders by a Swiss peasant named Jean Lanfray of his wife and two young daughters, on August 28th 1905. The murders were blamed on the two glasses of absinthe that Lanfray had drunk that day, rather than him being an alcoholic who regularly drank 5 liters of wine a day, as well as spirits. Switzerland banned absinthe in 1908, and other countries followed, culminating in a French ban in 1915. The ban remained in place until the European Union legalization allowed the manufacture and consumption of absinthe again in 1988.

Why Did Thujone Get the Blame for Absinthism?

Oil of wormwood was known to be toxic, and around 1900 the structure of thujone was discovered, and it was found to be present in wormwood, thuja oil and tansy oil. As the only component of absinthe to be identified, thujone was seen as a villain, though the evidence is now that absinthe, and thujone in particular, were blamed for large-scale alcoholism.

How Do We Know That?

People have claimed that the absinthe made a century ago, before it was banned, contained large amounts of thujone. Synthetic absinthe made using traditional recipes contains very little thujone, as do present-day commercial absinthe samples. In recent years, sealed bottles of pre-ban absinthe have been examined and they contain little thujone too. Of course, certain manufacturers of absinthe propagate the thujone myth, so that it is seen as an 'edgy' drink, whose consumers are doing something dangerous.

So, What Did Kill van Gogh?

It is generally accepted that he shot himself after suffering a lifetime of mental instability. Over the years, more than 150 psychiatrists have attempted

Self-portrait of Vincent van Gogh, showing his bandaged right ear, parts of which he had cut off in 1888 while in a delusional mental state to understand the exact nature of the mental illness, with around 30 different diagnoses being suggested. One of the most plausible is that by an American doctor, Wilfred Arnold, who, as a result of long study, believes that van Gogh suffered from acute intermittent porphyria. This disease affects the production of hemoglobin, and has a number of symptoms which match those reported for the artist, including anxiety, agitation, hallucinations, hysteria, delirium and depression. Absinthe was one of a number of things in van Gogh's lifestyle that could have brought on this complaint.

So It's Not True That Absinthe Makes the Mind Go Weaker?

Well, it still contains up to 70% ethanol (which makes it nearly twice as strong as most vodkas or whiskeys), so it wouldn't take many shots to make most people's minds go weak.

Chapter

63

TRIMETHYLAMINE
(AND FISH-BREATH ODOR)

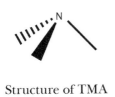

Structure of TMA

'What have we here? a man or a fish? dead or alive? A fish: he smells like a fish; a very ancient and fish-like smell; a kind of not of the newest Poor-John. A strange fish!'

William Shakespeare

The Tempest, Act 2, Scene 2 (Trinculo's monologue)

WHAT WAS TRINCULO ON ABOUT WHEN TALKING ABOUT CALIBAN?

Modern thinking inclines to the view that in describing Caliban he was giving an early documented example of *trimethylaminuria*.

WHICH IS?

People having a fish odor.

IS IT COMMON?

No.

AND IT IS CAUSED BY?

Trimethylamine (TMA), $(CH_3)_3N$.

Structure of TMA

Spacefill structure of TMA

How?

Fish contain a compound called trimethylamine N-oxide (TMAO). They are believed to use it to increase osmotic concentration and thus depress the freezing point of body fluids, an important detail in cold-water fish. It is also reckoned to counteract the perturbing effects of urea on proteins in enzyme systems.

So How Does Trimethylamine Get Formed?

The body contains enzymes called flavinmonoxygenases (FMOs) which reside in the liver and are thought to break down environmental toxins in the body. Toxic nitrogen compounds obtained from the diet are metabolized by enzymes and are ultimately converted into trimethylamine. Then, one of the FMOs called FMO3 oxidizes Me_3N to the odorless Me_3NO (TMAO), which can then be excreted. However, mutated copies of FMO3 lose this ability, and the Me_3N builds up. One source of trimethylamine is the compound choline ($Me_3N^+CH_2CH_2OH$), which is found in eggs, liver, legumes and some grains. It is broken down by anaerobic bacteria to form Me_3N. This mostly happens in the large intestine, although there are similar bacteria present in the oral cavity that perform this reaction too.

Also present in the human gut are methanogenic (methane making) bacteria, notably *Methanobrevibacter smithii*. The amount of them varies from person to person. These convert trimethylamine to methane, and it is said that this production of methane is responsible for flatulence (see p323). However, when the ability to decompose trimethylamine is impaired, such as when a person has a fault in the gene that makes FMO3 and/or if the number of methanogenic bacteria in the gut becomes greatly reduced for some reason, large amounts of the unchanged Me_3N are excreted, producing a highly unpleasant, fishy body odor. This has led to the colloquial terms 'fish-odor syndrome' or 'fish-breath syndrome' to describe the condition.

WHAT IS TRIMETHYLAMINE LIKE (APART FROM THE FISHY SMELL)?

It's a colorless liquid with a boiling point around 3.5°C, compared with the higher melting point of 224–226°C for the more polar Me_3NO, which presumably has dipole–dipole intermolecular forces. Trimethylamine is a base, like ammonia. Also like ammonia, it has a trigonal pyramidal structure. The C–N–C bond angle is 110.9°, compared with 107.2° in NH_3, presumably due to greater repulsions between the methyl groups. This angle is reduced to 109.0° in Me_3NO.

TMA   TMAO

HOW WIDESPREAD IS THE COMPLAINT?

Screening has indicated that the complaint is inherited as a recessive trait. Up to 1% of people are believed to contain a defective copy of the FMO3 gene, of which everyone has two copies, one inherited from each parent. The problem arises when *both* copies are defective. This means that possibly one person in 10,000 could suffer from fish-breath syndrome, which means that in the United States there are about 25,000 people with this disease.

"I ain't got no fish breath!"

Stinky

WHAT CAN SUFFERERS FROM FISH-BREATH SYNDROME DO ABOUT IT?

It is obviously a very unpleasant thing to live with. Sufferers get ostracized, children get called names at school. People start smoking to disguise the smell. Sufferers have even had to quit their jobs, and in extreme cases they have suffered from clinical depression and suicidal tendencies.

At present, no cure is known. Taking antibiotics can help (metronidazole, neomycin and lactulose have been used) and trying to control the amount of protein (and hence choline) in the diet, by eating less fish, eggs and liver, may be a good idea, though ingesting too little choline may lead to liver problems. A compound called indole-3-carbinole, found in broccoli and other dark green vegetables, can block the enzyme FMO3, so avoiding broccoli may also help. Things are complicated by the fact that indole-3-carbinole is now believed to help prevent cervical cancer.

ARE HUMANS THE ONLY SUFFERERS?

It seems that dogs sometimes smell of trimethylamine, while Swedish scientists have discovered the mutation in the FMO3 gene causing some Swedish cows to produce fish-smelling milk. The plant called Stinking Goosefoot (*Chenopodium olidum*) also produces trimethylamine.

Chapter

64

TNT

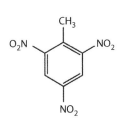

Detonation of 16 tonnes of TNT-equivalent
in Nevada

THE DELIVERY FIRM?

No, the high explosive which goes under the same abbreviation.

WHY CALL IT TNT?

You've got to admit that it is a whole lot easier than calling it 2,4,6-trinitrotoluene (let alone 2-methyl-1,3,5-trinitrobenzene), even though it doesn't tell you anything about the structure.

SO 'NITRO-' IS GOOD FOR EXPLOSIVES? BUT WHY?

The battleship *USS Iowa* test-firing its guns in 1984. The shells are filled with a mixture of TNT and another explosive, RDX

Hydrocarbon molecules, as found in natural gas or petrol, make good fuels. When they burn, they form two stable molecules, water and carbon dioxide, which owe their stability to strong C=O and O–H bonds. The formation of O–H bonds and C=O bonds gives out a lot of energy. There is also a good deal of hot gas which expands very rapidly, which creates the kind of explosive forces than can drive an engine. When nitro compounds like TNT are exploded, they also produce nitrogen (N_2) gas, forming the very strong N≡N triple bonds and releasing even more energy.

Although TNT produces less energy than burning petrol, it is the speed of the expanding shockwave (~6500 m/s) produced by exploding TNT that gives the explosion its blast.

Before TNT there were other nitro compounds in use as explosives, including nitroglycerine (see p207) and also picric acid (see p539). And before them there was gunpowder – that involves potassium nitrate. The NO_2 or NO_3 groups supply oxidizing oxygen atoms. When the nitro groups are part of the organic molecule, the oxygen atoms are in close contact with the carbon and hydrogen atoms they will oxidize. In other words, the explosive has the fuel and oxidizer built in. So when TNT explodes

$$2CH_3C_6H_2(NO_2)_3(s) \rightarrow 3N_2(g) + 5H_2O(g) + 7CO(g) + 7C(s)$$

As you can see, the products of the reaction include lots of gases, but it's also incomplete combustion. CO is formed instead of CO_2, and there's even some unburnt carbon remaining that forms a black cloud. During World War I, TNT was in short supply in England, hence a mixture of TNT with ammonium nitrate (see p21), known as *Amatol*, came into widespread use, to make the stock of TNT go further. By mixing TNT with ammonium nitrate, an oxygen-rich explosive, there was a further advantage in that the combustion is more efficient – Amatol explodes with a white or gray smoke instead of the black smoke of TNT.

$$2CH_3C_6H_2(NO_2)_3(s) + 7NH_4NO_3(s) \rightarrow 10N_2(g) + 19H_2O(g)$$
$$+ 14CO(g)$$

WHY IS TNT BETTER THAN NITROGLYCERINE?

Nitroglycerine is very shock-sensitive. (It was Alfred Nobel's great discovery that absorbing nitroglycerine in an inert mineral called *kieselguhr* made it much safer to handle. The product became known as dynamite, see p207.) Likewise, picric acid is very shock-sensitive when dry. In contrast, TNT was first made in 1863 by a German chemist named Joseph Wilbrand. He was trying to make

Flakes of yellow TNT

a dyestuff – TNT is yellow in color – and it was nearly 20 years before it was discovered to be an excellent high explosive. TNT was much safer to handle than the other explosives around at the time, and harder to detonate. It has quite a low melting point of 80°C – so shells can be filled safely with molten TNT.

AND IT'S A *HIGH* EXPLOSIVE, RIGHT?

Yes. High explosives are materials that require detonation and then produce a supersonic shockwave, giving them their power.

HOW DO YOU MAKE TNT?

It is made by nitrating toluene. For safety reasons, this has to be done in three stages, because nitrations are exothermic and there could be the risk of an uncontrolled reaction causing an explosion. It's also normally done in an ice-bath to prevent the reaction overheating. The nitro groups add on one at a time, first producing two different isomers (*ortho* and *para*, 2- and 4-) of nitrotoluene (NT), then 2,4-dinitrotoluene (DNT), then finally, if you're careful enough, trinitrotoluene. This first step (making NT) used to be a high-school practical experiment in the United Kingdom in the 1980s and 1990s until Health & Safety officials deemed it a bit too risky and spoilt all the fun!

The final product has to be carefully purified, to remove isomers of TNT (*e.g.* 2,3,4- and 2,4,5-TNT) as well as incomplete nitration

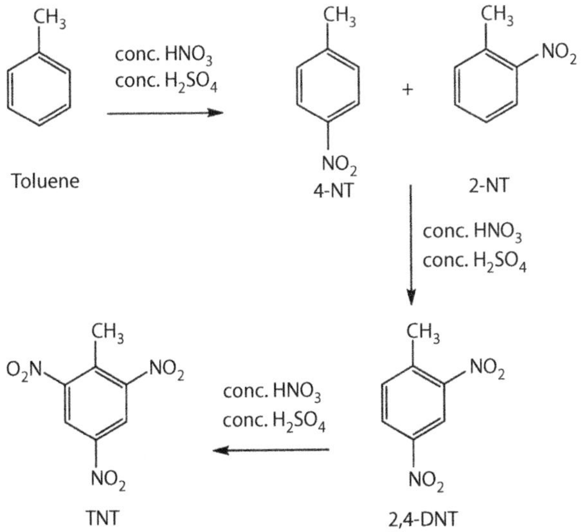

Synthesis of TNT by adding NO$_2$ groups sequentially
to toluene ... carefully

products like 2,4-DNT. Any impurities make the TNT product
less stable.

AND IT WAS BETTER THAN OTHER EXPLOSIVES AT THE TIME?

The Germans thought so, as they started using
TNT as their standard explosive for their armed
forces in 1902, while the British stuck with picric
acid. Picric acid is actually a stronger explosive,
but because it is more shock-sensitive, picric-
acid-filled shells don't penetrate armor so well,
as they tend to explode on contact. Partly as a
result of this, the Royal Navy suffered greater
losses than the German Navy at the Battle of Jutland in 1916.

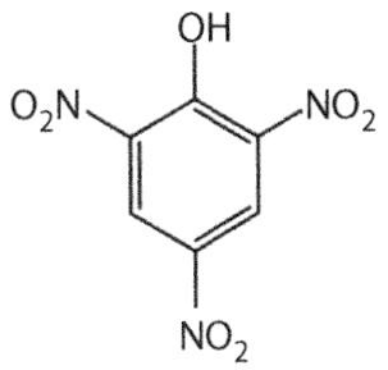

Picric acid

How Safe Is TNT?

As far as explosives go, pretty safe. But it has to be handled carefully, because of its toxicity. TNT is absorbed by the skin, so the skins of people handling it turn yellow. Over 17,000 cases of TNT poisoning were reported during World War I in the United States alone, causing around 500 deaths of munitions workers – mostly women – due to liver damage and aplastic anemia. TNT may also be a human carcinogen. Procedures for handling TNT were much tighter by the time of World War II. Even so, explosive factories have left their mark on the landscape, and cleaning up TNT-contaminated soil round factories or places where TNT has been used has become an important environmental issue. To try to solve this, scientists are studying plants that can absorb and detoxify TNT.

Is It Easy to Detect TNT?

A 'sniffer dog' helping US soldiers search for TNT and other explosives in Afghanistan in 2009

Given the increased awareness of possible terrorist incidents, this is obviously important. Scientists are working hard to develop chemical and electronic devices that are as sensitive as a dog's nose, but at the moment dogs are superior at detecting explosives. Although volatiles associated with the plastic casings of landmines can be detected, normally it is the explosive they are looking for. It's been shown that when dogs find TNT, they've actually responded to the presence of 2,4-DNT, an impurity in TNT. Although it is present at only a fraction of a percent level, the smaller and lighter DNT molecule is more volatile than TNT, so there are actually more DNT molecules present in the vapor above the TNT sample.

CAN YOU IMPROVE ON TNT?

In 1888, a German chemist named Albert Baur decided to introduce a butyl group into TNT to try to make it more explosive. He synthesized a variation of TNT with a tertiary butyl group replacing a hydrogen atom. Baur was disappointed when this did not have explosive properties, but things turned out well, as it had a marvelous musk smell, and was adopted by the perfume industry. In the ultimate compliment to a chemist, the compound came to be known as Musk Baur.

Musk Baur   Musk Xylene   Musk Ketone   Musk Ambrette

Baur went on to synthesize even better nitromusks, notably Musk Xylene, Musk Ketone and Musk Ambrette. These and similar molecules (*e.g.* Musk Moskene and Musk Tibetine) became widely used in the perfume industry. The famous French perfumer Ernest Beaux reportedly used over 10% nitromusks, especially Musk Ketone, in *Chanel No. 5* (1921), which reports indicate is still used in *No. 5* today (see p333). Musk Ambrette's additional floral note was used to advantage by the perfumer Francis Fabron for the perfume *L'Air du Temps* (Nina Ricci, 1948).

So Some Perfumes Contain a Derivative of TNT?

Yes, but it's probably best not to mention that when you take your duty free through airport security!

Chapter

65

VANCOMYCIN

It Looks Complicated ...

It's too complicated to design, and was discovered from soil samples taken from Borneo by a missionary.

It's an Antibiotic, Isn't It?

Yes, it was first isolated in 1953 from soil samples out of the jungles, from the microbe *Amycolatopsis orientalis*.

Why Were People Looking for Drugs in the Soil?

Researchers from the Eli Lilly drug company knew that soil fungi produce molecules for their defense, in the way that penicillin is made by a fungus (see p409), giving them an edge over their competitors. Therefore they searched soil samples from all over the world. They'd already hit on the first of the tetracycline family of antibiotics. When this new antibiotic was discovered it was given the name vancomycin (because it 'vanquished' bacteria!), and produced on an industrial scale by fermentation.

Isn't Penicillin Good Enough?

Penicillin did not become widely used until after World War II, but even just a decade later, when vancomycin was isolated, people were aware of bacteria that were resistant to penicillin, and the need for different antibiotics was evident. Vancomycin is regarded as the drug of last resort in hospital medicine cabinets, and is reserved for infections resulting from bacteria that are resistant to all other antibiotics, typically MRSA (methicillin-resistant *Staphyococcus aureus*, the best-known of the hospital 'superbugs', see p416).

How Do You Administer It?

When taken orally, vancomycin cannot enter the bloodstream efficiently, and therefore it is nearly always administered by injection. In fact, it is so potent, that the injection has to be diluted and given slowly over a period of about an hour to minimize pain, and to prevent 'red neck syndrome'.

'Redneck' Syndrome?

No, red neck syndrome. It's also called 'red man syndrome', and is an over-reaction by the cells in the face and neck to this potentially toxic drug, making these areas flush red. An oral version of vancomycin does exist, and this is used to treat stubborn infections of the intestines, such as those caused by another superbug, *C. difficile*. In this case, the fact that the vancomycin is not absorbed means it survives the trip though the digestive tract to reach the infected colon.

"Hey doc! That vanco-stuff ya'all gave me for my STD also done gave me 'red neck syndrome!'"

How Does Vancomycin Work?

It stops the bacteria growing and dividing properly, by inhibiting bacterial cell wall synthesis. These cell walls involve strings of sugar molecules which are cross-linked by short peptide chains. Vancomycin attaches itself to the D-alanyl-D-alanine groups at the end of the peptide chains, preventing the cross-links from forming and thus stopping the bacteria making the strong networks that form the cell walls. The attachment involves five hydrogen bonds (as well as various van der Waals' interactions).

DOESN'T PENICILLIN STOP CROSS-LINKS FORMING TOO?

Yes, but in a different way. Penicillin binds to the enzyme that makes the cross-links. The difference has been compared to penicillin acting as a saboteur of the machinery that builds the cell wall and vancomycin being a protester sitting in the way.

SO WHAT'S THE PROBLEM WITH VANCOMYCIN?

Unfortunately, bacteria are beginning to develop a resistance to vancomycin too. The bacteria that are resistant to vancomycin have a mutation in their DNA which means their proteins have a different terminal sequence. Instead of the last two amino acids being D-alanyl-D-alanine (D-Ala-D-Ala), they are D-Ala-D-Lac instead (where an alanine with an N–H group has been swapped for a lactate group with a C=O). This small change means that vancomycin can only form four hydrogen bonds to it, while extra repulsive lone pair interactions also occur. The binding is therefore a thousand times weaker, making vancomycin ineffective against these bacteria.

IS THERE AN ANSWER TO THE PROBLEM?

Dale Boger and his team at the Scripps Research Institute in California have created a simpler form of the vancomycin molecule. They have modified it by replacing an amide C=O oxygen with a C=NH group, so that it binds strongly to modified bacteria. This is confirmed by *in vitro* tests, which show that this molecule is very active against both normal and resistant bacteria. It now remains to be seen if this discovery can be applied to a real-life system.

AND THIS PROBLEM OF RESISTANCE TO DRUGS IS SOMETHING NEW?

Scientists recently investigated DNA from 30,000-year-old permafrost sediments, finding genes encoding resistance to β-lactam, tetracycline and glycopeptide antibiotics, showing that antibiotic resistance goes back much further than the problems caused by recent use of antibiotics.

Chapter

66

VX Gas

The Film-Star Molecule

VX, THAT'S A FUNNY FORMULA

It's not a formula; it is a shorthand for one of the nastiest compounds around. After all, would you call it *O*-ethyl *S*-[2-(diisopropylamino)ethyl]methylphosphonothiolate?

A NERVE GAS?

It's a liquid, not a gas at room temperature. Better to call it a 'nerve agent'.

WAS IT THE FIRST ONE TO BE MADE?

Gerhard Schrader in 1988 (© Press Association Images. With permission)

No, Gerhard Schrader, working for I.G. Farben at Frankfurt, was trying to make insecticides when he isolated Tabun on December 23rd 1936. The group of nerve agents he developed were called the G-agents, after Germany (although some sources say Gerhard). Tabun was abbreviated to 'GA' as it was the first one to be made. Schrader soon found that although it was a very potent insecticide, it was extremely toxic to mammals too, so toxic that it was manufactured in World War II as a potential weapon, though never used. When he spilt a drop in the laboratory, Schrader himself noticed one of the characteristic symptoms of nerve agent poisoning, *miosis* (contraction of the pupils of the eyes). Other notorious nerve agents, Sarin (GB) and Soman (GD), were also made before the end of the war.

After World War II, research on insecticides continued, and so did research on chemical warfare agents. Later agents were named

Tabun   Amiton

V-agents, supposedly after 'venom', due to the similarity in their mode of action to that of snake venom (although again, other reports say the 'V' stood for 'viscous' or even 'victory'). In 1952, Dr. Ranajit Ghosh was developing new pesticides at Imperial Chemical Industries (ICI). He synthesized a new and very potent insecticide, marketed as Amiton (VG) in 1954. Unfortunately, it was so potent that it was lethal to humans, with a toxicity similar to Sarin, and was rapidly removed from the market.

WHAT HAPPENED NEXT?

The Chemical Weapons Research Centre at Porton Down, near Salisbury (UK), was alerted. Further research took place, and a more deadly version called VX was developed. But by the time they were ready to make VX in any quantities in 1956, Britain took the step of renouncing chemical weapons, and instead traded knowledge of VX with the United States in exchange for information on thermonuclear weapons. The Americans built a plant for its manufacture at Newport, Indiana, and started to make VX there in 1961.

HOW DOES VX COMPARE WITH THE EARLIER NERVE AGENTS?

VX is a viscous, colorless liquid, with the consistency of motor-car engine oil. Its LD_{50}, the median lethal dose, is around

5–10 mg for a human. VX is less volatile but a good deal more toxic than agents like Sarin. The boiling point of VX is over 300°C, but it can be dispersed as a very dangerous aerosol and it is also absorbed through the skin. It is only hydrolyzed slowly, so it persists in soil for several weeks and does not evaporate. This means that it contaminates the ground for a long period, becoming a 'ground denial agent', turning land into a no-go area. Eventually, it breaks down to the detectable metabolites ethyl methylphosphonic acid (EMPA) and methylphosphonic acid (MPA). It's metabolites like these that groups such as the Organisation for the Prohibition of Chemical Weapons (OPCW), which received the Nobel Peace Prize in 2013, search for as evidence of the use of chemical weapons.

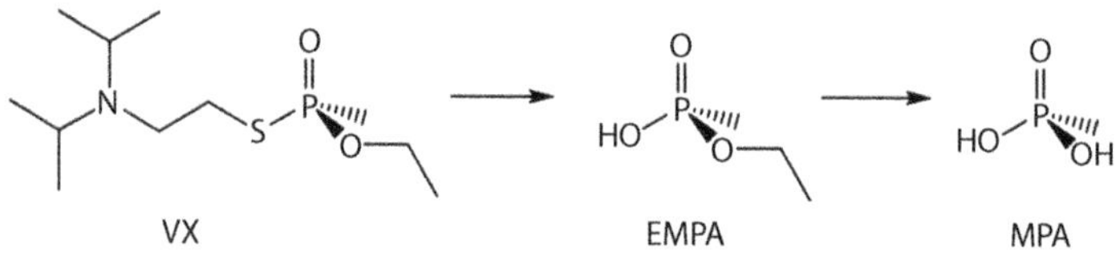

So the USA Had a Big Advantage over Other Countries?

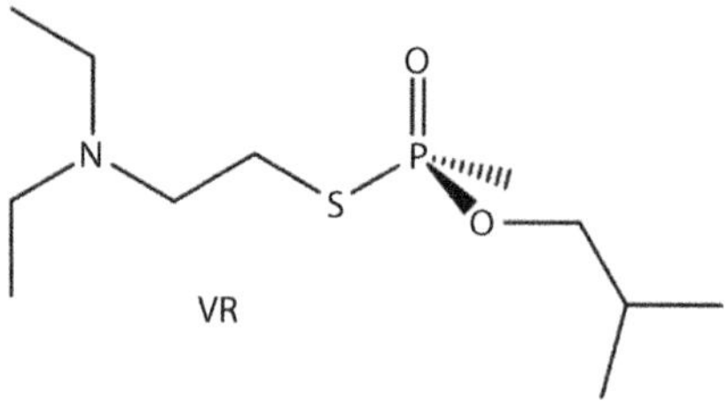

The Soviets prepared an isomeric version of VX (known as 'VR', 'Substance 33' or 'R-33'), maybe through incomplete intelligence. Possibly they had the molecular formula but not the structural formula, but it is thought more likely that the Russians did not have the technology to prepare American VX at that time.

HAS VX KILLED ANYONE?

The only known human fatality arose when two members of the *Aum Shinrikyo* cult (now known as *Aleph*) used VX to assassinate Tadahito Hamaguchi, a former member of their sect, in Osaka in December 1994. They sprinkled VX on his neck, he collapsed, and died 10 days later without recovering consciousness. This was the same cult that used Sarin nerve gas in the Tokyo subway attacks of 1995.

Victims of the chemical weapons attack on Halabja in 1988

It has been claimed that the Iraqis used VX during the Iran–Iraq war, particularly against Kurdish citizens in Halabja on March 16–17 1988, leaving an estimated 5000 people dead. The evidence indicates that both Tabun and Sarin were used in that war, but not VX, although it has been suggested that the Russian 'VR' may have been used. However, the Americans once used it successfully against sheep.

HOW DID THAT HAPPEN?

On March 13th 1968, trials were being carried out at the chemical-weapon proving ground at Dugway in the state of Utah. There was a malfunction when VX was released from a tank slung underneath a F4 Phantom jet, delaying the escape of some of the tank's

contents so that they fell upon sheep grazing 27 miles away in the
Skull Valley area, to the north of the test site. As a result of this over
3000 sheep were killed.

"Hostiles eliminated, over"

How Does VX Work?

Like other nerve agents, VX works by disrupting the central
nervous system. The neurotransmitter acetylcholine (ACh) passes
messages across synapses (gaps between neurons). When its job
is done, it is released by the receptor and has to be destroyed,
otherwise the ACh build up would lead to overstimulation of the
nervous system. This is done by the enzyme acetylcholinesterase
(AChE).

Nerve gases block the active site in AChE which stops the enzyme
from working. If nothing is done, the initial adduct 'ages',
eliminating a small molecule and permanently blocking the
active site of the enzyme, so that it cannot function. However,
if an antidote is rapidly administered (usually an oxime such as
pralidoxime, together with atropine), then the enzyme can be
reactivated.

So Why Call It a 'Film-Star' Molecule?

VX was very much a member of the cast in the Nicholas Cage–
Sean Connery movie, '*The Rock*' (1996), though it is shown as
a green liquid, rather than as a colorless oil. In the film, the
former prison island of Alcatraz, in San Francisco Bay ('*The
Rock*'), is seized by a group of renegade US Marines armed with
VX-carrying rockets. The FBI's chemical-weapon specialist
(Nicholas Cage) is accompanied by the one man to escape from
Alcatraz (Sean Connery) in a bid to retake the island. During the
final denouement, Cage uses a VX-containing pellet to kill the
man who is trying to strangle him. It's all good fun, but as usual
with Hollywood movies, scientifically dubious.

Chapter

67

WATER

70% of the Earth's surface is covered
in water

'Water, water, everywhere, and all the boards did shrink;
Water, water, everywhere, nor any drop to drink'.

Samuel Taylor Coleridge
The Rime of the Ancient Mariner

WHY WAS THERE NONE TO DRINK?

Because the Mariner was on a ship surrounded by salty seawater, which, if drunk, makes you sick, and then you become even thirstier.

STILL, WATER MUST BE ABOUT THE MOST BORING CHEMICAL EVER?

On the contrary, water is about the most remarkable chemical there is. Along with oxygen, it is vital to human life, and to all life, actually. In fact, scientists searching for evidence of life on other planets only bother to look at planets or moons where liquid water might exist. Its most remarkable property is the fact that it is about the only substance which exists as a solid, a liquid and a gas at normal temperatures on Earth. And for a molecule that size, that is not expected.

WHY NOT?

Just compare some properties of water with those of three similar molecules, and one element, which have the same number of electrons (10) and very similar masses (between 16 and 20 g).

	Methane (CH$_4$)	Ammonia (NH$_3$)	Water (H$_2$O)	Hydrogen Fluoride (HF)	Neon (Ne)
Covalent bonds per molecule	4	3	2	1	0
Melting point (°C)	−182.5	−77.7	0.0	−83.4	−248.6
Boiling point (°C)	−161.5	−33.4	100.0	19.5	−246.1

As you can see, water has melting and boiling points that are at much higher temperatures than the others.

SO WHY DOES THAT HAPPEN?

Melting and boiling points indicate the strength of the forces between molecules, as these have to be overcome when a change of state occurs. Water must therefore have unexpectedly strong intermolecular forces; these arise from what are called 'hydrogen bonds'.

YOU MEAN, LIKE O–H?

Hydrogen bonds are not strong covalent bonds between atoms like the ones you suggest. Hydrogen bonds arise because atoms of oxygen and hydrogen differ in their ability to attract the electron pair in the bond between them – oxygen has a significantly greater attraction ('electronegativity') than hydrogen. Because the negatively charged electrons in the O–H bond are on average nearer the oxygen atom, this means that the oxygen has a slight negative charge while the hydrogen has a slight positive charge. We normally depict this as $O^{\delta-}$–$H^{\delta+}$, indicating that the O–H bonds in a water molecule are polar, *i.e.* there is a separation of charge. This charge separation along a chemical bond (a dipole) occurs in most molecules, and its effect is to make molecules

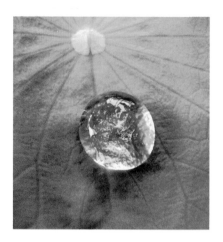

A water droplet on a leaf.
The leaf is hydrophobic so
the droplet forms a sphere
to minimize contact with
the surface

'sticky' toward their neighbors, because the positive regions of one molecule attract the negative parts of another. This mutual attraction is what causes gases to condense into liquids at low temperatures. For O–H bonds this force of attraction is particularly strong and so it has been given its own name – 'hydrogen bonds'. Molecules which contain OH groups, such as alcohols, carboxylic acids, sugars , *etc.*, often have high melting and boiling points, because higher temperatures are required to overcome the sticky intermolecular hydrogen bonds. For water, being a small, light molecule with two OH groups, this means that, proportionately, hydrogen bonding makes more difference to its properties than for any other molecule.

How Does This Affect the Structure and Properties of Water?

The oxygen atom in an isolated water molecule is surrounded by four electron pairs, two in O–H bonds and two non-bonding pairs. Because of repulsions, they keep apart as far as possible – if they were the same, they would be arranged tetrahedrally, at 109½° to each other (as in methane). But because the repulsions involving the non-bonding pairs are slightly stronger, the H–O–H bond angle is slightly reduced, to 104½°.

A water molecule

Owing to the polarity of the O–H bonds, hydrogen bonding creates an attraction between a hydrogen atom in one molecule and a non-bonding pair (a 'lone pair') of electrons in another molecule. Because a water molecule has two polar O–H bonds and two lone pairs, any water molecule can be involved in up to four hydrogen bonds. It is believed that in liquid water, the number is around 3.3.

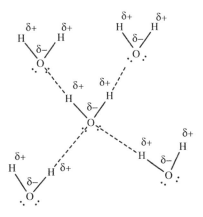

Each water molecule attracts several neighboring ones. The 'hydrogen bonds' are depicted as dashed lines

Although not as strong as a covalent bond, this attractive force is responsible for the unusually high melting and boiling points of water.

ARE THERE OTHER POLAR BONDS THAT CAN DO THIS?

Yes, N–H and F–H bonds.

SO WHY DON'T NH₃ AND HF HAVE SUCH HIGH MELTING AND BOILING POINTS AS WATER?

A molecule of NH_3 has got three polar N–H bonds but only one non-bonding pair. Conversely, a molecule of HF has got three non-bonding pairs of electrons, but only one polar H–F bond. Neither molecule can participate in as many hydrogen bonds per molecule as water – with two polar O–H bonds and two non-bonding pairs, it is 'just right'.

DO YOU GET HYDROGEN BONDS IN THE OTHER PHASES OF WATER?

In the gas phase, the molecules are too far apart, but hydrogen bonds are also found in ice. Ice has a very ordered structure, with the molecules lining up in an open, hexagonal structure to give four hydrogen bonds per water molecule. As a result, the molecules are slightly further apart in ice than in liquid water, which means that it is less dense than the liquid, and floats on it. Similarly, when you fill an ice cube tray with water, as it freezes, the ice cubes expand above the level of the tray.

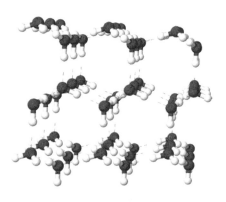

The open, porous structure of solid ice, showing why it has a low density

Icebergs float because ice is less dense than water

Water has its greatest density at about 4°C, which is why there is usually liquid water at the bottom of an apparently frozen lake, even in the depths of winter. Without this unusual property, life on

Earth would probably have died out during one of its various ice
ages because seas and oceans would have frozen solid right to the
bottom, killing all sea life.

ARE HYDROGEN BONDS RESPONSIBLE FOR ANY OTHER UNUSUAL PROPERTIES OF WATER?

Its high surface tension, for
one. This is why you can
float a needle on water, even
though steel is much denser.
And that is how insects like
the pond skater or water
strider can walk on water.
It also causes the capillary
action of water, which
enables water to rise up
the internal tubes (xylem)
within a plant to reach the
top of high trees.

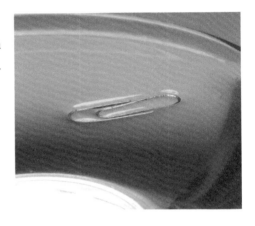

A metal paperclip floating due to the
surface tension of water

Water also has the very unusual property that when you increase
the pressure on it, its freezing point decreases. This is the opposite
of most materials, and is part of the reason why glaciers slide down
mountains. The pressure of the tonnes of ice pushing down on the
lower layers decreases the freezing point so much that the ice there
finds itself at a temperature above its new freezing point, and so
spontaneously melts. Thus, the glacier slides downhill on top of a
shallow river of melted water. It's also part of the reason why ice-
skaters can skate – the thin blades concentrate the skater's weight
to a huge local pressure on the ice surface, and this causes the ice
beneath the skate to briefly melt into water, long enough to let the
skate slide.

YOU KEEP SAYING 'PART OF THE REASON'. IS THERE MORE TO IT?

Sinead Kerr and John Kerr at the 2010 European Figure Skating Championships

Yes, for years people thought that ice melting under pressure was the only reason for ice being slippery. But it turns out that the story is more complex than that, because it doesn't explain why ice is still slippery at temperatures well below its freezing point, or when the pressure isn't enough to melt it (*e.g.* ice is still slippery if you wear flat shoes, not just ice-skates). It turns out that friction plays a big part – as the ice skates, skis, or glaciers slide, the friction generates heat, which melts the surface of the ice slightly. But that's still not the whole story, as it doesn't explain why ice is slippery even if you stand still. Modern analysis has shown that because the molecules at the surface of ice are inherently unstable due to the lack of molecules above them, the surface reconstructs to form a liquid-like layer. This verified the original hypothesis made by the famous physicist Michael Faraday in 1850 that all ice has an intrinsic thin layer of water at the surface. So ice is slippery due to its inherent surface water layer, which can be enhanced using pressure and friction.

IS THE GIANT STRUCTURE OF WATER ANYTHING TO DO WITH POLYWATER?

No, polywater was an exercise in scientific self-delusion. During the 1960s, a respected Russian chemist named Boris Derjaguin reported experiments in which water that had been condensed in narrow glass capillaries exhibited exceptionally unusual properties, with dramatically lowered freezing and elevated boiling points. He suggested this was due to water forming large clusters of around 10 linked water molecules, which he called 'polywater'. Most scientists doubted claims for a giant form of water, but some converts were made. Eventually it was shown that the material being studied was just impure water, containing dissolved impurities, like alkali metal ions as well as colloidal silica, and polywater was written off as fiction.

Recently, however, people that advocate homeopathic remedies have brought back the idea of polywater to try to explain their assertions that highly diluted medicines can affect or even cure certain ailments. They claim that because its structure changes depending on what molecules it has interacted with, polywater gives water a 'memory'; this is, they say, why homeopathic remedies supposedly work even though

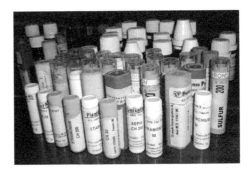

Various alleged homeopathic medicines, most of which contain nothing but water

there is not a single molecule of the active ingredient remaining in the administered dose. However, most real scientists think this is bunkum; there is no evidence for polywater even existing, or, if it does, nothing to show it would have a memory effect. Moreover, all rigorous scientific studies on homeopathic remedies have shown that any small effect they have (if any) can be explained simply by the placebo effect.

WHY IS WATER A GOOD SOLVENT?

Because many substances dissolve in it, water is often called a 'universal solvent'. But it doesn't dissolve everything, of course. It dissolves a lot of substances made of ions, because the polar water molecules are attracted to the charged ions ('solvation'). To dissolve a salt like sodium chloride, for example, the cost in energy is the average required to break a Na–Cl bond. But each Na^+ ion that's created is surrounded by as many as six water molecules, each with $Na^+–(OH_2)_n$ 'bonds' (with $n \approx 6$), while each chloride ion is also surrounded by many water molecules, each with $Cl^-–(H_2O)_m$ 'bonds' (with $m > 6$).

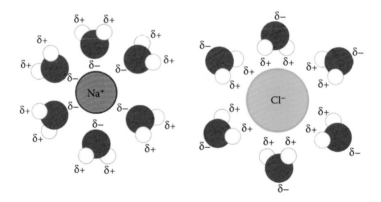

Solvation of NaCl. The Na^+ ions are surrounded by up to six polar water molecules. Even though the electrostatic attractions between the polar water molecules and the charged ions are relatively weak (compared to a normal chemical bond), there are so many of them that it is energetically favorable for the NaCl structure to break up into solvated ions rather than remain as a solid crystal

Water also dissolves other molecules containing polar O–H groups, such as ethanol and sugars, which can hydrogen bond to water. But it does not dissolve things that are not polar, including the kind of hydrocarbons found in oil and petrol, which explains the old saying that 'oil and water do not mix'. In the body, water dissolves and transports both molecules (*e.g.* glucose for 'fuel', see p193) and ions of metals like sodium and potassium that are responsible for transmission of nerve impulses. In fact, the whole of biochemistry can really be thought of as the chemistry of aqueous solutions!

Water also dissolves some gases. Oxygen dissolves in water slightly (which is how fish 'breathe' via their gills) but not on the scale where enough could dissolve in blood plasma for the needs of our body to support respiration in the cells. That is why our body needs a better oxygen carrier, in the form of hemoglobin (see p227).

IF WATER IS SO IMPORTANT, IS THERE ENOUGH TO GO AROUND?

Yes and no. It is estimated that of 1,386,000,000 km^3 of water on the planet, 97% is in the oceans, 2% is in glaciers and ice caps and just 1% fresh water (much of which is deep below the surface). If you divide this volume by the population of the Earth (7 billion), you get 0.2 km^3 of water each. This works out at around 200 billion liters each, of a resource that is recycled. This sounds a lot, but the

In undeveloped parts of the world, fresh water still has to be fetched daily from the local river or well

problem is that nearly all of it is undrinkable without expensive desalination (just ask the Ancient Mariner!) and not where it is needed (most people do not live next to an ocean). In some parts of the world, people walk several kilometers daily to access fresh water, while in the Western world, we have chlorinated water on tap and take it for granted. If current population trends continue to increase, in the future, wars may be fought over access to water, in the same way they are fought over access to oil nowadays.

BIBLIOGRAPHY

CHAPTER 1: ADENOSINE TRIPHOSPHATE (ATP)

Andersson, S. G., O. Karlberg, B. Canbäck and C. G. Kurland, *Philos. Trans. R. Soc. Lond., B, Biol. Sci.*, **358**, 2003, 165–177. (Mitochondrial DNA)

Bryant, D. E., K. E. R. Marriott, S. A. Macgregor, C. Kilner, M. A. Pasek and T. P. Kee, *Chem. Commun.*, **46**, 2010, 3726–3728. (Precursors to ATP)

Campbell, N. A., B. Williamson and R. J. Heyden, *Biology: Exploring Life* (Pearson Prentice Hall, Boston, MA, 2006). (Chloroplasts)

Henze, K. and W. Martin, *Nature*, **426**, 2003, 127–128. (Mitochondria)

Knowles, J. R. *Annu. Rev. Biochem.*, **49**, 1980, 877–919. (Mechanism of action)

Lodish, H., A. Berk, P. Matsudaira, C. A. Kaiser, M. Krieger, S. L. Zipursky and J. Darnell, *Molecular Cell Biology* (W. H. Freeman, New York, 2004).

Mereschkowsky, C. *Biol. Centralbl.*, **25**, 1905, 593–604. (Original theory of endosymbiosis)

Smith, A. M. *Biomacromolecules*, **2**, 2001, 335–341. (Biosynthesis of starch)

Törnroth-Horsefield, S. and R. Neutze, *Proc. Natl. Acad. Sci. USA*, **105**, 2008, 19565–19566. (Review of ATP)

Wernegreen, J. J. *PLoS Biol.*, **2**, 2004, e68. (Review of endosymbiosis)

Images

Chloroplasts – (Photographer Kristian Peters) – http://commons.wikimedia.org/wiki/File:Plagiomnium_affine_laminazellen.jpeg

Mitochondria – (Photographer: Louisa Howard) – http://commons.wikimedia.org/wiki/File:Mitochondria,_mammalian_lung_-_TEM.jpg

CHAPTER 2: ADRENALINE/EPINEPHRINE (NORADRENALINE/NOREPINEPHRINE)

Aronson, J. K. *Brit. Med. J.,* **320**, 2000, 506–509. (Naming controversy)

Ballantyne, C., Can a person be scared to death?, *Scientific American,* Jan 30th 2009.

Berecek, K. H. and B. M., Brody, *Am. J. Physiol.,* **242**, 1982, H593–H601. (Epinephrine as a neurotransmitter)

Devauges, V. and S. J. Sara, *Behav. Brain Res.,* **39**, 1990, 19–28. (Norepinephrine and attention)

Hoffman, B. B. *Adrenaline* (Harvard University Press, Cambridge, MA, 2013).

http://www.classicrockmagazine.com/news/motley-crue-nikki-sixx-marks-overdose-anniversary/ (Nikki Sixx overdose)

Lyrics to 'Kickstart my Heart', Motley Crüe, Dr Feelgood album, written by Nikki Sixx, 1989, copyright Downtown Music Publishing International Inc., Tommyland Music, WB Music Corp.

Nieuwenhuis, S., G. Aston-Jones and J. D. Cohen, *Psychol. Bull.,* **131**, 2005, 510–532. (Norepinephrine and decision making)

Sicherer, S. H. *Understanding and Managing Your Child's Food Allergy*
(The Johns Hopkins University Press, Baltimore, 2006).
(Epinephrine used to treat allergies)

Simons, F. E. *Ann. of Allergy, Asthma & Immunol.*, **104**, 2010, 405–412.
(Anaphylaxis and epinephrine)

Simons, F. E. *J. Allergy Clin. Immunol.*, **124**, 2009, 625–636.
(Anaphylaxis and epinephrine)

Sixx, N. and I. Gittins, *The Heroin Diaries: A Year in the Life of a
Shattered Rock Star* (Pocket Books, London, 2007).

Wilbert-Lampen, U. *et al.*, *New Engl. J. Med.*, **358**, 2008, 475–483.
(German soccer team and heart attacks).

Yamashima, T. *J. Med. Biogr.*, **11**, 2003, 95–102. (Jokichi Takamine)

Yu, A. J., P. Dayan, *Neuron*, **46**, 2005, 681–692. (Norepinephrine and
attention)

Images

Asthma inhaler – http://commons.wikimedia.org/wiki/
File:AsthmaInhaler.jpg

Bungee Jumper – http://commons.wikimedia.org/wiki/File:Jump_
from_nevis_bungee_platform.jpg

Epi-pen – http://commons.wikimedia.org/wiki/File:Epipen.jpg

Motley Crüe – http://commons.wikimedia.org/wiki/
File:M%C3%B6tley_Cr%C3%BCe_-_2005.jpg

CHAPTER 3: AMMONIUM NITRATE

Akhavan, J. *Chemistry of Explosives* (Royal Society of Chemistry, Cambridge, 3rd edition, 2011, pp. 4–7, 42–43, 135–137).

Brown, G. I. *The Big Bang!: History of Explosives* (Sutton Publishing Ltd, Stroud, 1998, pp. 87, 89–91, 234).

Cotton, F. A. and G. Wilkinson, *Advanced Inorganic Chemistry* (John Wiley, New York, 5th edition, 1988, p. 315).

Davis, T. L. *The Chemistry of Powder and Explosives* (Angriff Press, Las Vegas, 2012, reprint of 1943 edition).

http://www.texascity-library.org/disaster/ (Texas City Disaster)

Leigh, G. J. *The World's Greatest Fix* (Oxford University Press, Oxford, 2004, pp. 160–161).

Images

Oklahoma City bomb – http://commons.wikimedia.org/wiki/File:Oklahomacitybombing-DF-ST-98-01356.jpg

Texas City explosion – (Photo used with permission of Moore Memorial Library and the City of Texas City, TX, USA): http://www.texascity-library.org/disaster/

CHAPTER 4: ARTEMISININ

Chaturvedi, D., A. Goswami, P. P. Saikia, N. C. Barua and P. G. Rao, *Chem. Soc. Rev.*, **39**, 2010, 435–454. (Artemisinin and its derivatives)

Cheeseman, H. *et al. Science*, **336**, 2012, 79–82. (A major genome region underlying artemisinin resistance in malaria)

Copple, M. *et al.*, *Mol. Med.*, **18**, 2012, 1045–1055. (RKA 182)

Dondorp, M. *et al.*, *New England Journal of Medicine*, **361**, 2009, 5. (Artemisinin resistance)

Eckstein-Ludwig, U., R. J. Webb, I. D. A.Van Goethem, J. M. East, A. G. Lee, M. Kimura, P. M. O'Neill, P. G. Bray, S. A. Ward and S. Krishna, *Nature*, **424**, 2003, 957–961. (Artemisinin and PfATP6)

http://www.laskerfoundation.org/awards/2011clinical.htm (Youyou Tu's award for discovering artimisin)

Klayman, D. L. *Science*, **228**, 1985, 1049–1055. (Qinghaosu (artemisinin): An antimalarial drug from China)

Klonis, N., M. P. Crespo-Ortiz, I. Bottova, N. Abu-Bakar, S. Kenny, P. J. Rosenthal and L. Tilley, *Proc. Natl. Acad. Sci.*, **108**, 2011, 11405–11410. (Artemisinin activity against *Plasmodium falciparum* requires hemoglobin uptake and digestion)

Krishna, S., L. Bustamante, R. K. Haynes and H. M. Staines, *Trends in Pharmacol. Sci.*, **29**, 2008, 520–527. (Review, including artemisone)

Lévesque, F. and P. H. Seeberger, *Angew. Chem. Int. Ed.*, **51**, 2012, 1706–1709. (Continuous flow process for converting dihydroartemisinic acid to artemisinin)

Moehrle, J. J., S. Duparc, C. Siethoff, P. L. M van Giersbergen, J. C. Craft, S. Arbe-Barnes, S. A Charman, M. Gutierrez, S. Wittlin and J. L. Vennerstrom, *Br. J. Clin. Pharmacol.* **75**, 2013, 524–537.

Nilsen A. *et al.*, *Sci. Transl. Med.*, **5**, 2013, 177. (ELQ-300 and new classes of malarial drugs)

O'Neill, P. M. *et al.*, *Angew. Chem. Int. Ed.*, **49**, 2010, 5693–5697. (RKA182)

O'Neill, P. M., V. E. Barton and S. A. Ward, *Molecules*, **15**, 2010, 1705–1721. (Mechanism of action of artemisinin)

Paddon C. J. *et al.*, *Nature*, **496**, 2013, 528–532. (Making artemisinin in bulk using yeast fermentation)

Sherman, W. *Magic Bullets to Conquer Malaria* (ASM Press, Washington, 2011).

Youyou Tu, *Nat. Med.* **17**, 2011, 1217–1220. (Discovery of artemisinin)

Images

Sweet wormwood – http://commons.wikimedia.org/wiki/File:Natural_anti-malarial_(4738072658).jpg

Youyou Tu – http://www.laskerfoundation.org/awards/2011clinical.htm (Courtesy of the Albert and Mary Lasker Foundation)

CHAPTER 5: ASPIRIN

Bayer's website about aspirin: http://www.aspirin.com

Houlton, S. *Chemistry World*, **4**, December 2007, 56–59. (Vioxx withdrawal)

Jeffreys, D. *Aspirin, The Story of a Wonder Drug* (Bloomsbury, London, 2004).

Klein, B. E. K., K. P. Howard, R. E. Gangnon, J. O. Dreyer, K. E. Lee and R. Klein, *J. Am. Med. Assoc.*, **308**, 2012, 2469–2478. (Aspirin and macular degeneration)

Kurumbail, R. G. *et al.*, *Nature*, **384**, 1996, 644–648. (Structural basis for COX-2 inhibitors)

Loll, P. J., D. Picot and R. M. Garavito, *Nat. Struct. Biol.*, **2**, 1995, 637–643. (How aspirin deactivates prostaglandin H2 synthase)

Mann, C. C. and M. L. Plummer, *The Aspirin Wars* (A. A. Knopf, New York, 1991).

Nesi, T. *Poison Pills: The Untold Story of the Vioxx Drug Scandal* (Thomas Dunne Books, New York, 2008).

Osborn, C. (ed), *Aspirin* (Royal Society of Chemistry, London, 1998).

Picot, D., P. J. Loll and R. M. Garavito, *Nature*, **367**, 1994, 243–249.

Rainsford, K. D. (ed), *Aspirin and Related Drugs* (Taylor & Francis, London, 2004).

Sheldon, P. *The Fall and Rise of Aspirin – the Wonder Drug* (Brewin Books, Studley, 2007).

Sneader, W. *Brit. Med. J.*, **321**, 2000, 1591–1594. (Discovery of aspirin: A reappraisal)

Starko, K. M., *Clin. Infectious Dis.*, **49**, 2009, 1405–1410. (Salicylates and 1918-9 flu pandemic)

Starko, K. M., C. G. Ray, L. B. Dominguez, W. L. Stromberg and D. F. Woodall, *Pediatrics*, **66**, 1980, 859–864. (Reye's Syndrome and salicylate use)

Stone, E. *Philos. Trans. Roy. Soc.*, **53**, 1763, 195.

Vane, J. R. and R. M. Botting (eds), *Therapeutic Roles of Selective COX-2 Inhibitors* (William Harvey Press, London, 2001).

Vane, J. R. and R. M. Botting, *Thrombosis Res.*, **110**, 2003, 255–258. (How aspirin acts).

Images

Aspirin cartoon – copyright Randy Glasbergen, used with permission (www.glasbergen.com)

Aspirin tablets – http://commons.wikimedia.org/wiki/File:Regular_strength_enteric_coated_aspirin_tablets.jpg

Felix Hoffman – http://commons.wikimedia.org/wiki/File:Felix_Hoffman.jpg

CHAPTER 6: CAFFEINE

Arnaud, M. J. *Progress in Drug Research*, **31**, 1987, 273–313. (Pharmacology of caffeine)

Ashihara, H., H. Sano and A. Crozier, *Phytochemistry*, **69**, 2008, 841–856. (Biosynthesis)

Berger, A. J. and K. Alford, *Med. J. Aust.*, **190**, 2009, 41–43. (Cardiac arrest attributed to energy drinks)

Childs, E., C. Hohoff, J. Deckert, K. Xu, J. Badner and H. de Wit, *Neuropsychopharmacology*, **33**, 2008, 2791–2800. (*ADORA2A* polymorphism and caffeine-induced anxiety)

Cornelis, M. C. *et al.*, *PLoS Genetics*, **7**, 2011, e1002033. (Genetic factors for caffeine consumption)

Crozier, T. W. M., A. Stalmach, M. E. Lean and A. Crozier, *Food Funct.*, **3**, 2012, 30–33. (Potential health implications)

Czerny, M. and W. Grosch, *J. Agric. Food Chem.*, **48**, 2000, 868–872. (Odorants in raw arabica)

Donnerstein, R. L., D. Zhu, R. Samson, A. M. Bender and S. J. Goldberg, *Am. Heart J.*, **136**, 1998, 643–646. (Caffeine ingestion and ECGs)

Guerreiro, S., D. Toulorge, E. Hirsch, M. Marien, P. Sokoloff and P. P. Michel, *Mol. Pharmacol.*, **74**, 2008, 980–989. (Paraxanthine and Parkinson's)

Heckman, M. A., J. Weil and E. G. Demeija, *J. Food Sci.*, **75**, 2010, R77–R87. (Review on consumption, functionality, safety)

Illy, E., The Complexity of Coffee, *Scientific American*, June 2002, 86–91.

Kim, Y.-S., H. Uefuji, S. Ogita and H. Sano, *Transgenic Res.*, **15**, 2006, 667–672. (Caffeine-producing tobacco plants)

Klebanoff, M. A., R. J. Levine, R. DerSimonian, J. D. Clemens and D. G. Wilkins, *New. Engl. J. Med.*, **341**, 1999, 1639–1644. (Paraxanthine and risk of spontaneous abortion)

Lauren Wolf, *Chem. Eng. News*, Feb 4, 2013, 9–12. (Caffeine and health)

Mazzafera, P., T. W. Baumann, M. M. Shimizu and M. B. Silvarolla, *Tropical Plant Biology*, **2**, 2009, 63–76. (Coffee from caffeine-free arabica plants)

Nathanson, J. A. *Science*, **226**, 1984, 184–187. (Methylxanthine pesticide activity)

Noever, R., J. Cronise and R. A. Relwani, *NASA Tech Briefs*, **19**, 1995, 82. [Published in *New Scientist* magazine, 27 April 1995]. (Using spider-web patterns to determine toxicity)

Orrú M. *et al.*, *Neuropharmacology*, **67**, 2013, 476–484. (Caffeine, paraxanthine and locomotor activation)

Perrine, D. M. *The Chemistry of Mind-Altering Drugs* (American Chemical Society, Washington, DC, 1996, esp. pp. 7–9, 179–181).

Silvarolla, M. B., P. Mazzafera and L. C. Fazuoli, *Nature*, **429**, 2004, 826. (Naturally decaffeinated arabica coffee)

Stoffelen, P., M. Noirot, E. Couturon and F. Anthony, *Bot. J. Linn. Soc.*, **158**, 2008, 67–72. (Caffeine-free coffee)

Weinberg, B. A. and B. K. Bealer, *The World of Caffeine* (Routledge, New York, 2001).

Xu, K., Y.-H. Xu, J.-F. Chen and M. A. Schwarzschild, *Neuroscience*, **167**, 2010, 475–481. (Paraxanthine and Parkinson's)

Images

Café-au-lait – http://commons.wikimedia.org/wiki/File:Café _au_ lait.jpg

Red Bull mini – Courtesy of © Red Bull Media House

Red Catucaí coffee – http://commons.wikimedia.org/wiki/ File:FruitColors.jpg?uselang=en-gb

Roasted coffee beans – http://commons.wikimedia.org/wiki/ File:Roasted_coffee_beans.jpg

Spiderweb on caffeine – http://en.wikipedia.org/wiki/ File:Caffeinated_spiderwebs_modified.jpg

CHAPTER 7: CAPSAICIN

Acworth, N., J. Waraska and P. Gamache, *Am. Lab.*, **11**, 1997, 25–32. (HPLC determination of capsaicin content)

Billing, J. and P. W. Sherman, *Quart. Rev. Biol.*, **73**, 1998, 3–49. (Antimicrobial functions of spices)

Binshtok, M., B. P. Bean and C. J. Woolf, *Nature*, **449**, 2007, 607. (Capsaicin and anaesthetic)

Caterina, M. J., M. A. Schumacher, M. Tominaga, T. A. Rosen, J. D. Levine and D. Julius, *Nature*, **389**, 1997, 816. (Discovery of TRPV1 receptor)

Chapman and Hall Combined Chemical Dictionary (Compound code number: HGT21-K).

Collingham, L. *Curry: A Biography of a Dish* (Chatto and Windus, London, 2005).

Conway, S. J. *Chem. Soc. Rev.*, **37**, 2008, 1530–1545. (Review of capsaicin and TRPV1)

Gavva, N. R. *et al.*, *J. Neurosci.*, **27**, 2007, 3366–3374. (TRPV1 receptor and body temperature regulation)

Iwai, K., T. Suzuki, H. Fujiwake and S. Oka, *J. Chromatogr.*, **172**, 1979, 303–311. (HPLC determination of capsaicin content)

Kachoosangi, R. T., G. G. Wildgoose and R. G. Compton, *Analyst*, **133**, 2008, 888–895. (Carbon nanotube-based sensors for capsaicin)

Kashyap, A. and S. Weber, Starch grains from farmana give new insights into harappan plant use. *Antiquity Project Gallery*, 84:326. Online at http://www.antiquity.ac.uk/projgall/kashyap326/

Katritzky, R., Y.-J. Xu and A. V. Vakulenko, *J. Org. Chem.*, **68**, 2003, 9100–9104. (Essential structural features of capsaicinoids)

Lawler, A. *Science*, **337**, 2012, 288. (Spices in the ancient Indus civilisation)

McGee, H. *McGee on Food and Cooking* (Hodder, London, 2004, esp. pp. 418–430).

Reilly, A., D. J. Crouch and G. S. Yost, *J. Forensic Sci.*, **46**, 2001, 502–509. (Quantitative analysis of capsaicinoids in peppers and pepper spray products)

Reyes-Escogido, M. de L., E. G. Gonzalez-Mondragon and E. Vazquez-Tzompantzi, *Molecules*, **16**, 2011, 1253–1270. (Capsaicin review)

Sherman, P. W. and J. Billing, *BioScience*, **49**, 1999, 453. (Darwinian gastronomy: Why we use spices)

Story, G. M. and L. Cruz-Orengo, *Am. Sci.*, **95**, 2007, 326. (Hot and cold sensors)

Tewksbury, J. J. and G. P. Nabhan, *Nature*, **412**, 2001, 203–204. (Squirrel deterrent)

Tewksbury, J. J., K. M. Reagan, N. J. Machnicki, T. A. Carlo, D. C. Haak, A. L. Calderon-Penaloza and D. J. Levey, *Proc. Natl. Acad. Sci.*, **105**, 2008, 11808–11811. (Evolutionary ecology of wild chilies)

Willoughby, H., R. L. Jinks, G. W. Morgan, H. Pepper, J. Budd and B. Mayle, *Eur. J. Forest Res.*, **130**, 2001, 601–618. (Woodmice and squirrel deterrent)

Images

Capsicum green chilli plant – http://commons.wikimedia.org/wiki/File:Capsicum_-_green_chili_chilli_-_piment_verte_-_gr%C3%BCner_Chili_02.jpg

Dorset naga – http://en.wikipedia.org/wiki/File:Dorset_Naga_Peppers.jpg

Police pepper spray – http://en.wikipedia.org/wiki/File:Pepper_spray_Demonstration.jpg

Red chillies – http://upload.wikimedia.org/wikipedia/
commons/d/d8/Red_Chilli.jpg?uselang = en-gb

Wilbur Scoville – http://en.wikipedia.org/wiki/File:Scoville,_
Wilbur_Prof_med.jpg

CHAPTER 8: CARBON DIOXIDE

Christianson, G. E. *Greenhouse: The 200-Year Story of Global Warming*
(Walker & Co., New York, 1999).

Fagan, B. M. *The Long Summer: How Climate Changed Civilization*
(Granta Books, London, 2004).

Gattuso, J.-P., D. Allemand and M. Frankignoulle, *Am.
Zoologist*, **39**, 1999, 160–183. (Photosynthesis and
calcification at cellular, organismal and community levels
in coral reefs)

Greenwood, N. N. and A. Earnshaw, *Chemistry of the Elements*
(Butterworth-Heinemann, Oxford, 2nd edition, 1997,
pp. 310–314).

http://www.esrl.noaa.gov/gmd/ccgg/trends/#mlo (Mauna Loa
records)

Langdon, C. and M. J. Atkinson, *J. Geophys. Res.*, **110**, 2005, C09S07.
(Effect of elevated pCO$_2$ on photosynthesis and calcification of
corals)

Martini, M., CO$_2$ emissions in volcanic areas: Case histories and
hazards, in A. Raschi, F. Miglietta, R. Tognetti and P. R. van

Gardingen (eds), *Plant Responses to Elevated CO2: Evidence from Natural Springs* (Cambridge University Press, Cambridge, 1997, pp. 69–86).

Nienhuis, S., A. R. Palmer and C. D. G. Harley, *Proc. R. Soc. B*, **277**, 2010, 2553–2558. (Elevated CO2 affects shell dissolution rate but not calcification rate in a marine snail)

Petit, J. R. *et al.*, *Nature*, **399**, 1999, 429–436. (Climate and atmospheric history of the past 420,000 years from the Vostok ice core, Antarctica)

Roston, E. *The Carbon Age: How Life's Core Element Has Become Civilization's Greatest Threat* (Walker & Company, New York, 2008).

Volk, T. *CO2 Rising* (MIT Press, Cambridge, MA, 2008).

Ward, P. D., *Under a Green Sky, Global Warming, the Mass Extinctions of the Past, and What They Can Tell Us About Our Future* (Harper Collins, New York, 2007).

Weart, S. R. *The Discovery of Global Warming* (Harvard University Press, Cambridge, MA, 2nd edition, 2003).

Images

Carbon dioxide beer bubbles—(Photographer: Ildar Sagdejev)—http://commons.wikimedia.org/wiki/File:2009-03-21_Beer_brewing_bubbles.jpg

Carbon emissions since 1850 – http://commons.wikimedia.org/wiki/File:Global_Carbon_Emissions.svg

Fire-extingisher training – http://upload.wikimedia.org/wikipedia/commons/2/26/Flickr_-_The_U.S._Army_-_Fire_training.jpg

Greenhouse effect – http://commons.wikimedia.org/wiki/
File:Greenhouse_effect_diagram.png

Solid dry ice (Photo: MarkS at the German language Wikipedia) –
http://commons.wikimedia.org/wiki/File:Trockeneis.jpg

Statue of Liberty – http://commons.wikimedia.org/wiki/
File:Statue_of_Liberty_7.jpg

Statues damaged by acid rain (Photo: Nino Barbieritpolla)
– http://commons.wikimedia.org/wiki/File:Pollution_-_
Damaged_by_acid_rain.jpg

Chapter 9: β-Carotene

Coultate, T. P. *Food: The Chemistry of its Components* (RSC,
Cambridge, 3rd edition, 1996, pp. 142–148).

Cunningham, F. X. and E. Gantt, *Annu. Rev. Plant Physiol. Plant Mol.
Biol.*, **49**, 1998, 557–583. (Genes and enzymes of carotenoid
biosynthesis in plants)

Golley, J. *John 'Cat's-Eyes' Cunningham: The Aviation Legend* (Airlife
Press, Shrewsbury, England, 1999).

Litwack, G. (ed), *Vitamins and Hormones 75: Vitamin A* (Academic
Press, San Diego, 2007).

Online lipid library – http://lipidlibrary.aocs.org/plantbio/
carotenoids/index.htm

Paine J. A. *et al.*, *Nat. Biotech.*, **23**, 2005, 482–487. (Golden rice)

Rawnsley, C. F. and R. Wright, *Night Fighter* (Elmfield Press, London,
1976) [and earlier editions].

Rodahl, K. and T. Moore, *Biochem. J.*, **37**, 1943, 166–168. (The vitamin A content and toxicity of bear and seal liver)

Tang, G., J. Qin, G. G. Dolnikowski, R. M. Russell and M. A. Grusak, *Am. J. Clin. Nutr.* **89**, 2009, 1776–1783. (Golden rice as a source of Vitamin A)

Ye, X., S. Al-Babili, A. Klöti, J. Zhang, P. Lucca, P. Beyer and I. Potrykus, *Science*, **287**, 2000, 303–305. (Golden rice)

Images

Autumn colours – http://commons.wikimedia.org/wiki/File:Autumn_colour_-_geograph.org.uk_-_1041221.jpg

Bristol Beaufighter – http://en.wikipedia.org/wiki/File:415th_Night_Fighter_Squadron_-_Bristol_Beaufighter.jpg

Carrots – http://commons.wikimedia.org/wiki/File:Carrots.JPG

John Cunningham – Copyright © Press Association.

CHAPTER 10: CHLOROPHYLL

Amesz, J. *Photosynthesis* (Elsevier, Amsterdam, 1987).

Barber, J. (ed) *The Photosystems: Structure, Function and Molecular Biology* (Elsevier, Amsterdam, 1992).

Feyer, G., J. P. Allen, M. Y. Okamura and D. C. Rees, *Nature*, **339**, 1989, 111–116. (Mechanism of photosynthesis)

Fleming, I. *Nature*, **216**, 1967, 151–152. (Structure of chlorophyll)

Hall D. O. and K. K. Rao, *Photosynthesis* (Cambridge University Press, Boca Raton Fl, USA, 1994).

Hughton R. A. and G. M. Woodwell, *Sci. Am.*, **260**, 1989, 36–44. (CO_2 and global climactic change).

Huzisige H. and B. Ke, *Photosyn. Res.*, **38**, 1993, 185–209. (History of discovery of photosynthesis)

Larkum A. W. D. and M. Kühl, *Trends in Plant Sci.*, **10**, 2005, 355–357. (Different types of chlorophyll)

Marcus, R. A. and N. Sutin, *Biochim. Biophys. Acta*, **811**, 1985, 265–322. (Details of electron transfer in photosynthesis)

Scheer, H. (ed), *Chlorophylls* (CRC Press, Boca Raton, FL, USA, 1991).

Whitmarsh J. and Govindjee, The photosynthetic process, in G. S. Singhal, G. Renger, S. K. Sopory, K. D. Irrgang and Govindjee, (eds), *Concepts in Photobiology: Photosynthesis and Photomorphogenesis* (Kluwer Academic, Boston, 1999)

Witt, H. T. *Photosyn. Res.*, **29**, 1991, 55–77. (Mechanism of photosynthesis)

Woodward, R. B. *et al.*, *J. Am. Chem. Soc.*, **82**, 1960, 3800–3802. (Total synthesis of chlorophyll)

Images

Aerial view of the Amazon rainforest – http://commons.
wikimedia.org/wiki/File:Aerial_view_of_the_Amazon_
Rainforest.jpg

Antoine Lavoisier – http://commons.wikimedia.org/wiki/
File:Antoine_lavoisier.jpg

Autumn leaf colors – http://commons.wikimedia.org/wiki/File:Mo
miji_%E7%B4%85%E8%91%89%E3%81%99%E3%82%8B%
E3%83%A4%E3%83%9E%E3%83%A2%E3%83%9F%E3%8
2%B8_B221212.JPG

Chlorophyll absorption spectra – http://commons.wikimedia.org/
wiki/File:Chlorophyll_ab_spectra-en.svg

Chloroplasts – (Photographer: Wilfredo R. Rodriguez H.) – http://
commons.wikimedia.org/wiki/File:Clorofila_3.jpg

Jan Ingenhousz – Modified from the image at – http://commons.
wikimedia.org/wiki/File:Jan_Ingenhousz.jpg

Jean Senebier – http://commons.wikimedia.org/wiki/File:Jean_
Senebier.jpg

Joseph Priestley – http://commons.wikimedia.org/wiki/File:PSM_
V05_D400_Joseph_Priestley.jpg

Julius Robert Mayer – http://commons.wikimedia.org/wiki/
File:Julius_Robert_Mayer_von_Friedrich_Berrer.jpg

Photosynthesis – http://commons.wikimedia.org/wiki/
File:Photosynthesis.gif

Theodore de Saussure – http://commons.wikimedia.org/wiki/
    File:Nicolas-Th%C3%A9odore_de_Saussure.jpg

CHAPTER 11: CHOLESTEROL

Bloch, K. *Science,* **150**, 1965, 19–29. (Biosynthesis of cholesterol)

Drexel, H. *Euro. Heart J. Suppl.,* **8**, 2006 F23. (Reducing cardiac risk
    by raising HDL levels)

Endo, A. *Atheroscl. Suppl.,* **5**, 2004, 125–130. (History of statin
    development).

Endo, A., M. Kuroda and Y. Tsujita, *J. Antibiotics,* **29**, 1976, 1346–1348.
    (Discovery of mevastatin – the first statin)

Hausenloy, D. J. and D. M. Yellon, *Heart,* **94** 2008 706–714.
    (Affect of weight loss on LDL levels)

Lecerf, J. M. and M. de Lorgeril, *Br. J. Nutr.,* **106**, 2011, 6–14. (Link
    between cholesterol and heart disease)

Lemay, P. and R. Oesper, *J. Chem. Educ.,* **25**, 1948, 62–70. (Michel
    Chevreaul)

Mensink, R. P., P. L. Zock, A. D. M. Kester and M. B. Katan, *Am. J.
    Clin. Nutr.,* **77**, 2003 1146–1155. (Affect of eating fatty acids on
    LDL levels)

Olson, R. E. *J. Nutr.,* **128** (2 Suppl), 1998, 439S–443S. (Discovery of
    cholesterol)

Pate-Douglas, T. and R.E Keyser, *Arch. Phys. Med. Rehab.*, **80**, 1999, 691–695. (Affect of exercise on LDL levels)

Rosanoff, A. and M. S. Seelig, *J. Am. Coll. Nutr.,* **23**, 2004, 501S–505S. (Statins)

Thijssen, M. A. and R. P. Mensink, in A. von Eckardstein (ed), *Atherosclerosis: Diet and Drugs.* Springer, Berlin, pp. 171–172, 2005. (Fatty acids and atherosclerotic risk)

Images

Cholesterol cartoon – copyright Randy Glasbergen, used with permission (www.glasbergen.com)

Michel Chevreul – http://commons.wikimedia.org/wiki/File:Michel_Eug%C3%A8ne_Chevreul.jpg

CHAPTER 12: CISPLATIN

Alderden, R. A., M. D. Hall and T. W. Hambley, *J. Chem. Educ.*, **83**, 2006, 728. (Discovery)

Armstrong, L. *It's Not about the Bike* (Yellow Jersey Press, London, 2000, esp. p. 86).

Champion, R. and J. Powell, *Champion's Story* (Victor Gollancz, London, 1981).

Dhara, S. C. *Indian J. Chem.*, **8**, 1970, 193–194. (Best synthesis)

http://www.history.qmul.ac.uk/research/modbiomed/Publications/wit_vols/44855.pdf (History of cisplatin)

Kauffman, G. B., R. Pentimalli, S. Dolti and M. B. Hall, *Platinum Metals Rev.*, **54**, 2010, 250–256. (Peyrone and cisplatin)

Kelland, L. *Nat. Rev. Cancer*, **7**, 2007, 573–584.

Lippert, B. *Cisplatin: Chemistry and Biochemistry of a Leading Anticancer Drug* (Wiley-VCH, Weinheim, 1999).

Peyrone, M. *Justus Liebigs Annalen der Chemie*, **51**, 1844, 1–29. (Discovery)

Rosenberg, B., E. B. Grimley. L. Van Camp and A. J. Thomson, *J. Biol. Chem.*, **242**, 1967, 1347–1352. (Testing on mice with cancer)

Rosenberg, B., L. Van Camp and T. Krigas, *Nature*, **205**, 1965, 698–699. (Discovery of inhibited cell division)

Rosenberg, B., L. Van Camp, J. E. Trosko and V. H. Mansour, *Nature*, **222**, 1969, 385–386. (Testing on mice with cancer)

Todd, R. C. and S. J. Lippard, *Metallomics*, **1**, 2009, 280–291. (Pt anti-tumor compounds)

Wang, D. and S. J. Lippard, *Nat. Rev. Drug Discov.*, **4**, 2005, 307–320.

Images

Barnett Rosenberg – http://commons.wikimedia.org/wiki/File:Nci-vol-8173-300_barnett_rosenberg.jpg

Bob Champion – © Press Association Images.

Lance Armstrong – http://commons.wikimedia.org/wiki/File:Lance_Armstrong_Tour_2010_team_presentation.jpg

CHAPTER 13: COCAINE

Barron, B., R. Maleszka, P. G. Helliwell and G. E. Robinson, *J. Exp. Biol.*, **212**, 2009, 163–168. (Cocaine and bees)

Bencharit, S., C. L. Morton, Y. Xue, P. M. Potter and M. R. Redinbo, *Nat. Struct. Biol.*, **10**, 2003, 349–356. (Cocaine metabolism)

Bones, J., K. V. Thomas and B. Paull, *J. Environ. Monit.*, **9**, 2007, 701–707. (Cocaine in waste water)

Bowden, M. *Killing Pablo: The Hunt for the World's Richest, Most Powerful Criminal in History* (Atlantic Books, New York, 2001).

Calatayud, J. and A. González, *Anesthesiology*, **98**, 2003, 1503–1508. (Anesthesia)

Casale, J. F. and R. F. X. Klein, *Forensic Sci. Rev.*, **5**, 1993, 95–107. (Illicit production of cocaine)

Casale, J. F., J. R. Ehleringer, D. R. Morello and M. J. Lott, *J. Forensic Sci.*, **50**, 2005, 1315–1321. (Isotopic fractionation of carbon and nitrogen in cocaine processing)

Cecinato, A., C. Balducci and G. Nervegna, *Sci. Total Environ.*, **407**, 2009, 1683–1690. (Cocaine in city air)

Dietz, D. M. *et al.*, *Nature Neuroscience*, **15**, 2012, 891–896. (Cocaine and brain damage)

Dillehay, T. D., J. Rossen, D. Ugent, A. Karathanasis, V. Vasquez and P. J. Netherly, *Antiquity*, **84**, 2010, 939–953. (Early Holocene coca chewing in northern Peru)

Dronsfield, A., P. M. Ellis and K. Pooley, *Educ. Chem.*, **44**, 2007, 183–186. (Anaesthesia)

Ehleringer, J. R., J. F. Casale, M. J. Lott and V. L. Ford, *Nature*, **408**, 2000, 311. (Isotopic fingerprint from growing region)

Ersche, K. D., A. Barnes, P. S. Jones, S. Morein-Zamir, T. W. Robbins and E. T. Bullmore, *Brain*, **134**, 2011, 2013–2024. (Cocaine and the brain)

Hazarika, P., S. M. Jickells, K. Wolff and D. A. Russell, *Angew. Chem. Int. Ed.*, **47**, 2008, 10167–10170. (Imaging of cocaine in latent fingerprints)

Lobo, M. K. *et al.*,*Science*, **330**, 2010, 385–390. (Cocaine and brain neurons)

Mendelson, J. E. *et al.*, *Forensic Toxicol.*, **28**, 2010, 33–37. (Capsaicin potentiates cocaine)

Natanson, J. A., E. J. Hunnicutt, L. Kantham and C. Scavone, *Proc. Natl. Acad. Sci. USA*, **90**, 1993, 9645–9648. (Cocaine as plant defence)

Perrine, D. M. *The Chemistry of Mind-Altering Drugs: History, Pharmacology, and Cultural Context* (American Chemical Society, Washington, 1996. pp. 181–192).

Scheidweiler, K. B., M. A. Plessinger, J. Shojaie, R. W. Wood and T. C. Kwong, *J. Pharm. Exp. Ther.*, **307**, 2003, 1179–1187. (Methylecgonidine biomarker for crack cocaine)

Sleeman, R., I. F. A. Burton, J. F. Carter and D. J. Roberts, *Analyst*, **124**, 1999, 103–108. (Screening of banknotes for cocaine)

Streatfeild, D. *Cocaine: An Unauthorized Biography* (Virgin, London, 2001).

Tannu, N. S., L. L. Howell and S. E. Hemby, *Molecular Psychiatry*, **15**, 2010, 185–203. (Cocaine and brain damage)

Images

Coca bush – http://commons.wikimedia.org/wiki/File:Colcoca04.jpg

Cocaine hydrochloride powder – http://commons.wikimedia.org/wiki/File:CocaineHydrochloridePowder.jpg

Pablo Escobar – © Press Association Images.

Poster for *Vin Mariani* – http://commons.wikimedia.org/wiki/File:Vin_mariani_publicite156.jpg

Robin Williams – modified from http://commons.wikimedia.org/wiki/File:Williams,_Robin_(USGov)_crop.jpg

Sherlock Holmes tiles – http://commons.wikimedia.org/wiki/File:Sherlock_Holmes,_No_Smoking.jpg

CHAPTER 14: DEET

American Lyme disease foundation – http://www.aldf.com

Ditzen, M., M. Pellegrino and L. B. Vosshall, *Science*, **319**, 2008, 1838. (Mosquito's homing-in ability)

Fradin, M. S. and J. F. Day, *New Engl. J. Med.*, **347**, 2002, 347. (Comparison of effectiveness of different insect repellents)

Handwerk, B. *DEET Blocks Bugs From Smelling Humans as "Food"*, National Geographic News, March 13th 2008; http://news. nationalgeographic.co.uk/news/2008/03/080313-deet-smell. html

http://www.bangkokpost.com/news/local/310348/cocktail-lethal-for-sisters (Deaths due to DEET in Thai cocktails)

Kain, P., S. M. Boyle, S. K. Tharadra, T. Guda, C. Pham, A. Dahanukar and A. Ray, *Nature*, **502**, 2013, 507–514.

Katritzky, A. R., Z. Wang, S. Slavov, M. Tsikolia, D. Dobchev, N. G. Akhmedov, C. D. Hall, U. R. Bernier, G. G. Clark and K. J. Linthicum, *Proc. Nat. Acad. Sci.,* **105**, 2008, 7359–7364. (Design of new insect repellent molecules)

Metcalf, R. L. *Insect Control* in *Ullmann's Encyclopedia of Industrial Chemistry* (Wiley-VCH, Weinheim, 2002) (pyrethrins).

Moore, S. J. and A. Lenglet, in M. Wilcox, G. Bodeker and P. Rasoanaivo (eds), *Traditional Medicinal Plants and Malaria,* (CRC Press, Taylor & Francis, London, 2004), pp. 343–363. (Citronella and eucalyptus as mosquito repellents)

Pellegrino, M., N. Steinbach, M. C. Stensmyr, W. S. Hansson and L. B. Vosshall, *Nature*, **478**, 2011, 511–516. (DEET's mechanism of action)

Stanczyk, N. M., J. F. Y. Brookfield, L. M. Field and J. G. Logan, *PLoS ONE,* **8**, 2013, e54438. (Mosquitoes becoming less sensitive to DEET)

Syed, Z. and W. S. Leal, *Proc. Nat. Acad. Sci. USA*, **105**, 2008, 13195–13196. (Mosquitoes dislike DEET smell)

Takken, W. and B. G. Knols, *Annu. Rev. Entomol.*, **44**, 1999, 131. (Mosquitoes are attracted by CO_2, octenol and lactic acid)

Vertellus – manufacturers of DEET – http://www.deet.com

Images

Care-Plus DEET lotion – courtesy of Care-Plus, used with permission.

Mosquito coil – http://commons.wikimedia.org/wiki/ File:Katorisenkou.jpg

Mosquitoes on hands – image used with permission of USDA – ARS, with thanks to Greg Allen.

CHAPTER 15: DIFLUORODICHLOROETHANE, CF_2Cl_2: (*FREON*-12, CFC-12 OR R-12) AND RELATED COMPOUNDS

Brown, A. R. and D. W. George, *Pharm. Res.*, **14**, 1997, 1542–1547. (HFC-134a driven aerosols)

Cagin, S. and P. Dray, *Between Earth and Sky* (Pantheon Books, New York, 2003).

Emmen, H. H. *et al.*, *Regul. Toxicol. Pharm.*, **32**, 2000, 22–35. (Human safety and pharmacokinetics of HFC 134a)

Forster, P. M. de F. *et al.*, *J. Quant. Spectrosc. Radiat. Transfer*, **93**, 2004, 447–460. (Radiative forcing of HFC-134a)

Garrett, A. B. *J. Chem. Educ.*, **39**, 1962, 1361–1362. (Midgley's account of the discovery of CF_2Cl_2)

Giunta, C. J. *Bull. Hist. Chem.*, **31**, 2006, 66–74. (Thomas Midgley Jr. and the Invention of CFC Refrigerants); online at http://www. scs.illinois.edu/~ mainzv/HIST/bulletin_open_access/v31-2/ v31-2%20p66-74.pdf

Greenwood, N. N. and A. Earnshaw, *Chemistry of the Elements* (Pergamon, Oxford, 1984, 1st edition, p. 323). (Synthesis of CFCs)

Han, Y., Q. Ma, L. Wang and C. Xue, *J. Ocean Univ. China*, **11**, 2012, 562–568. (HFC 134a solvent in extraction of astaxanthin)

Lapkin, A. A., P. K. Plucinski and M. Cutler, *J. Nat. Prod.* **69**, 2006, 1653–1664. (HFC 134a solvent in extraction of artemisinin)

Lee, Y., D.-G. Kang and D. Jung, *Int. J. Refrig.*, **36**, 2013, 1203–1207. (HFO1234yf/HFC134a mixtures as refrigerant)

Levin, P. D., D. Levin and A. Avidan, *Br. J. Anaesth.*, **92**, 2004, 865–869. (Interference of HFC-134a with infrared anaesthetic gas monitors)

Lu, J., H. Yang, S. Chen, L. Shi, J. Ren, H. Li and S. Peng, *Cat. Lett.*, **41**, 1996, 221–224. (Synthesis of HFC-134a)

Midgley, T. IV, *From the Periodic Table to Production: The Life of Thomas Midgley, Jr., the Inventor of Ethyl Gasoline and Freon Refrigerants* (Stargazer Publ. Co., Corona, CA, 2001).

Molina, M. J. and F. S. Rowland, *Nature*, **249**, 1974, 810–812. (Destruction of ozone)

Rattigan, O. V., D. M. Rowley, O. Wild, R. L. Jones and A. Cox, *J. Chem. Soc. Faraday Trans.*, **90**, 1994, 1819–1829. (Mechanism of atmospheric oxidation of HFC-134a)

Wayne, R. P. *Chemistry of Atmospheres* (Oxford University Press, Oxford, 2nd edition, 1991).

Williams, K. R. *J. Chem. Educ.*, **77**, 2000, 1540–1541. (Refrigeration)

Images

Aerosol spray (Photo: PiccoloNamek) – http://commons. wikimedia.org/wiki/File:Aerosol.png

Air conditioner (Photo: Piotrus) – http://commons.wikimedia. org/wiki/File:Northeastern_University_-_air_conditioner.JPG

Expanded polystyrene coffee cup (Photo: Cvilletomorrow) – Extracted from image at – http://commons.wikimedia.org/ wiki/File:Spudnuts_sampler.jpg

Ozone hole 1979 & 2008 – http://commons.wikimedia.org/wiki/ File:Agujero_en_la_capa_de_ozono_2008.jpg

Refrigerator cycle – http://commons.wikimedia.org/wiki/ File:Refrigerator-cycle.svg

Thomas Midgley, Jr. – http://commons.wikimedia.org/wiki/ File:Thomas_Midgley,_Jr.jpg

CHAPTER 16: DDT

Beatty, R. G. *The DDT Myth: Triumph of the Amateurs*, John Day Co, New York, 1973.

Carson, R. *Silent Spring* (Houghton Mifflin, Cambridge, 1962).

Gribble, G. *Organochlorines – Natural and Anthropogenic*, in Mooney and Bate, pp. 161–176.

Holm, L., A. Blomqvist, I. Brandt, B. Brunström, Y. Ridderstråle and C. Berg, *Environ. Toxicol. Chem.*, **25**, 2006, 2787–2793. (Eggshell thinning due to o,p'-DDT)

Kemm, K. *Malaria and the DDT Story*, in Mooney and Bate, pp. 1–16.

Kinkela, D. *DDT and the American Century: Global Health, Environmental Politics, and the Pesticide That Changed the World* (The University of North Carolina Press, Chapel Hill, 2011).

Lundholm, C. E. *Comp. Biochem. Physiol. C: Pharmacol. Toxicol. Endocrinol.*, **118,** 1997, 113–128. (DDE-Induced eggshell thinning in birds)

Mooney, L. and R. Bate (eds), *Environmental Health: Third World Problems, First World Preoccupations* (Butterworth-Heinemann, Oxford, 1999).

Roberts, D. R., W. D. Alecrim, J. M. Heller, S. R. Ehrhardt and J. B. Lima, *Nature*, **297**, 1982, 62–63. (*Eufriesia purpurata* and DDT-collecting)

Silinskas, K. C. and A. B. Okey, *J. Natl. Cancer Inst.*, **55**, 1975, 653–657. (DDT protects female rats against mammary tumours and leukemia)

Terzolo, M. *et al.*, *N. Engl. J. Med.*, **356**, 2007, 2372–2380. (Mitotane for adrenal cancer)

Tren, R. and R. Bate, *The DDT Story* (IEA, London, 2001).

Vetter, W. and D. Roberts, *Sci. Total Environ.*, **377**, 2007, 371–377. (DDT, organohalogens and *Eufriesea purpurata*)

West, T. F. and G. A. Campbell, *DDT and Newer Persistent Insecticides* (Chapman & Hall, London, 2nd edition, 1950).

Whelan, E. M. *Toxic Terror* (Buffalo, Prometheus Books, Buffalo, NY, 1993).

Images

A WW2 soldier being sprayed with DDT – http://commons. wikimedia.org/wiki/File:DDT_WWII_soldier.jpg

DDT spraying – http://commons.wikimedia.org/wiki/File:DDT_ spray_1958.jpg

Rachel Carson – US Fish & Wildlife Service National Digital Library – http://digitalmedia.fws.gov/cdm/singleitem/collection/ natdiglib/id/2810/rec/1

CHAPTER 17: DIGITALIS

Aronson, J. K. *An account of the Foxglove and its Medical Uses, 1785– 1985* (Oxford University Press, Oxford; 1985), an annotated version of Withering's original book, available from: http:// www.botanicus.org/title/b12023139.

Krikler, D. M., The foxglove, *J. Am. Coll. Cardiol.*, **5** (5 Suppl A) 1985, 3A–9A. (The old woman from Shropshire and William Withering)

Lee, M. R. *Proc. R. Coll. Physicians Edinb.*, **31**, 2001, 77–83. (William Withering (1741–1799): A Birmingham Lunatic)

Peck, T. W. and K. D. Wilkinson, *William Withering of Birmingham MD, FRS, FLS* (John Wright and Sons Ltd, Bristol, 1950)

Images

Purple foxglove – http://commons.wikimedia.org/wiki/File:Foxgloves_in_Farningham_-_geograph.org.uk_-_191176.jpg?uselang=en-gb

William Withering – http://commons.wikimedia.org/wiki/File:William_Withering.jpg?uselang=en-gb

Withering's memorial plaque – http://en.wikipedia.org/wiki/File:WW_memorial_plaque.jpg

CHAPTER 18: DIMETHYLMERCURY

Blayney, M., J. Winn and D. Nierenberg, *Chem. Eng. News*, 12 May 1997, **75**(19) 7. (Toxicity of $(CH_3)_2Hg$)

Dartmouth Toxic metals Tribute to Karen Wetterhahn – http://www.dartmouth.edu/~toxmetal/about/tribute-to-karen-wetterhahn.html

Eccles, C. U. and Z. Aman, (eds), *The Toxicity of Methylmercury* (The John Hopkins University Press, Baltimore, 1987).

Friberg, L. and J. Vostal, *Mercury in the Environment* (CRC Press, Cleveland, Ohio, 1972).

Goldwater, L. J. *Mercury; a History of Quicksilver* (York Press, Baltimore, 1972).

Kashiwabawa, K., S. Konaka, T. Iijima and M. Kimura, *Bull. Chem. Soc. Japan*, **46**, 1973, 307. (Structure of $(CH_3)_2Hg$)

Krishnamurthy, S. *J. Chem. Educ.*, **69**, 1992, 347. (Vitamin B12 and methylation of mercury)

National Research Council Committee on the Toxicological Effects of Methylmercury, *Toxicological effects of Methylmercury* (National Academy Press, Washington, 2000).

Nierenberg, D. W., R. E. Nordgren, M. B. Chang, R. W. Siegler, M. B. Blayney, F. Hochberg, T. Y. Toribara, E. Cernichiari and T. Clarkson, *New Engl. J. Med.*, **338**, 1998, 1672–1676. (Fatality)

Pazderova, J., A. Jirasek, M. Mraz and J. Pechan, *Int. Arch. of Arbeitsmedizin*, **33**, 1974, 323–328. (Fatality)

Smith, W. E. and A. M. Smith, *Minamata* (Holt, Rinehart and Winston, New York, 1975).

Images

Biomagnification – adapted from: http://commons.wikimedia.org/wiki/File:MercuryFoodChainMercureBioconcentration-frFL.png

Cinnabar – http://commons.wikimedia.org/wiki/File:Cinnabarit_01.jpg

Karen Wetterhahn – http://commons.wikimedia.org/wiki/File:KarenWetterhahn.jpg

Latex gloves – http://commons.wikimedia.org/wiki/
File:Disposable_gloves_09.JPG

Minimata map – http://commons.wikimedia.org/wiki/
File:Minamata_map_illustrating_Chisso_factory_effluent_
routes2.png

The Mad Hatter's teaparty – http://commons.wikimedia.org/wiki/
File:Teaparty.svg

CHAPTER 19: DIMETHYLSULPHIDE

Aprea, E., F. Biasioli, S. Carlin, G. Versini, T. D. Märk and F. Gasperi, *Rapid Commun. Mass Spectrom.*, **21**, 2007, 2564–2572. (Me_2S in white truffles).

Bills, D. D. and T. W. Keenan, *J. Agric. Food Chem.*, **16**, 1968, 643–645. (Me_2S in sweet corn)

Brillat-Savarin's quote about Italian white truffles is actually: "*On trouve en Piémont les truffes blanches qui sont très estimées: elles ont un petit goût d'ail qui ne nuit point à leur perfection, parce qu'il ne donne lieu à aucun retour désagréable. Les meilleures truffes de France viennent du Périgord et de la Haute Provence; c'est vers le mois de janvier qu'elles ont tout leur parfum.*"

Brillat-Savarin, J. *The Physiology of Taste*, English translation: http://ebooks.adelaide.edu.au/b/brillat/savarin/b85p/

Chasteen, T. G. and R. Bentley, *J. Chem. Educ.*, **81**, 2004, 1524–1528. (Me_2S in sulfur cycle).

Giles MacDonogh, *Brillat-Savarin: The Judge and His Stomach*, John Murray, 1992 (Brillat-Savarin biography); Text of the book at: http://etext.library.adelaide.edu.au/b/brillat/savarin/b85p/complete.html

Hall, I. R., G. Brown and A. Zambonelli, *Taming the Truffle: The History, Lore, and Science of the Ultimate Mushroom* (Timber Press, Portland, 2008).

Hansen, J. *Appl. Environ. Microbiol.*, **65**, 1999, 3915–3919. (Me_2S in brewing)

Henriksson, A. S., M. Sarnthein, G. Eglinton and J. Poynter, *Geology*, **28**, 2000, 499–502. (Me_2S over 200 000 years in equatorial Atlantic)

Lovelock, J. *Philos. Trans. R. Soc. Lond. B*, **352**, 1997, 143–147. (Me_2S and climate)

Lovelock, J., R. A. Maggs and R. A. Rasmussen, *Nature*, **237**, 1972, 452–453. (Me_2S and climate)

March, R. E., D. S. Richards and R. W. Ryan, *Int. J. Mass Spectrometry*, 2006, **249–250**, 60–67. (Me_2S in black truffles)

McSweeney, P. L. H. and M. J. Sousa, *Lait*, **80**, 2000, 293–324. (Me_2S in cheese)

Pawlik, J. R., G. McFall and S. Zea, *J. Chem. Ecol.*, **28**, 2002, 1103–1115. (Me_2S in sponges)

Pripis-Nicolau, L., G. Revel, A. Bertrand and A. Lonvaud-Funel, *J. Appl. Microbiol*, **96**, 2004, 1176–1184. (Me_2S in plants and vegetables)

Schäfer, H., N. Myronova and R. Boden, *J. Exp. Bot.*, **61**, 2010, 315–334. (Me$_2$S in the biosphere)

Stensmeyr, M. C., I. Urru, I. Collu, M. Celander, B. S. Hansson and A. M. Angioy, *Nature*, **420**, 2002, 625–626. (Me$_2$S in arum lily)

Suarez, F. L., J. Springfield and M. D. Levitt, *Gut*, **43**, 1998, 100–104. (Me$_2$S in human flatulence)

Talou, T., A. Gaset, M. Delmas, M. Kulifaj and C. Montant, *Mycol. Res.*, **94**, 1990, 277–278. (Dogs and pigs detect Me$_2$S in black truffles).

Todd, J. D., R. Rogers, Y. G. Li, M. Wexler, P. L. Bond, L. Sun, A. R. J. Curson, G. Malin, M. Steinke and A. W. B. Johnston, *Science*, **315**, 2007, 666–669. (Genes that make Me$_2$S in bacteria and the S cycle)

Images

Black truffles – http://commons.wikimedia.org/wiki/File:Truffles_black_Croatia.jpg?uselang=en-gb

Pig rooting for truffles—(Photographer: Evelyn Simak) – http://commons.wikimedia.org/wiki/File:A_favourite_pastime_..._-_geograph.org.uk_-_578188.jpg?uselang=en-gb

Savarin and his book – http://commons.wikimedia.org/wiki/File:Jean_Anthelme_Brillat-Savarin.jpg

White truffles – http://commons.wikimedia.org/wiki/File:Truffles_white_Croatia.jpg?uselang=en-gb

Chapter 20: Dopamine

Arias-Carrión O. and E. Pöppel, *Act. Neurobiol. Exp.*, **67**, 2007, 481–488. (Dopamine and reward-seeking behaviour)

Barron, A. B., R. Maleszka, R. K. Vander Meer and G. E. Robinson, *Proc. Natl. Acad. Sci. U.S.A.*, **104**, 2007, 1703–1707. (Octopamine in honey bees)

Benes, F. M. *Trends in Pharmacol. Sci.*, **22**, 2001, 46–47. (Discovery of dopamine).

Bjorklund, A. and S. B. Dunnett, *Trends in Neurosci.*, **30**, 2007, 194–202. (Review)

Brefel-Courbon, C., P. Payoux, C. Thalamas, F. Ory, I. Quelven, F. Chollet, J. L. Montastruc and O. Rascol, *Movement Disord.*, **20**, 2005, 1557–1563. (Effect of levodopa on pain threshold in Parkinson's disease)

Clemens, S., D. Rye and S. Hochman, *Neurology,* **67**, 2006, 125–130. (Dopamine and restless legs syndrome)

Cotzias, G. C., P. S. Papavasiliou and R. Gellene, *New Engl. J. Med.* **281**, 1969, 272. (L-dopa in Parkinson's syndrome)

Davis, K. L., R. S. Kahn, G, Ko and M. Davidson, *Am. J. Psychiatry,* **148**, 1991, 1474–1486. (Dopamine in schizophrenia)

Dent, J. Y. *Brit. J. of Addiction to Alcohol & Other Drugs,* **46**, 1949, 15–28. (Apomorphine treatment of addiction)

Depue, R. A. and P. F. Collins, *Behavioral and Brain Sci.*, **22**, 1999, 491–517. (How dopamine affects mood and personality)

Flaherty, A. W. *J. Comp. Neur.*, **493**, 2005, 147–153. (Dopamine's role in ides and creative drive)

Fusar-Poli, P., K. Rubia, G. Rossi, G. Sartori and U. Balottin, *Am. J. Psychiatry*, **169**, 2012, 264–272. (Dopamine and ADHD)

Hsieh, G. C. *et al.*, *J. Pharmacol. Exp. Therapeutics,* **308**, 2004, 330–338. (Dopamine and apomorphine in erectile dysfunction)

Moran, M. *Medical News Today, MediLexicon, Intl.*, 2008. http://www.medicalnewstoday.com/releases/94023.php (Dopamine linked to aggression)

Nestler, E. J. *Nature Neurosci.*, **8**, 2005, 1445–1449. (Dopamine and addiction)

Parkinson's disease website – http://www.parkinsons.org.uk/

Previc, F. *The Dopaminergic Mind in Human Evolution and History* (Cambridge University Press, Cambridge, 2009).

Schultz, W. *Neuron,* **36**, 2002, 241–263. (Dopamine and reward behaviour)

Schwab, R., L. Amador and J. Lettvin, *Trans. Am. Neurol. Assoc.*, **56**, 1951, 251–253. (Apomorphine in Parkinson's disease)

Schwaerzel, M., M. Monastirioti, H. Scholz, F. Friggi-Grelin, S. Birman and M. Heisenberg, *J. Neurosci.*, **23**, 2003, 10495–10502. (Dopamine and octapomine in insects)

Simuni, T. and H. Hurtig. Levadopa: A pharmaologic miracle four decades later, in S.A. Factor and W.J. Weiner. (eds),

Parkinson's Disease: Diagnosis and Clinical Management (Google eBook). (Demos Medical Publishing, 2008)

The Michael J. Fox Foundation for Parkinson's Research – https://www.michaeljfox.org/

Wood, P. B. *Expert Rev. Neurother.*, **8**, 2008, 781–797. (Dopamine and pain)

Images

Michael J. Fox – (Cropped from a photo by Alan Light) http://commons.wikimedia.org/wiki/File:Michael_J_Fox_Tracy_Pollan2.jpeg

Muhammad Ali – http://commons.wikimedia.org/wiki/File:Ali.jpg

CHAPTER 21: EPIBATIDINE

Albuquerque, E. X., J. W. Daly and B. Witkop, *Science*, **172**, 1971, 995–1002. (Batrachotoxin review)

Cestèle, S. and W. A. Catterall, *Biochimie*, **82**, 2000, 883–892. (Batrachotoxin and the voltage-gated sodium channel)

Daly, J. W., H. M. Garraffo, T. F. Spande, M. W. Decker, J. P. Sullivan and M. Williams, *Natural Product Rep.*, **17**, 2000, 131–135. (Epibatidine discovery and other frog alkaloids review)

Daly, J. W., T. F. Spande and H. M. Garraffo, *J. Nat. Prod.*, **68**, 2005, 1556–1575. (Review on amphibian alkaloids)

Darst, C. R., P. A. Menéndez-Guerrero, L. A. Coloma and D. C. Cannatella, *Am. Naturalist*, **165**, 2005, 56–69. (Dietary specialization and chemical defence in poison frogs)

Devlin, S. and J. Du Bois, *Chem. Sci.*, **4**, 2013, 1059–1063. (Synthesis)

Du, Y., D. Garden, B. Khambay, B. S. Zhorov and K. Dong, *Mol. Pharmacol.*, **80**, 2011, 426–433. (Batrachotoxin and insect sodium channels)

Dumbacher, J. P., A. Wako, S. R. Derrickson, A. Samuelson, T. F. Spande and J. W. Daly, *Proc. Natl. Acad. Sci. USA*, **101**, 2004, 15857–15860. (Beetles and batrachotoxin)

Dumbacher, J. P., G. K. Menon and J. W. Daly, *The Auk*, **126**, 2009, 520–530. (Batrachotoxin storage in birds)

Dumbacher, J. P., T. F. Spande and J. W. Daly, *Proc. Natl. Acad. Sci. USA*, **97**, 2000, 12970–12975. (Batrachotoxins in birds)

Evans, D. A., K. A. Scheidt and C. W. Downey, *Org. Lett.*, **3**, 2001, 3009–3012. (Epibatidine synthesis)

Fitch, R. W., H. M. Garraffo, T. F. Spande, H. J. C. Yeh and J. W. Daly, *J. Nat. Prod.*, **66**, 2003, 1345–1350. (Epibatidine)

Fitch, R. W., T. F. Spande, H. M. Garraffo, H. J. C. Yeh and J. W. Daly, *J. Nat. Prod.*, **73**, 2010, 331–337. (Phantasmidine)

Prince, R. J. and S. M. Sine, *Biophys. J.*, **75**, 1998, 1817–1827. (Epibatidine and muscle acetylcholine receptors)

Saporito, R. A., H. M. Garraffo, M. A. Donnelly, A. L. Edwards, J. T. Longino and J. W. Daly, *Proc. Natl. Acad. Sci. USA.*, **101**, 2004, 8045–8050. (Pumiliotoxins)

Wang, S.-Y., J.Mitchell, D. B.Tikhonov, B. S. Zhorov and G. K. Wang, *Mol. Pharmacol.*, **69**, 2006, 788–795. (Batrachotoxin and the voltage-gated sodium channel)

Images

Phantasmal frog – http://commons.wikimedia.org/wiki/ File:Epipedobates-tricolor-dreistreifen-baumsteiger.jpg

Phyllobates terribilis poison-dart frog – (Photographer: Leon Weber) – http://commons.wikimedia.org/wiki/File:Schrecklicherpfeilgiftf rosch-01.jpg

Hooded Pitohui – Photo: copyright Jack Dumbacher, California Academy of Science, with permission.

CHAPTER 22: ESTRADIOL

Allen, W. M., *Science*, **82**, 1935, 89–93; *South. Med. J.*, **63**, 1970, 1151–1155. (Discovery and isolation of progesterone)

Asbell, B. *The Pill: A Biography of the Drug That Changed the World* (Random House, New York, 1995)

Bentley, G. R. and C. G. N. Mascie-Taylor (eds), *Infertility in the Modern World: Present and Future Prospects* (Cambridge University Press, Cambridge, UK, 2000). (Estrogenic compounds in the environment)

Boron, W. F. and E. L. Boulpaep, *Medical Physiology: A Cellular And Molecular Approach* (Elsevier/Saunders, Philadelphia, 2003. p. 1300). (Biosynthesis of estradiol)

Brunner, R. L. *et al.*, *Arch. Inter. Med.*, **165**, 2005, 1976–1986. (Effectiveness of horse estrogens in HRT)

Goldzieher, J. W. and H. W. Rudel, *J. Am. Med. Assoc.*, **230**, 1974, 421–425. (Review of development of oral contraceptives)

Massart, F., V. Meucci, G. Saggese and G. Soldani, *J. Pediatr.*, **152**, 2000, 690–695. (Early puberty onset due to xenoestrogens)

Rolland, M., J. Le Moal, V. Wagner, D. Royére and J. De Mouzzon, *Human Reprod.*, **28**, 2012, 452–470. (Decline in sperm rates)

Rossouw, J. E. *et al.*, *J. Am. Med. Assoc.*, **288**, 2002, 321–333. (Review of HRT issues)

Sharpe, R. M. and N. E. Skakkebaek, *Lancet*, **341**, 1993, 1392–1395. (Estrogen link to male infertility)

Speroff, L. and P. D. Darney, *Oral Contraception. A Clinical Guide for Contraception* (Lippincott Williams & Wilkins, Philadelphia, 2011). (The Pill and its mechanism of action)

Walsh, B., The perils of plastic, in *Time Magazine*, April 2010 – http://www.time.com/time/specials/packages/ article/0,28804,1976909_1976908_1976938-1,00.html. (Estrogenic compounds in plastic)

Wang-Cheng R. and J. M. Neuner, *Menopause* (ACP Press, Philadelphia, 2007. p. 96).

Wise, A. and K. O'Brien, T. Woodruff, *Environ. Sci. Tech.*, **1**, 2011, 51–60. (Estrogens in drinking water)

Images

Different types of contraceptive pill – http://commons.wikimedia. org/wiki/File:Plaquettes_de_pilule.jpg

CHAPTER 23: GLUCOSE

Ball S. G. and M. K. Morell, *Ann. Rev. Plant Biol.*, **54**, 2003, 207–233. (Glycogen and starch)

Campbell, N. A. *Biology* (Benjamin Cummings, New York, 1996). (Chitin)

Dahl, R. *Charlie and the Chocolate Factory* (Penguin Books, London, 2007). Text copyright Roald Dahl Nominee Ltd, 1964.

Emil Fischer – Nobel prize – http://www.nobelprize.org/nobel_ prizes/chemistry/laureates/1902/fischer-bio.html

Gebel, E. *Diabetes Forecast*, Feb. 2010. (Diabetes and glucose levels)

Klemm, D., B. Heublein, H.-P. Fink and A. Bohn, *Angew. Chem. Int. Ed.*, **44**, 2005, 3358–3393. (Review on cellulose)

Perry, G. H. *et al.*, *Nature Genetics*, **39**, 2007, 1256–1260. (Amylase and human evolution)

Revedin, A. *et al.*, *Proc. Nat. Acad. Sci.*, **107**, 2010, 18815–18819. (Prehistoric foods, including starch)

Sarkanen, K. V. and C. H. Ludwig (eds), *Lignins: Occurrence, Formation, Structure, and Reactions* (Wiley Interscience, New York, 1971).

Schenck, F. W. Glucose and glucose-containing syrups, in *Ullmann's Encyclopedia of Industrial Chemistry* (Wiley-VCH, Weinheim, 2006). (Structure of glucose & isomers)

Sjöström, E. *Wood Chemistry: Fundamentals and Applications* (Academic Press, San Diego, CA, 1993). (Lignin)

Smith, A. M. *Biomacromol.*, **2**, 2001, 335–341. (Biosynthesis of starch)

Smither, R. and C. Surowiec (eds), *This Film is Dangerous: A Celebration of Nitrate Film* (FIAF, Brussels, 2002). (Celluloid films)

Varki, A., R. Cummings, J. Esko, H. Freeze, P. Stanley, C. Bertozzi, G. Hart and M. Etzler, *Essentials of Glycobiology* (Cold Spring Harbor Laboratory Press, Cold Spring Harbour, NY, 2008). (Biological uses for glucose)

Young, R. *Cellulose Structure Modification and Hydrolysis* (Wiley, New York, 1986).

Images

Amylose – modified from the version at – http://commons. wikimedia.org/wiki/File:Amylose_3Dprojection.corrected.png

Amylopectin chains – http://commons.wikimedia.org/wiki/ File:Amylopectin_chains.jpg

Bowl of rice – http://commons.wikimedia.org/wiki/File:Steamed_ rice_in_bowl_01.jpg

Lobster – http://commons.wikimedia.org/wiki/File:2005-12-31_-_ Hummer2.jpg

Poatoes – http://commons.wikimedia.org/wiki/File:India_-_Koyambedu_Market_-_Potatoes_03_(3986298003).jpg

Redwood tree (Photographer: Allie Caulfield from Germany) – http://commons.wikimedia.org/wiki/File:Sequoia_sempervirens_Big_Basin_Redwoods_State_Park_8.jpg

Rhinoceros bettle – http://species.wikimedia.org/wiki/File:Dynastes_hercules_ecuatorianus_MHNT.jpg

Women hurdlers (London Grand Prix, 2012, 100 m hurdles), Photographer: Robbie Dale – http://commons.wikimedia.org/wiki/File:London_Grand_Prix_2012_100m_Hurdles.jpg

CHAPTER 24: GLYCEROL

Berry, E. M., J. Hirsch, J. Most, D. J. McNamara and J. Thornton, *Am. J. Clin. Nutr.*, **44**, 1986, 220–231. (Adipose composition depends on what fats you eat).

Cheitlin, M. D., A. M. Hutter, R. G. Brindis, P. Ganz, S. Kaul, R. O. Russell and R. M. Zusman, *Circulation*, **99**, 1999, 168–177. (Risk of taking Sildenafil with nitroglycerine)

Chen, Z., M. W. Foster, J. Zhang, L. Mao, H. A. Rockman, T. Kawamoto, K. Kitagawa, K. I. Nakayama, D. T. Hess and J. S. Stamler, *Proc. Nat. Acad. Sci.*, **102**, 2005, 12159–12164. (Mitochondrial aldehyde dehydrogenase activates nitroglycerine)

Chen, Z., M. W. Foster, J. Zhang, L. Mao, H. A. Rockman, T. Kawamoto, K. Kitagawa, K. I. Nakayama, D. T. Hess and J. S. Stamler, *Proc. Natl. Acad. Sci. USA*, **102**, 2005, 12159. (How GTN works to widen blood vessels)

Fight Club at the Internet Movie database – http://www.imdb.com/title/tt0137523/

Graboys, T. B. and B. Lown, *Circulation*, **108**, 2003, e78–e79. (Nitroglycerine and angina)

Grosso, A. *The Everything Soapmaking Book: Recipes and Techniques for Creating Colorful and Fragrant Soaps* (Adams Media, Avon, MA, 2007). (Soaps and glycerine soaps)

Koch, H. *Deutsche Zeitsch. für Chir.*, **186**, 1924, 273–278. (Humanol)

Kokatnur, M. G., M. C. Oalmann, W. D. Johnson, G. T. Malcom and J. P. Strong, *Am. J. of Clin. Nutr.*, **32**, 1979, 2198–2205. (Composition of human fat)

Krut, L. H. and B. Brontestewart, *J. Lipid Res.*, **5**, 1964, 343 (Composition of human fat)

National Research Council, *Fat Content and Composition of Animal Products* (Printing and Publishing Office, National Academy of Science, Washington, DC, 1976).

Nutter, M. K., E. E. Lockhart and R. S. Harris, *Oil & Soap*, **20**, 1943, 231. (Chemical composition of fats in chickens and turkeys)

Parker, J. O. *Am. J. Managed Care*, October 2004, S332–338. (Therapies for angina, including nitroglycerine)

Schück, H. and R. Sohlman, *The Life of Alfred Nobel* (William Heinemann Ltd., London, 1929)

Sneader, W. *Drug Discovery; A History* (John Wiley & Sons, Chichester, 2005). (Use of GTN for heart ailments)

The Nobel Foundation – http://www.nobelprize.org

The Tallini Tales of Destruction – http://www.logwell.com/ tales/menu/ (A collection of historical stories involving nitroglycerine use and abuse in the oilfields)

Thomas, A., Fats and Fatty Oils, *Ullmann's Encyclopedia of Industrial Chemistry* (Weinheim, Wiley-VCH, 2002).

Images

Alfred Nobel – http://commons.wikimedia.org/wiki/ File:AlfredNobel_adjusted.jpg

Humanol ampules – http://commons.wikimedia.org/wiki/ File:HUMANOL_Sterile.jpg

CHAPTER 25: HEAVY WATER: DEUTERIUM OXIDE, D_2O

Andreev, B. M. *Separation Science and Technology*, **36**, 2001, 1949–1989. (Girdler process)

Buteau, K. C. *J. High Technol. Law*, **10**(2), 2010, 62–77. (Deuterated drugs) www.jhtl.org/docs/pdf/buteau_10jhtll.pdf

Cassidy, D. *Uncertainty: The Life and Science of Werner Heisenberg*, (W. H. Freeman, New York, 1992).

Coffey, P. *Cathedrals of Science* (Oxford University Press, Oxford, 2008, pp. 208–223). (Urey and Lewis).

Cook, R. L., F. C. De Lucia and P. Helminger, *J. Mol. Spectrosc.*, **53**, 1974, 62–76. (Structure of isolated H_2O and D_2O molecules)

Dahl, P. F. *Heavy Water and the Wartime Race for Nuclear Energy* (Institute of Physics Publishing, Bristol, 1999).

Goldsmith, M. *Frederick Joliot-Curie: A Biography* (Lawrence & Wishart, London, 1976).

Kurzman, D. *Blood and Water* (Henry Holt, New York, 1997). (Rjukan)

Lewis, G. N. and R. E. Cornish, *J. Am. Chem. Soc.*, **55**, 1933, 2616–2617. (Separation by fractional distillation)

Lewis, G. N. and R. T. MacDonald, *J. Chem. Phys.*, **1**, 1933, 341–344. (Separation by electrolysis)

Mears, R. *The Real Heroes of Telemark* (Hodder and Stoughton, London, 2003). (Ryukan)

Miller, A. I. *Can. Nucl. Soc. Bull.*, **22**(1), 2001, 14 pp. at http://www.media.cns-snc.ca/Bulletin/A_Miller_Heavy_Water.pdf (Girdler process)

O'Driscoll, C. *Chemistry & Industry*, 9 March 2009, pp. 24–26. (Deuterated drugs)

Powers, T. *Heisenberg's War* (Knopf, New York, 1993).

Reader, J. and C. W. Clark, *Physics Today*, **66**, 2013, 44–49. (Neutrons and deuterium)

Soper, A. K. and C. J. Benmore, *Phys. Rev. Lett.*, **101**, 2008, 065502. (Quantum differences between heavy and light water)

Waltham, C. *An Early History of Heavy Water* (Department of Physics and Astronomy, UBC). August 1998, revised June 2002, at http://arxiv.org/abs/physics/0206076

Washburn, E. W. and H. C. Urey, *Proc. Natl. Acad. Sci. USA,* **18**, 1932, 496–498. (Separation by electrolysis)

Images

Gilbert Lewis – copyright © Sciencephoto.com. Used with permission.

Harold Urey – http://commons.wikimedia.org/wiki/File:Harold_Urey.jpg

Homer Simpson – modified from the version at: http://www.simpsoncrazy.com/pictures/homer

Menthol NMR – Glenn A. Facey, University of Ottawa NMR Facility BLOG, October 3, 2007. http://u-of-o-nmr-facility.blogspot.ca/2007/10/proton-nmr-assignment-tools-d2o-shake.html

Rjukan hydroelectric power plant – http://commons.wikimedia.org/wiki/File:Vemork_Hydroelectric_Plant_1935.jpg

CHAPTER 26: HEME

Bielig, H. J., E. Bayer, L. Califano and L. Wirth, *Publ. Staz. Zool. Napoli,* **25**, 1954, 26–66. (Hemovanadin)

Campbell, M. K. *Biochemistry* (Harcourt, Philadelphia, PA, 1999). (Review)

Collman, J. P., J. I. Brauman, T.R. Halbert and K. S. Suslick, *Proc. Natl. Acad. Sci. U.S.A.*, **73**, 1976, 3333–3337. (How O_2 and CO bond to Hb)

Greenhall, A. M., Feeding behavior, in: A. M. Greenhall and U. Schmidt (eds), *Natural History of Vampire Bats* (CRC Press, Boca Raton, FL, 1988). (Vampire bats)

Lodish, H., A. Berk, L. S. Zipursky, P. Matsudaira, D. Baltimore and J. Darnell, *Molecular Cell Biology* (W. H. Freeman, New York, 2000). (Evolution of Mb and Hb)

McGee, H. *On Food and Cooking: The Science and Lore of the Kitchen* (Scribner, New York, 2004). (Meat color and myoglobin)

Ordway and G. A., D. J. Garry, *J. Exp. Biol.*, **207**, 2004, 3441–3446. (Myoglobin overview)

Perutz, M., *New Scientist Sci. J.*, **56**, 1971, 676. (Hemoglobin review)

Shuster, C. N., Jr, R. B. Barlow and H. J. Brockmann, *The American Horseshoe Crab* (Harvard University Press. 2004). (Blood of crabs and molluscs)

Stoker, B. *Dracula* (Archibald Constable and Co., London, 1897).

Images

Bleeding finger (Photographer: Crystal (Crystl) from Bloomington, USA- http://commons.wikimedia.org/wiki/File:Bleeding_finger.jpg

Blood drops – http://commons.wikimedia.org/wiki/File:NIK_3232-Drops_of_blood_medium.JPG

CHAPTER 27: HEXENAL

Aou, S., M. Mizuno, Y. Matsunaga, K. Kubo, X-L. Li and A. Hatanaka, *Chem. Senses*, **30** (suppl 1), 2005, i262–i263. (Green odor reduces pain sensation)

Baldwin, E. A., J. W. Scott, C. K. Shewmaker and W. Schuch, *HortScience*, **35**, 2000, 1013–1022. (Key odorants in tomato)

Bisignano, G., M. G. Laganà, D. Trombetta, S. Arena, A. Nostro, N. Uccella, G. Mazzanti and A. Saija, *FEMS Microbiol Lett.*, **198**, 2001, 9–13. (Antibacterial activity of (*E*)-2-alkenals from olives)

Brilli, F., L. Hörtnagl, I. Bamberger, R. Schnitzhofer, T. M. Ruuskanen, A. Hansel, F. Loreto and G. Wohlfahrt, *Environ. Sci. Technol.*, **46**, 2012, 3859–3865. (Green leaf volatiles from cut grass)

Bult, J. H. F., H. N. J. Schifferstein, J. P. Roozen, E. D. Boronat, A. G. J. Voragen and J. H. A. Kroeze, *Chem. Senses*, **27**, 2002, 485–494. (Apples)

Chalke, S. *The Wisden Cricketer*, October 2004, p. 61. (Alan Dixon)

Engelberth, J., H. T. Alborn, E. A. Schmelz and J. H. Tumlinson, *Proc. Natl. Acad. Sci. USA*, **101**, 2004, 1781–1785. ((*Z*)-3-hexenal to stimulate production of jasmonic acid in plants)

Hansson, B. S., M. C. Larsson and W. S. Leal, *Physiological Entomology*, **24**, 1999, 121–126. (Scarab beetle)

Hatanaka, A. *Phytochemistry*, **34**, 1993, 1201–1218. (Review of biogeneration of green odor)

Holldobler, B. and E. O. Wilson, *The Ants* (Belknap Press, Harvard, 1990, p. 263). (Hexenal in ants)

http://www.cf.ac.uk/biosi/staff/jacob/teaching/sensory/olfact1. html (smell and memory)

http://www.leffingwell.com/olfaction.htm (smell and memory)

Kilpinen, O., D. Liu and A. P. S. Adamsen, *PLoS ONE*, **7**, 2012, e50981. (*E*-2-hexenal released by bed bugs)

Kubo, I., K. Fujita, A. Kubo, K. Nihei and T. Ogura, *J. Agric. Food Chem.*, **52**, 2004, 3329–3332. (Antibacterial activity of (*E*)-2-alkenals from cinnamon against *Salmonella choleraesuis*)

Moraes, M. C. B., M. Pareja, R. A. Laumann and M. Borges, *Neotrop. Entomol.*, **37**, 2008, 489–505. (Hexenal in neotropical stink bugs)

Nikaido, Y., S, Miyata and T. Nakashima, *Physiol. Behav.*, **103**, 2011, 547–556. (A mixture of *cis*-3-hexenol and *trans*-2-hexenal reduces stress in rats)

Noge, K., K. L. Prudic and J. X. Becerra, *J. Chem. Ecol.*, **38**, 2012, 1050–1056. (Defensive roles of (*E*)-2-hexenal in *Heteroptera*)

Sasabe, T. *et al.*, *Chem. Senses*, **28**, 2003, 565–572. (Green grass smell is good for you)

Song: *Green, Green Grass of Home* – http://en.wikipedia.org/wiki/ Green,_Green_Grass_of_Home

Song: *Whispering Grass* – http://en.wikipedia.org/wiki/ Whispering_Grass

Tadie, J.-Y. *Marcel Proust: A Life* (Viking, New York, 2000).

Images

Apple juice – http://commons.wikimedia.org/wiki/File:Apple_ juice_with_2apples.jpg

Bed bug – http://commons.wikimedia.org/wiki/File:Bed_bug,_ Cimex_lectularius.jpg

Cricket – http://commons.wikimedia.org/wiki/File:Close_fielders. jpg

Marcel Proust – http://commons.wikimedia.org/wiki/File:Marcel_ Proust_1900.jpg

Tom Jones – http://commons.wikimedia.org/wiki/File:Tom_ Jones_concert.jpg

Tomatoes – http://commons.wikimedia.org/wiki/ File:Tomate_2008-2-20.JPG

CHAPTER 28: HYDROGEN PEROXIDE

Activation energies for the decomposition of H_2O_2; adapted from E. A. Moelwyn-Hughes, *The Kinetics of Reactions in Solution* (OUP, Oxford, 2nd edition, 1947, p. 299), and J. G. Stark and H. G. Wallace, *Chemistry Data Book* (John Murray, London, 1975, p. 85).

Cotton, F. A., G. Wilkinson, C. A. Murillo and M. Bochmann, *Advanced Inorganic Chemistry* (John Wiley, New York, 6th Edition, 1999, pp. 457ff). (Properties)

Eisner, T. *For Love of Insects* (Belknap Press, Harvard, 2003, pp. 9–43). (Bombardier beetle)

Eisner, T. and D. J. Aneshansley, *Proc. Natl. Acad. Sci. USA.*, **96**, 1999, 9705–9709. (Bombardier beetle)

Eisner, T., M. Eisner and M. Siegler, *Secret Weapons: Defenses of Insects, Spiders, Scorpions, and Other Many-Legged Creatures* (Belknap Press, Harvard, 2005, pp. 157–162). (Bombardier beetle)

Greenwood, N. N. and A. Earnshaw, *Chemistry of the Elements* (Butterworth-Heinemann, Oxford, 2nd edition, 1997, p. 633ff.) (Properties)

http://boingboing.net/2010/05/26/bombardier-beetle-up.html (video of bombardier beetle)

http://www.esa.org/esablog/research/ballistics-experts-of-the-bug-world/

http://www.rafmuseum.org.uk/messerschmitt-me-163b-1a-komet.htm

Johnson, B. *The Secret War* (BBC, London, 1978, pp. 278–282). (Me-163)

Milas, N. A., A. Golubović, *J. Am. Chem. Soc.*, **81**, 1959, 6461–6462. (Acetone peroxide)

Moore, R. *A Time to Die: The Kursk Disaster* (Doubleday, London, 2002).

Myers, R. L., *The 100 Most Important Chemical Compounds: A Reference Guide* (Greenwood Press, Westport, CT, 2007, pp. 144–146). (Uses of hydrogen peroxide)

Peroxide propulsion: http://www.peroxidepropulsion.com/article/2

Truscott, P. *Kursk: Russia's Lost Pride* (Simon & Schuster, London, 2002).

Images

Bombardier beetle – Copyright © 1999 National Academy of Sciences, U.S.A. Used with permission. Taken from: T. Eisner and D. J. Aneshansley, *Proc. Natl. Acad. Sci. USA.*, **96**, 1999, 9705.

Debbie Harry in 1980 – © Press Association Images.

Jean Harlow – http://commons.wikimedia.org/wiki/File:Jean_Harlow,_Black_and_White,_Photograph.jpg

London bus bombing in 2005 —© Press Association Images.

Marilyn Monroe – http://commons.wikimedia.org/wiki/File:Marilyn_Monroe_in_Gentlemen_Prefer_Blondes_trailer.jpg

Me163 – http://commons.wikimedia.org/wiki/File:Messerschmitt_Me_163B_USAF.jpg

Soviet Oscar-class submarine – http://commons.wikimedia.org/wiki/File:Oscar_class_submarine_2.JPG

CHAPTER 29: INSULIN

Aggarwal, S. R. *Nat. Biotechnol.*, **30**, 2012, 1191–1197. (Modern synthesis of insulin using DNA technology)

Banting, F. G., C. H. Best, J. B. Collip, W. R. Campbell and A. A. Fletcher, *Can. Med. Assoc. J.*, **12**, 1922, 141–146. (Discovery of insulin)

Bliss, M. *J. Hist. Med. Allied Sci.*, **48**, 1993, 253–274. (Story of the discovery of insulin)

Discovery and Early Development of Insulin (University of Toronto library): http://link.library.utoronto.ca/insulin/

Kjeldsen, T. *Appl. Microbiol. and Biotechnol.*, **54**, 2000, 277–286. (Insulin made by yeast)

Lestradet, H. *Diabetes & Metabolism*, **23**, 1977, 112. (Story of discovery of insulin)

Sanger, F. and E. O. Thompson, *Biochem. J.*, **53**, 1953, 353–366 and 366–374. (Structure of insulin)

Sanger, F. and H. Tuppy, *Biochem. J.*, **49**, 1951, 463–481 and 481–490. (Structure of insulin)

Images

Best and Banting – http://commons.wikimedia.org/wiki/File:C._H._Best_and_F._G._Banting_ca._1924.png

Insulin hexamer (created by Isaac Yonemoto) – http://commons.wikimedia.org/wiki/File:InsulinHexamer.jpg

Islet of Langerhans from a mouse (taken by Jakob Suckale) – http://commons.wikimedia.org/wiki/File:Mouse_pancreatic_islet.jpg

James Collip – http://commons.wikimedia.org/wiki/File:J._B._Collip_in_his_office_at_McGill_University_ca._1930.png

McLeod, J. J. R. – http://commons.wikimedia.org/wiki/File:J. J. R._Macleod_ca._1928.png

Mechanism of Insulin action – modified from the
version at: http://commons.wikimedia.org/wiki/
File:Insulin_glucose_metabolism_ZP.svg

Paul Langerhans – http://commons.wikimedia.org/wiki/File:Paul_
Langerhans.jpg

CHAPTER 30: KISSPEPTIN

Dhillo, W. S. *J. Neuroendicrin.*, **20**, 2008, 963.

Dungan, H. M., D. K. Clifton and R. A. Steiner , *Endocrinology*, **147**,
2006, 1154–1158. (Review of role in reproduction)

Gottsch, M. L., D. K. Clifton and R. A. Steiner, *Peptides*, **30**, 2009,
4–9. (Review of kisspeptin and Kiss gene, and naming)

Harms, J. F., D. R. Welch and M. E. Miele, *Clin. Exp. Metastasis*, **20**,
2003, 11–18. (Kisspeptin as an anti-cancer agent)

Kaiser U. B. and W. Kuohung, *Endocrine*, **26**, 2006, 277–284. (Role
in puberty)

Lee, J. H., M. E. Miele, D. J. Hicks, K. K. Phillips, J. M. Trent, B.
E. Weissman and D. R. Welch, *J. Natl. Cancer Inst.*, **88**, 1996,
1731–1737. (Discovery of the KiSS-1 gene)

Web portal for kisspeptin researchers – http://http://www.
kisspeptin.org/

Images

Ewe and lamb – (Modified from an original photo by Derek
Harper) – http://commons.wikimedia.org/wiki/File:Ewe_
and_lamb,_Parke_-_geograph.org.uk_-_761127.jpg

Kiss (the rock band) – http://commons.wikimedia.org/wiki/
File:KISS_in_concert_Boston_2004.jpg

CHAPTER 31: LAURIC ACID

Anneken, D. J., S. Both, R. Christoph, G. Fieg, U. Steinberner and
A. Westfechtel, Fatty acids in *Ullmann's Encyclopedia of Industrial Chemistry* (Wiley-VCH, Weinheim, 2006).

Baert, J. H. and R. J. Veys, *J. Oral Pathol. Med.*, **24**, 1997, 181.
(Toxicity results)

Barker, G. *Surfactants in Cosmetics* (Marcel Dekker, New York, 1985).
(Foaming in detergents)

Beare-Rogers, J., A. Dieffenbacher and J. V. Holm, *Pure Appl. Chem.*
73, 2001, 685. (Fatty acids in milks)

Brady, G. S., H. R. Clauser and J. A. Vaccari, *Materials Handbook – An
Encyclopedia for Managers, Technical Professionals, Purchasing and
Production Managers, Technicians, and Supervisors* (McGraw-Hill,
15th edition, 2002, pp. 250–251). (Cochin oil)

Carpo, B. G., V. M. Verallo-Rowell and J. Kabara, *J. Drugs Derm.*, **6**,
2007, 991–998. (Antibacterial activity of monolaurin)

Cascorbi, H. F., F. G. Rudo and G. G. Lu, *J. Pharm. Sci.*, **52**, 1963,
803. (Toxicity studies)

Isaacs, C. E. *Adv. Nutrit. Res.* **10**, 2001, 271–285. (Monolaurin as an
antibiotic)

James, T. K. and A. Rahman, *New Zealand Plant Protect.*, **58**, 2005,
157. (Coconut-oil-derived herbicides)

Mensink, R. P., P. L. Zock, A. D. M. Kester and M. B. Katan, *Am. J. of Clin. Nutr.* **77**, 2003, 1146. (Role of fatty acids in heart disease).

Smulders, E., W. Rybinski, E. Sung, W. Rähse, J. Steber, F. Wiebel and A. Nordskog, Laundry detergents in *Ullmann's Encyclopedia of Industrial Chemistry* (Wiley-VCH, Weinheim, 2002). (SLS in detergents)

Suikho, C. and J. Serup, *Skin. Res. Technol.*, **14**, 2008, 498. (Irritant studies)

Thijssen, M. A. and R. P. Mensink, Fatty acids and atherosclerotic risk, in A. von Eckardstein (ed.) *Atherosclerosis: Diet and Drugs* (Springer, Berlin, 2005, p. 171).

Turro, N. J. and A. Yekta, *J. Am. Chem. Soc.*, **100**, 1978, 5951–5952. (Surfactant properties of SLS)

Images

Oil-palm plantation – http://commons.wikimedia.org/wiki/File:Oilpalm_malaysia.jpg

Palm trees on a beach – http://commons.wikimedia.org/wiki/File:Palm_trees.jpg

CHAPTER 32: LIMONENE

Bouwmeester, H. J., J. Gershenzon, M. C. J. M. Konings and R. Croteau, *Plant Physiol.*, **117**, 1998, 901–912. (Biosynthesis)

Friedman, L. and J. G. Miller, *Science*, **172**, 1971, 1044–1046. (Incorrect smells of limonene enantiomers)

http://www.leffingwell.com/chirality/limonene.htm (Smell of isomers)

Hyatt, D. C., B. Youn, Y. Zhao, B. Santhamma, R. M. Coates, R. B. Croteau and C.-H. Kang, *Proc. Natl. Acad. Sci.*, **104**, 2007, 5360–5365.

Laszlo, P. *Citrus. A History* (University of Chicago Press, Chicago, 2007).

Ohloff, G., W. Pickenhagen and P. Kraft, *Smell and Chemistry* (Wiley-VCH, Zurich, 2012).

Orange Guard website: http://www.orangeguard.com

Sell, C. S., Scent through the looking glass, in P. Kraft and K. A. D. Swift (eds), *Perspectives in Flavor and Fragrance Research* (Wiley-VCH, Zurich, 2005, pp. 67–88), and in *Chemistry and Biodiversity*, **1**, 2004, 1899–1920. (Smells of limonene enantiomers).

Thomas, A. F. and Y. Bessière, *Nat. Prod. Rep.*, **6**, 1989, 292–309. (Review)

Images

Lemons – http://commons.wikimedia.org/wiki/File:Lemon.jpg

Orange Guard™ bottle– used with permission, courtesy of Orange Guard Inc.

Zesting an orange – http://commons.wikimedia.org/wiki/File:Zesting_an_orange.jpg

Chapter 33: Linoleic Acid

Burr, G. O., M. M. Burr and E. Miller, *J. Biol. Chem.*, **86**, 1930, 1–9. (EFAs in nutrition)

Chow, C. K. *Fatty Acids in Foods and Their Health Implications* (Routledge Publishing, New York, 2001).

Hayakawa, K., Y. Y. Linko and P. Linko, *J. Lipid Sci. Technol.*, **102**, 2000, 419–425. (TFAs in human nutrition)

Hu, F. B., J. E. Manson and W. C. Willett, *J. Am. Coll. Nutr.*, **20**, 2001, 5–19. (Review of dietary fat and risk of heart disease)

Jim Clark. *The Hydrogenation of Alkenes: Margarine Manufacture; Chemguide: Helping you to understand Chemistry* (http://www.chemguide.co.uk/organicprops/alkenes/hydrogenation.html).

Michel, R. S. *Lancet* **343**, 1994, 8911. (Mediterranean diet and heart diseases)

O'Connor, A. *The Claim: Margarine Is Healthier Than Butter*, 2007 *New York Times* (http://www.nytimes.com/2007/10/16/science/16real.html).

Simopoulos, A. P. *Biomed. Pharmacoth*, **56**, 2002, 365–379. (ω–3 and ω–6 EFAs)

Images

Corn oil – http://commons.wikimedia.org/wiki/File:Corn_oil_%28mais%29.jpg?uselang=en-gb

Margarine – http://commons.wikimedia.org/wiki/File:Margarine_
BMK.jpg (Credit: BMK, Germany)

CHAPTER 34: LYSERGIC ACID DIETHYLAMIDE (LSD)

Albarelli, H. P. *Terrible Mistake: The Murder of Frank Olson and the
CIA's Secret Cold War Experiments* (Trine Day, Waterville, 2008).

Caporael, L. R. *Science,* **192**, 1976, 21–26; N. P. Spanos and J.
Gottlieb, *Science,* **194,** 1976, 1390–1394. (Ergotism and Salem)

González-Maeso, J. *et al., Neuron,* **53**, 2007, 439–452 (LSD and
signalling pathways).

Hofmann, A. *LSD: My Problem Child* (Multidisciplinary Association
for Psychedelic Studies (MAPS), Saratosa, USA. ISBN:
0-9660019-8-2, p48), and The Beckley Foundation, Oxford UK
(ISBN: 978-0-19-963941-0, p. 19).

Lee, M. A. *Acid Dreams: The Complete Social History of LSD* (Grove
Press, New York, 2000).

Lyrics to '*Lucy in the Sky with Diamonds*', written by John Lennon,
1966, copyright EMI records, Ltd.

Mann, J. *Turn on and Tune in: Psychedelics, Narcotics and Euphoriants*
(Royal Society of Chemistry, Cambridge, 2009, pp. 1–25).

Matossian, M. K., *Poisons of the Past: Molds, Epidemics, and History* (Yale
University Press, New Haven, 1989). (Ergotism and history)

Perrine, D. M. *The Chemistry of Mind-Altering Drugs* (American
Chemical Society, Washington DC, 1996, esp. pp. 259–278).

Roberts, A. *Albion Dreaming: A Popular History of LSD in Britain* (Marshall Cavendish, London, 2008).

Schardl, L., D. G. Panaccione and P. Tudzynski, *The Alkaloids*, **63**, 2006, 45–86. (Ergot alkaloids rev.)

Smith, D. E. *California Medicine*, **110**, 1969, 472–476. (LSD and Haight-Ashbury)

Stevens, J. *Storming Heaven: LSD and The American Dream* (Heinemann, London, 1988).

Images

Albert Hofmann – http://commons.wikimedia.org/wiki/File:Albert_Hofmann_Oct_1993.jpg

Ergot (*Claviceps purpurea*) – http://commons.wikimedia.org/wiki/File:Claviceps_purpurea.JPG

LSD strips – http://commons.wikimedia.org/wiki/File:10_strip.jpg

Leary being arrested in 1972 – http://commons.wikimedia.org/wiki/File:Leary-DEA.jpg

CHAPTER 35: MEDROXYPROGESTERONE ACETATE

http://www.independent.co.uk/news/uk/this-britain/the-turing-enigma-campaigners-demand-pardon-for-mathematics-genius-1773480.html (Campaign to get Turing pardoned)

http://news.bbc.co.uk/1/hi/scotland/north_east/8636891.stm (Ryan Yates).

Leavitt, D. *The Man Who Knew Too Much: Alan Turing and the Invention of the Computer* (W. W.Norton & Co., London, 2006). (Turing's life story)

Light, S. A. and S. Holroyd, *J. Psychiatry Neurosci.*, **31**, 2006, 132–134 (Use of MPA for dementia).

Murray, M. A., J. H. Bancroft, D. C. Anderson, T. G. Tennent and P. J. Carr, *J. Endocrinol.* **67**, 1975, 179–188 (Benperidol usage).

Neumann, F. and M. Töpert, *J. Steroid Biochem.* **25**, 1986, 885–895. (Cyproterone and other antiandrogens)

Rondeaux, C., *Can castration be a solution for sex offenders?*, *Washington Post*, 2006: http://www.washingtonpost.com/wp-dyn/content/article/2006/07/04/AR2006070400960_pf.html

Scott, C. L. and T. Holmberg, *J. Am. Acad. Psychiatry Law*, **31**, 2003, 502. (Castration of sex offenders: Prisoners' rights versus public safety)

Images

Statue of Turing – http://commons.wikimedia.org/wiki/File:Alan_Turing.jpg

CHAPTER 36: METHAMPHETAMINE

Armstrong, D. W., K. L. Rundett, U. B. Nair and G. L. Reid, *Curr. Sep.*, **15**, 1996, 57. (Separation of isomers)

Chouvy, P.-A. and J. Meissonnier, *Yaa Baa: Production, Traffic and Consumption of Methamphetamines in Mainland Southeast Asia*, (Singapore University Press, Singapore, 2004).

Fotheringham, W. *Put Me Back on My Bike: In Search of Tom Simpson* (Yellow Jersey Press, London, 2002).

Friscolanti, M. *Friendly Fire: The Untold Story of the U.S. Bombing that Killed Four Canadian Soldiers in Afghanistan* (John Wiley, Ontario, 2006).

Green, A. R., A. O. Mechan, J. M. Elliott, E. O'Shea and M. I. Colado , *Pharmacol. Rev.* **55**, 2003, 463–508. (MDMA pharmacology)

Guillot, C. and D. Greenway *J. Psychopharmacol.* **20**, 2006, 411–416. (Recreational use of MDA linked to depression)

Holley, M. F. *Crystal Meth* (Tate Publishing, Mustang, OK, 2005).

http://jurisprudence.tas-cas.org/sites/CaseLaw/Shared%20 Documents/376.pdf (Court of Arbitration for Sport judgment on Baxter's Appeal)

Iversen, L. *Speed, Ecstasy, Ritalin* (OUP, Oxford, 2006). (The best single source)

Jin, H. L. and T. E. Beesley, *Chromatographia,* **38**, 1994, 595. (Separation of isomers)

Johnny Cash, *Cash: The Autobiography* (Harper Collins, New York, 1997).

Lyrics to '*Here Come the Nice*' by *The Small Faces*, off the album '*Itchycoo Park*' written by Steve Marriott and Ronnie Lane, copyright EMI Music publishing 1967.

Mottram, D. R. (ed), *Drugs In Sport* (E and F. N. Spon, London, 2nd edition, 1996, esp. pp. 19, 86–94).

Ogata, A. *J. Pharm. Soc. Jpn.,* **445**, 1919, 193; *Chem. Abs.*, **13**, 1919, 1709. (Synthesis)

Otto Snow, *Amphetamine Syntheses: Industrial* (Thoth Press, 2002).

Rasmussen, N. *On Speed: The Many Lives of Amphetamine,* (New York University Press, NY, 2008).

Rosen, D. M. *Dope: A History of Performance Enhancement in Sports From the Nineteenth Century to Today* (Praeger, Westport, CT, 2008).

Ross, A. *Unfinished Business: Alain Baxter* (Dewi Lewis Media, Stockport, 2005).

Roussotte, F., L. Soderberg and E. Sowell, *Neuropsychol. Rev.*, **20**, 2010, 376–397. (Effect of meth exposure on foetus)

Shulgin, A. and A. Shulgin, *Pihkal: A Chemical Love Story* (Transform Press, Berkeley, CA, 1995).

Smith, D. E. and C. M. Fischer, *Clinical Toxicology,* **3**, 1970, 117–124. (Speed and Haight-Ashbury)

The Horrors of Methamphetamines – http://www.rehabs.com/explore/meth-before-and-after-drugs/(US Police mugshots of meth users taken over the course of a few years showing their deterioration due to use of the drug).

Thompson, P. M. *et al.*, *J. Neurosci.*, **24**, 2004, 6028. (Structural abnormalities in the brains of methamphetamine users)

Uncle Fester, *Secrets of Methamphetamine Manufacture* (Loompanics, Port Townsend, WA, 5th edition, 1994). (Synthesis)

Weisheit, R. and W. L. White, *Methamphetamine: Its History, Pharmacology, and Treatment* (Hazelden Publishing, Center City, MN, 2009).

Weisheit, R., *Southern Rural Sociol.*, **23**, 2008, 78. (Synthesis of methamphetamine)

Images

Alain Baxter – ©Press Association Images.

'Breaking Bad' Poster – 'All Hail the King'. 'Breaking Bad' © 2013 Sony Pictures Television, all rights reserved.

Crystal meth – http://commons.wikimedia.org/wiki/File:Crystal_Meth.jpg

Ecstasy pills – http://commons.wikimedia.org/wiki/File:Ecstasy_monogram.jpg

CHAPTER 37: METHANE

Atreya, S. K. *Sci. Am.*, May 2007, 42–51. (Planetary methane)

Callow, C. *Power from the Sea* (Victor Gollancz, London, 1973). (North Sea Gas)

Darley, J. *High Noon for Natural Gas* (Chelsea Green Publishing, Vermont, 2004). (Natural gas supplies)

Dueck, T. A. *et al.*, *New Phytol.*, **175**, 2007, 29–35. (Evidence against methane from plants)

Fraser, P. J., R. A. Rasmussen, J. W. Creffield, J. R. French and M. A. K. Khalil, *J. Atmosp. Chem.*, **4**, 1986, 295–310. (Emission by termites)

Global Climate Change and Environmental Stewardship by Ruminant Livestock Producers is a useful summary: http://www.epa.gov/methane/pdfs/ffa.pdf

Grainger, C., T. Clarke, S. M. McGinn, M. J. Auldis, K. A. Beauchemin, M. C. Hanna, G. C. Waghorn, H. Clark and R. J. Eckard, *J. Dairy Sci.*, **90**, 2007, 2755–2766. (Emission by cows)

Johnson, K. A. and D. E. Johnson, *J. Anim. Sci.*, **73**, 1995, 2483–2492. (Emission by cows)

Keppler, F., J. T. G. Hamilton, M. Brass and T. Röckmann, *Nature*, **439**, 2006, 187–191. (Evidence for methane from plants)

Lenhart, K. *et al.*, *Nat. Comm.*, **3**, 2012, 1046. (Methane from fungi).

Li, C., J. Qiu, S. Frolking, X. Xiao, W. Salas, B. Moore, S. Boles, Y. Huang and R. Sass, *Geophys. Res. Lett.*, **29**, 2002, 1972. (Emission by rice)

Lu, Y. and R. Conrad, *Science*, **309**, 2005, 1088–1090. (Emission by rice)

Max, M. and W. Dillon, *Chem. and Ind*, 10 January 2000, 16. (Methane clathrates)

McGeer, P. and E. Durbin, (eds), *Methane. Fuel for the Future* (Plenum Press, New York, 1982).

National Academy of Sciences, *Methane Generation from Human, Animal, and Agricultural Wastes* (Books for Business, New York, 2001).

Nisbet, R. E. R. *et al.*, *Proc. R. Soc. B*, **276**, 2009, 1347–1354. (Evidence against methane from plants)

Ruppel, C. D. *Nature Educ. Know*, **3**, 2011, 29 (Methane clathrates and climate change); online at http://www.nature.com/scitable/knowledge/library/methane-hydrates-and-contemporary-climate-change-24314790

Seuss, E., G. Bohrmann, J. Greinert and E. Lausch, *Flammable Ice, Scientific American*, November 1999, 52. (Methane clathrates)

Tulk, C. A., D. D. Klug, A. M. dos Santos, G. Karotis, M. Guthrie, J. J. Molaison and N. Pradhan, *J. Chem. Phys.*, **136**, 2012, 054502 and refs therein. (Structures of clathrates)

US EPA report, 2010 – http://www.epa.gov/outreach/pdfs/Methane-and-Nitrous-Oxide-Emissions-From-Natural-Sources.pdf (Methane emissions)

Zimmerman, P. R., J. P. Greenberg, S. O. Wandiga and P. J. Crutzen, *Science*, **218**, 1982, 563–565. (Emission by termites)

Zoback, M., S. Kitasei and B. Copithorne, *Addressing the Environmental Risks from Shale Gas Development*, Worldwatch

Institute, 2010. http://efdsystems.org/Portals/25/
Hydraulic%20Fracturing%20Paper%20-%20World%20Watch.
pdf (Fracking and the environment)

Images

Burning methane clathrate – http://commons.wikimedia.org/wiki/
File:Burning_hydrate_inlay_US_Office_Naval_Research.jpg

Davy lamp – http://commons.wikimedia.org/wiki/File:Pieler_
safety_lamp.jpg

Gas cooker flame – http://commons.wikimedia.org/wiki/File:Gas_
flame.jpg

North Sea oil-rig (Photographer: Jarle Vines) – http://commons.
wikimedia.org/wiki/File:StatfjordA(Jarvin1982).jpg

Surface of Titan (Courtesy NASA/JPL-Caltech) – http://commons.
wikimedia.org/wiki/File:Huygens_surface_color_sr.jpg

CHAPTER 38: 2-METHYLUNDECANAL

Arctander, S. *Perfume and Flavor Chemicals II*, Montclair, NJ, 1969,
compound 2132. (2-methylundecanal)

Charles-Roux, E. *The World of Coco Chanel: Friends, Fashion, Fame*
(Thames & Hudson, London, 2005).

de Barry, N. M., M. Turonnet and G. Vindry, *L'abcdaire du Parfum*
(Flammarion, Paris, 1999, pp. 38–39). (Chanel No. 5).

Emsley, J. *Chem. Ind.*, **72**, (22 December, 2008), 17–19.

Enders, D. and H. Dyker, *Liebigs Ann. Chem.*, 1990, 1107–1110. (Synthesis of the enantiomers of 2-methylundecanal)

Gibka, J. and M. Glinski, *Flav. Fragr. J.*, **11**, 2006, 480–483. (Odors of aldehydes)

Koyasako, A. and R. A. Bernhard, *J. Food Sci.*, **48**, 1983, 1807–1812. (2-MNA in kumquat oil)

Kraft, P., C. Leard and P. Goutell, From *Rallet No.1* to *Chanel No.5* via Mademoiselle Chanel No.1, *Perfumer and Flavorist*, October 2007, pp. 36–48.

Laszlo, P. *Bull. Hist. Chem.*, **15/16**, 1994, 59–64. (Georges Darzens' biography). Online at http://www.scs.illinois.edu/~mainzv/HIST/bulletin_open_access/num15-16/num15-16%20p59-64.pdf

Mazzeo, T. J. *The Secret of Chanel No. 5* (Harper Collins, New York, 2010).

Picardie, J. *Coco Chanel: The Legend and The Life* (Harper Collins, London, 2010).

Turin, L. and T. Sanchez, *Perfumes: The Guide* (Viking Adult, New York, 2008, pp. 59–261).

Veuillet-Gallot, D. *Le Guide du Parfum*, Éditions Hors Collection, 1996, pp. 59–61. (Chanel No.5).

Wallis, J. *Coco Chanel* (Reed Educational, Oxford, 2001).

Image

Ernest Beaux – http://commons.wikimedia.org/wiki/File:Ernest_Beaux.jpg

CHAPTER 39: MONOSODIUM GLUTAMATE

Chandrashekar, J., M. A. Hoon, N. J. Ryba and C. S. Zuker, *Nature*, **444**, 2006, 288–294. (Mammalian taste receptors)

Freeman, M. *J. Am. Acad. Nurse Pract.*, **18**, 2006, 482–486. (Health studies).

International glutamate information service – http://www. glutamate.org/ (FAQs about MSG)

Kawamura, Y. and M. R. Kare (eds), *Umami: A Basic Taste* (Marcel Dekker Inc., New York, 1987).

Kurihara, K. *Am. J. Clin. Nutr.*, **90**, 2009, 719S–722S. (Umami/MSG as a food additive)

Lindemann, B., Y. Ogiwara and Y, Ninomiya, *Chem. Senses*, **27**, 2002, 843–844. (Discovery of umami)

Loliger, J. *J. Nutr.*, **130** (4s Suppl) 2000, 915s–920s. (Glutamate as savory taste)

Maragakis, N. J. and J. D. Rothstein, *JAMA Neurol.*, **58**, 2001, 365–370. (Glutamate and neurological illness)

McEntee, W. J. and T. H. Crook, *Psychopharmacology*, **111**, 1993, 391–401. (Role in learning, memory, and aging)

Meldrum, B. S. *J. Nutr.*, **130**, 2000 1007S–1015S. (Glutamate as a neurotransmitter)

Mosby, I. *Social History of Med.* **22**, 2009, 133–151. (Chinese Restaurant Syndrome)

Tarasoff, L. and M. F. Kelly, *Food Chem. Toxicol.*, **31**, 1993, 1019–1035. (Double-blind health studies).

Willams, A. N. and K. M. Woessner, *Clin. Exp. Allergy*, **39**, 2009, 640–646. (Review of health studies)

Yamaguchi, S. *Physiol. Behavior*, **49**, 1991, 833–841. (Umami)

Yamaguchi, S. and C. Takahashi, *J. Food Sci.*, **49**, 1984, 82–85. (Using MSG means you can reduce salt in foods)

Images

Human body outline – http://commons.wikimedia.org/wiki/File:Man_shadow_with_organs.png

Kikunae Ikeda – http://commons.wikimedia.org/wiki/File:Kikunae_Ikeda.jpg

Packet of MSG – http://commons.wikimedia.org/wiki/File:Ajinomoto_pro.jpg

CHAPTER 40: MORPHINE, CODEINE AND HEROIN

Booth, M., Opium. *A History* (Simon & Schuster, London, 1996).

Chouvy, P.-A. *Opium: Uncovering the Politics of the Poppy* (I. B. Tauris, London, 2009).

Dormandy, T. *Opium: Reality's Dark Dream* (Yale University Press, New Haven, CT, 2012).

Fillingim, R. B. and R. W. Gear, *Euro. J. of Pain*, 2004, 413–425. (Sex differences and analgesia)

Gear, R. W., N. C. Gordon, M. Hossaini-Zadeh, J. S. Lee, C. Miaskowski, S. M. Paul and J. D. Levine, *J. Pain*, **9**, 2008, 337–341. (Morphine and analgesia)

Hodgson, B. *In the Arms of Morpheus: The Tragic History of Laudanum, Morphine, and Patent Medicines* (Firefly Books, Buffalo, NY, 2001).

Hodgson, B. *Opium: A Portrait of the Heavenly Demon* (Souvenir Press, Oxford, 2000.)

Jay, M. *High Society. Mind-Altering Drugs in History and Culture* (Thames and Hudson, London, 2010).

Lachenmeier, D. W., C. Sproll and F. Musshoff, *Ther. Drug Monit.*, **32**, 2010, 11–18. (Poppy seed foods and drug testing)

Mann, J. *Turn On and Tune In: Psychedelics, Narcotics and Euphoriants* (Royal Society of Chemistry, Cambridge, 2009, pp. 26–59).

Perrine, D. M. *The Chemistry of Mind-Altering Drugs* (American Chemical Society 1996, Washington, DC, esp. pp. 43–112).

Schiff, P. L. *Am. J. Pharm. Educ.*, **66**, 2002, 186–194. (Opium and its alkaloids, review)

The French Connection: Internet Movie Database – http://www.imdb.com/title/tt0067116/?ref_=fn_al_tt_1

Trainspotting – Internet Movie Database: http://www.imdb.com/title/tt0117951/

Yamaguchi, K., M. Hayashida, H. Hayakawa, M. Nihira and Y. Ohno, *Forensic Toxicol.*, **29**, 2011, 69–71. (Poppy seed foods and drug testing)

Images

Bayer heroin bottle – http://commons.wikimedia.org/wiki/File:Bayer_Heroin_bottle.jpg

Janis Joplin – http://commons.wikimedia.org/wiki/File:Janis_Joplin_seated_1970.JPG

Jim Morrison – cropped from: http://commons.wikimedia.org/wiki/File:Doors_electra_publicity_photo.JPG

Laudanum bottle – http://commons.wikimedia.org/wiki/File:Laudanum_poison_100ml_flasche.jpg

Poppy and seedpod – http://commons.wikimedia.org/wiki/File:Papaver_somniferum_flowers.jpg

Poppy seedpod oozing opium – http://commons.wikimedia.org/wiki/File:Slaapbol_R0017601.JPG

Sid Vicious (Photo: Chicago Art Dept, John Schorr) – http://commons.wikimedia.org/wiki/File:Vicious.jpg

CHAPTER 41: NANDROLONE

Ayotte, *Br. J. Sports. Med.*, **40**, 2006, 25–29. (Norandrosterone in athletes' urine samples)

Birch, J. *J. Chem. Soc.*, 1950, 367–368; A. L. Wilds and N. A. Nelson, *J. Am. Chem. Soc.*, **75**, 1953, 5366–5369. (Synthesis)

Catlin, D. H., B. Z. Leder, B. Ahrens, B. Starcevic, C. K. Hatton, G. A. Green and J. S. Finkelstein, *J. Am. Med. Assoc.*, **284**, 2000, 2618–2621. (Contamination of OTC androstenedione)

Gambelunghe, C., M. Sommavilla and R. Rossi, *Biomed. Chromatogr,* **16**, 2002, 508–512. (Urine testing for nandrolone metabolites)

Grosse, J., P. Anielski, P. Hemmersbach, H. Lund, R. K. Mueller, C. Rautenberg and D. Thieme, *Steroids,* **70**, 2005, 499–506. (Norandrosterone formation in stored urine samples)

Hemmersbach, P. and J. Große, Nandrolone: A multi-faceted doping agent, in D. Thieme and P. Hemmersbach (eds), *Handbook of Experimental Pharmacology,* **195**, 2010, 127–154 (Review)

Kintz, P., V. Cirimele, V. Dumestre-Toulet and B. Ludes, *J. Pharmaceut. Biomed. Anal.,* **24**, 2001, 1125–1130. (Detecting nandrolone using hair analysis)

Rob Kingston, Catching the drug runners, *Chemistry in Britain,* September **36**, 2000, 26–29.

Steven Ungerleider, *Faust's Gold: Inside the East German Doping Machine* (Thomas Dunne Books, New York, 2001).

Images

East German Olympic swimmers – http://commons.wikimedia. org/wiki/File:Bundesarchiv_Bild_183-W0727-138,_Moskau,_ Olympiade,_Siegerinnen_%C3%BCber_200_m_R%C3%BCc ken.jpg

Linford Christie – © Press Association Images.

CHAPTER 42: NICOTINE

BBC News (29th April 2013) – http://www.bbc.co.uk/news/world-europe-22335520 (EU to ban neonicotinoids)

Blum, A. P., H. A. Lester and D. A. Dougherty, *Proc. Natl. Acad. Sci.*, **107**, 2010, 13206–13211. (Nicotinic pharmacophore)

Brandt, A. *Cigarette Century: The Rise, Fall, and Deadly Persistence of the Product That Defined America* (Basic Books, New York, 2007).

Buccafusco, J. J. and A. V. Terry, *Life Science*, **72**, 2003, 2931–2942. (Cotinine and Alzheimer's)

Buccafusco, J. J., J. W. Beach and A. V. Terry, *J. Pharmacol. Exp. Ther.*, **328**, 2009, 364–370. (Nicotine and cotinine improve working memory in macaques).

EFSA report *EFSA identifies risks to bees from neonicotinoids* (16th Jan 2013) – http://www.efsa.europa.eu/en/press/news/130116.htm

Faulkner, J. M. *J. Am. Med. Assoc.*, **100**, 1933, 1664–1665. (Nicotine poisoning of florist)

Gately, I. *La Diva Nicotina: The Story of How Tobacco Seduced the World* (Simon & Schuster, London, 2001).

Goodman, J. *Tobacco in History: The Cultures of Dependence* (Routledge, London, 1993).

Goodman, S. L. and Z. Xun (eds), *Smoke: A Global History of Smoking* (Reaktion Books, London, 2004).

Gorrod, J. W. and M-C. Tsai, in R. W. Waring, G. B. Steventon and S. C. Mitchell (eds), *Molecules of Death* (Imperial College Press, London, 2nd ed., 2007, pp. 233–252).

Kessler, D., K. Gase and I. T. Baldwin, *Science*, **321**, 2008, 1200–1202. (Nicotine in floral scents)

Lockhart, L. P. *Br. Med. J.*, **1**, 1933, 246–247. (Nicotine poisoning of insecticide worker)

Melton, L. *Chemistry World*, **4**, July 2007, 44–48. (Antismoking drugs)

Minematsu, N., H. Nakamura, M. Furuuchi, T. Nakajima, S. Takahashi, H. Tateno and A. Ishizaka, *Eur. Respir. J.*, **27**, 2006, 289–292. (Genetics of CYP2A6 and smoking)

Mineur, Y. S. *et al. Science*, **332**, 2011, 1330–1332. (Nicotine and appetite)

Oreskes, N. and E. M. Conway, *Merchants of Doubt: How a Handful of Scientists Obscured the Truth on Issues from Tobacco Smoke to Global Warming* (Bloomsbury Press, New York, 2010).

Parker-Pope, T. *Cigarettes: Anatomy of an Industry from Seed to Smoke* (New Press, New York, 2000).

Picciotto, M. R. and P. J. Kenny, *Cold Spring Harb. Perspect. Med.*, **3**, 2013, a012112. (Desensitisation of nicotinic acetylcholine receptors)

Steppuhn, A., K. Gase, B. Krock, R. Halitschke and I. T. Baldwin, *PLoS Biology*, **2**, 2004, 1074–1080. (Nicotine's defensive function in nature)

Stratton, K., P. Shetty, R. Wallace and S. Bondurant (eds), *Clearing the Smoke: Assessing the Science Base for Tobacco Harm Reduction* (National Academies Press, Washington, DC, 2001).

Xiu, X., N. L. Puskar, J. A. P. Shanata, H. A. Lester and D. A. Dougherty, *Nature*, **458**, 2009, 534–537. (Nicotine binding to brain receptors)

Yamamoto, I., Nicotine to nicotinoids: 1962 to 1997, in I. Yamamoto and J. Casida (eds), *Nicotinoid Insecticides and the Nicotinic Acetylcholine Receptor* (Springer-Verlag, Tokyo, 1999, pp. 3–27).

Zagorevski, D. V. and J. A. Loughmiller-Newman, *Rapid Commun. Mass Spectrom.*, **26**, 2012, 403–411. (Nicotine in a Mayan flask)

Images

Beehive full of dead bees – http://commons.wikimedia.org/wiki/File:Abeilles-mortes-dead-bees.JPG

Nicotine patch – http://en.wikipedia.org/wiki/File:Nicoderm.JPG

No Smoking sign – http://commons.wikimedia.org/wiki/File:English_No_Smoking_sign.JPG

CHAPTER 43: NITROUS OXIDE, N_2O

Beaulieu, J. J. *et al.*, *Proc. Natl. Acad. Sci. USA*, **108**, 2011, 214–219. (Greenhouse gas)

Eger, E. I. (ed), *Nitrous Oxide*, N_2O (Elsevier, New York, 1985).

Greenwood, N. N. and A. Earnshaw, *Chemistry of the Elements* (Pergamon, Oxford, 1st edition, 1984, pp. 508–511).

http://www.general-anaesthesia.com/index.html

Jevtović-Todorović, V., S. M. Todorović, S. Mennerick, S. Powell, K. Dikranian, N. Benshoff, C. F. Zorumski and J. W. Olney, *Nat. Med.*, **4**, 1998, 460–463. (Nitrous oxide as an NMDA antagonist, neuroprotectant and neurotoxin)

Knight, D. *Humphry Davy: Science and Power* (Cambridge University Press, Cambridge, 1992).

Lamont-Brown, R. *Humphry Davy: Life Beyond the Lamp* (Alan Sutton, Stroud, 2004).

Partington, J. R. *A Short History of Chemistry* (Macmillan, London, 1939).

Ravishankara, A. R., J. S. Daniel and R. W. Portmann, *Science*, **326**, 2009, 123–125. (Ozone depletion)

Zuck, D., P. Ellis and A. Dronsfield, *Educ. Chem.*, **49**, March 2012, 26–29. (Anaesthetic)

Images

Davy's Royal Institution lecture – http://commons.wikimedia.org/wiki/File:Royal_Institution_-_Humphry_Davy.jpg

First etherized surgical operation – http://commons.wikimedia.org/wiki/File:Southworth_%26_Hawes_-_First_etherized_operation_(re-enactment).jpg

Grand prix racing cars – http://commons.wikimedia.org/wiki/
    File:Formula_one.jpg

Joseph Priestley – http://commons.wikimedia.org/wiki/
    File:Priestley.jpg

Sir Humphry Davy – http://commons.wikimedia.org/wiki/File:Sir_
    Humphry_Davy,_Bt_by_Thomas_Phillips.jpg

CHAPTER 44: 1-OCTEN-3-OL

Combet, E., J. Henderson, D. C. Eastwood and K. S. Burton,
    Mycosci., **47**, 2006, 317–326. (Review of C_8 volatiles in
    mushrooms and fungi and their formation)

Glindemann, D., A. Dietrich, H.-J. Staerk and P. Kuschk, *Angew.
    Chem. Int. Ed.*, **45**, 2006, 7006–7009. (Odour of iron objects
    when touched)

Grant, A. J. and J. C. Dickens, *PLoS ONE* **6**, 2011, e21785.
    (Functional characterization of the octenol-receptor neuron of
    the yellow-fever mosquito)

Högnadóttir, Á. and R. L. Rouseff, *J. Chromatogr. A*, **998**, 2003,
    201–211. (In orange juice oil)

http://www.leffingwell.com/chirality/octenol.htm (Smells of
    isomers)

Klesk, K., M. Qian and R. R. Martin, *J. Agric. Food Chem.*, **52**, 2004,
    5155–5161. (In raspberry)

Logan, J. G. *et al.*, *J. Med. Entomol.*, **46**, 2009, 208–219. (Scottish
    midge)

Maga, J. A., *J. Agric. Food Chem.*, **29**, 1981, 1–4. (Review, mushroom flavor)

Maggi, F., T. Bílek, D. Lucarini, F. Papa, G. Sagratini and S. Vittori, *Food Chem.*, **113**, 2009, 216–221. (Synthesis from *Melittis melissophyllum*)

Marilley, L. and M. G. Casey, *Int. J. Food Microbiol.*, **90**, 2004, 139–159. (Cheese flavour review)

Mosandl, A., G. Heusinger and M. Gessner, *J. Agric. Food Chem.* **34**, 1986, 119–122. (Smells of the isomers)

Noordermeer, M. A., G. A. Veldink and J. F. G. Vliegenthart, *Chembiochem.*, **2**, 2001, 494–504. (Biosynthesis)

Pérès, C., C. Denoyer, P. Tournayre and J.-L. Berdagué, *Anal. Chem.*, **74**, 2002, 1386–1392. (In Camembert)

Ramoni, R. *et al.*, *J. Biol. Chem.*, **276**, 2001, 7150–7155. (1-Octen-3-ol as the natural ligand of bovine odorant-binding protein)

Whitfield, F. B., D. J. Freeman, J. H. Last, P. A. Bannister and B. H. Kennett, *Austr. J. Chem.*, **35**, 1982, 373–384. (In prawns and sand-lobsters)

Wood, W. F., C. L. Archer and D. L. Largent, *Biochem. Syst. Ecol.*, **29**, 2001, 531–533. (Banana slug)

Images

Camembert cheese – http://commons.wikimedia.org/wiki/File:Camembert_cheese.jpg

Melittis melissophyllum – http://commons.wikimedia.org/wiki/
　　File:Melittis_melissophyllum_230506.jpg

Mushrooms – http://commons.wikimedia.org/wiki/
　　File:ChampignonMushroom.jpg

Chapter 45: Oxygen (and Ozone)

Benedick, R. E. *Ozone Diplomacy – New Directions in Safeguarding the
　　Planet* (Harvard University Press, revised edition, Cambridge,
　　MA, 1998).

Booth, N. *How Soon is Now?: The Truth About the Ozone Layer* (Simon
　　& Schuster, London, 1999).

Cagin, S. and P. Dray, *Between Earth and Sky* (Pantheon Books, New
　　York, 2003).

Jackson, J. *A World on Fire* (Viking, New York, 2005).

Lane, N. *Oxygen. The Molecule that Made the World* (OUP, Oxford,
　　2002).

Litfin, K. T., *Ozone Discourses: Science and Politics in Global
　　Environmental Cooperation* (Columbia University Press, New
　　York, 1995).

Morgenthaler, G. W., D. A. Fester and C. G. Cooley, *Acta Astronaut*
　　32, 1994, 39–49. (Oxygen use in spacesuits)

Ogryzlo, E. A. *J. Chem. Educ.*, **42**, 1965, 647–648. (Spectrum of O_2
　　and Liquid O_2, and why it's blue)

Parson, E. A. *Protecting the Ozone Layer* (OUP, Oxford, 2003).

Roan, S. L. *Ozone Crisis: The 15-Year Evolution of a. Sudden Global Emergency* (Wiley-Interscience, New York, 1989).

Walker, G. *An Ocean of Air* (Bloomsbury, London, 2007).

Wayne, R. P. *Chemistry of Atmospheres* (2nd edition, OUP, Oxford, 1991).

Images

Cyanobacteria (Photo: Matthew J Parker) – http://commons. wikimedia.org/wiki/File:Tolypothrix_(Cyanobacteria).JPG

Flask of liquid oxygen (Photo: Dr Warwick Hillier – http:// commons.wikimedia.org/wiki/File:Liquid_Oxygen.gif

Ozone hole – http://commons.wikimedia.org/wiki/File:Largest_ ever_Ozone_hole_sept2000_with_scale.jpg

Scuba diver (Photo: Soljaguar) – http://commons.wikimedia.org/ wiki/File:Buzo.jpg

CHAPTER 46: OXYTOCIN

Bartz, J. A., D. Simeon, H. Hamilton, S. Kim, S. Crystal, A. Braun, V. Vicens and E. Hollander, *Social Cog. Aff. Neurosci.*, **6**, 2011, 556–563. (Oxytocin and borderline personality disorder).

Bartz, J. A., J. Zaki, N. Bolger and K. N. Ochsner, *Trends Cogn. Sci.*, **15**, 2011, 301–309. (Review: Social effects of oxytocin in humans)

Baumgartner, T., M. Heinrichs, A. Vonlanthen, U. Fischbacher and E. Fehr, *Neuron*, **58**, 2008, 639–650. (Trust)

Carter, C. S. *Psychoneuroendocrinology*, **23**, 1998, 778. (Prairie vole)

Carter, C. S. and L. L. Getz, *Sci. Am.*, **268**, 1993, 100–106. (Monogamy and the prairie vole).

De Dreu, C. K. W., L. L. Greer, G. A. Van Kleef, S. Shalvi and M. J. J. Handgraaf, *Proc. Natl. Acad. Sci. USA*, **108**, 2011, 1262–1266. (Oxytocin and human ethnocentrism)

Domes, G., M. Heinrichs, A. Michel, C. Berger and S. C. Herpertz, *Biol. Psychiatry*, **61**, 2007, 731–733. (Mind-reading)

Fanelli, F., P. Barbier, D. Zanchetta, P. G. de Benedetti, B. Chini, *Mol. Pharmacol.*, **56**, 1999, 214–225. (Activation mechanism of the oxytocin receptor)

Fink, S., L. Excoffier and G. Heckel, *Proc. Natl. Acad. Sci. USA*, **103**, 2006, 10956–10960. (Monogamy)

Gimpl, G. and F. Fahrenholz, *Physiolog. Rev.*, **81**, 2001, 629–683. (Oxytocin receptor review).

Kosfeld, M., M. Heinrichs, P. J. Zak, U. Fischbacher and E. Fehr, *Nature*, **435**, 2005, 673–676.

Lee, A. G., D. R. Cool, W. C. Grunwald, D. E. Neal, C. L. Buckmaster, M. Y. Cheng, S. A. Hyde, D. M. Lyons and K. J. Parker, *Biol. Lett.*, **7**, 2011, 584–587. (Amino acid sequence)

Lee, H.-J., A. H. Macbeth, J. H. Pagani and W. S. Young, *Prog. in Neurobiol.*, **88,** 2009, 127–151. (Review: Oxytocin, the great facilitator of life)

Lim, M. M., Z. Wang, D. E. Olazabal, X. Ren, E. F. Terwilliger and L. J. Young, *Nature*, **429**, 2004, 754–757. (Genetics and monogamy).

MacKenzie, I. Z. *Reproduction*, **131**, 2006, 989–998. (Induction of labour)

Miller, G. *Science*, **339**, 2013, 267–269. (The promise and perils of oxytocin)

Ophir, A. G., S. M. Phelps, A. B. Sorin and J. O. Wolff, *Anim. Behav.*, **75**, 2008, 1143. (Social but not genetic monogamy and prairie voles)

Rose, J. P., C. K. Wu, C. H. Hsaio, E. Breslow and B. C. Wang, *Nat. Struc. Biol.*, **3**, 1996, 163–169. (Structure of the neurophysin-oxytocin complex)

Ross, H. E., S. M. Freeman, L. L. Spiegel, X. Ren, E. F. Terwilliger and L. J. Young, *J. Neurosci.*, **29**, 2009, 1312–1318. (Oxytocin receptor density and behaviour)

Scheele, D., N. Striepens, O. Güntürkün, S. Deutschländer, W. Maier, K. M. Kendrick and R. Hurlemann, *J. Neurosci.*, **32**, 2012, 16074–16079. (Oxytocin modulates social distance)

Shamay-Tsoory, S. G., M. Fischer, J. Dvash, H. Harari, N. Perach-Bloom and Y. Levkovitz, *J. Biopsych.*, **66**, 2009, 864–870. (Oxytocin increases *schadenfreude*)

Thackare, H., H. D. Nicholson and K. Whittington, *Hum. Reprod. Update* **12**, 2006, 437–448. (Reproduction and other uses)

Uvnas-Moberg, K. *The Hormone of Closeness: The Role of Oxytocin in Relationships* (Pinter & Martin, London, 2013).

Uvnas-Moberg, K. *The Oxytocin Factor: Tapping the Hormone of Calm, Love and Healing* (Da Capo Press, Cambridge, MA, 2003).

Young, L. J. and E. A. Hammock, *Trends Genet.*, **23**, 2007, 209–212. (Monogamy)

Young, L. J. and Z. Wang, *Nat. Neurosci.*, **7**, 2004, 1048–1054. (Pair bonding)

Zak, P. J. *The Moral Molecule* (Bantam Press, London, 2012).

Zak, P. J., A. A. Stanton and S. Ahmadi, *PLoS ONE*, **2**, 2007, e1128. (Generosity)

Zak, P. J., R. Kurzban and W. T. Matzner, *Hormones Behav.*, **48**, 2005, 522–527.

Images

Prairie vole – (Used with permission of the Photographer Ryan Rehmeier) – Source: http://www.konza.ksu.edu/gallery/prairie_vole.JPG

CHAPTER 47: PARACETAMOL/ACETAMINOPHEN

Anderson, B. J. *Pediatric Anesthesia,* **18**, 2008, 915–921. (How paracetamol acts)

Andersson, D. A. *et al.*, *Nat. Commun.*, **2**, 2011, 551. (Paracetamol metabolites activate receptors in spinal cord)

Bartlett, D. *J. Emerg. Nurs.*, **30**, 2004, 281–283. (Treatment for overdoses)

Bertolini, A. A. Ferrari, A.Ottani, S. Guerzoni, R. Tacchi and S. Leone, *CNS Drug Reviews*, **12**, 2006, 250–275. (Review)

Högestätt, E. D., B. A. G. Jönsson, A. Ermund, D. A. Andersson, H. Björk, J. P. Alexander, B. F. Cravatt, A. I. Basbaum and P. M. Zygmunt, *J. Biol. Chem.*, **280**, 2005, 31405–31412. (AM404)

http://cen.acs.org/articles/91/i9/Chemical-Characters-Dead-Mouse-Trap.html (Recent update on Guam)

Johnston, J. J., P. J. Savarie, T. M. Primus, J. D. Eisemann, J. C. Hurley and D. J. Kohler, *Environ. Sci. Technol.*, **36**, 2002, 3827–3833. (Paracetamol and brown snakes)

Jones, A. L., P. C. Hayes, A. T. Proudfoot, J. A. Vale and L. F. Prescott, *Brit. Med. J. (Clin. Res. Ed.)*, **315**, 1997, 301–303. (Should methionine be added to every paracetamol tablet?)

Ottani, A. S. Leone, M. Sandrini, A. Ferrari and A. Bertolini, *Euro. J. Pharmacol.*, **531**, 2006, 280–281. (Paracetamol and cannabinoid CB1 receptors)

Steventon, G. B. and A. Hutt, in R. H. Waring, G. B. Steventon and S. C. Mitchell (eds), *Molecules of Death* (2nd edition, Imperial College Press, London, 2007, pp. 253–263).

Images

Brown tree snake – image modified from the one at: http://commons.wikimedia.org/wiki/File:Brown_tree_snake_Boiga_irregularis_2_USGS_Photograph.jpg

Packet of Tylenol – Permission for use granted by McNeil
Consumer Healthcare Division of McNeil-PPC, Inc., maker of
Tylenol®.

Name Game Answers

Acetaminophen = 6 + 3 + 4, Tylenol = 2 + 9,
Panadol = 1 + 8 + d + 10,
A.P.A.P. (acetyl-*para*-aminophenol) = 6 + 2 + 1 + 3 + 4 + 10.

CHAPTER 48: PENICILLINS

Bud, R. *Penicillin: Triumph and Tragedy* (Oxford University Press,
Oxford, 2007).

Butler, D. *Nature,* **438**, 2005, 6. (Probenecid plus penicillin)

Dalhoff, A., N. Janjic and R. Echols, *Biochem. Pharmacol.* **71**, 2006,
1085–1095. (Penams)

Diggins, F. W. *Biologist* (London), **47**, 2000, 115–119. (Discovery of
penicillin)

Diggins, F. W. *Brit. J. Biomed. Sci.,* **56**, 1999, 83–93. (Discovery of
penicillin)

Lax, E. *The Mould in Dr. Florey's Coat: How Penicillin Began the Age of
Miracle Cures* (Little, Brown, London, 2004).

Mann, J. *Life Saving Drugs: The Elusive Magic Bullet* (Royal Society of
Chemistry, Cambridge, 2nd edition, 2004, pp. 35–60).

Patrick, G. L. *An Introduction to Medicinal Chemistry* (Oxford
University Press, Oxford, 4th edition, 2009, pp 387–403).

Penicillin and Beyond. The Betalactam Antibiotics: Case Study 4 (Open University Worldwide, Milton Keynes, 2nd edition, 1998).

Queener, S. W. *Antimicrobial Agents and Chemotherapy*, **34**, 1990, 943–948. (Penicillin and cephalosporin biosynthesis)

Roach, P. L., I. J. Clifton, C. M. H. Hensgens, N. Shibata, C. J. Schofield, J. Hajdu and J. E. Baldwin, *Nature*, **387**, 1997, 827–830. (Isopenicillin N synthase and penicillin formation)

Sandanayaka, V. P. and A. S. Prashad, *Curr. Med. Chem.*, **9**, 2002, 1145–1165. (Resistance to beta-lactam antibiotics)

Selwyn, S. *The Beta-lactam Antibiotics: Penicillins and Cephalosporins in Perspective* (Hodder, London, 1980).

Sheehan, J. C. *Enchanted Ring: Untold Story of Penicillin* (MIT Press, Cambridge, MA, 1982).

Trehan, I., F. Morandi, L. C. Blaszczak and B. K. Shoichet, *Chem. & Biol.*, **9**, 2002, 971–980. (Structure of amoxycillin bound to β-lactamase)

Images

Alexander Fleming – http://commons.wikimedia.org/wiki/File:Alexander_Fleming.jpg

Worker making peniciilin in 1943 – http://commons.wikimedia.org/wiki/File:Penicillin_Past,_Present_and_Future-_the_Development_and_Production_of_Penicillin,_England,_1943_D16958.jpg

CHAPTER 49: PROSTANOIC ACID AND PROSTAGLANDINS

Bergström, S., H. Danielsson and B. Samuelsson, *Biochim. Biophys. Acta* **90**, 1964, 207–210. (Formation of PGE_2 from arachidonic acid).

DeCaterina, R. and G. Basta, *Euro. Heart J. Suppl.* **3**, 2001, D42–D49. (General review)

Funk, C. D. *Science*, **294**, 2001, 1871–1875.

Kawabata, A. *Biol. Pharm. Bull.*, **34**, 2011, 1170.

Nelson, R. F. *An Introduction to Behavioral Endocrinology* (Sinauer Associates, Sunderland, MA, 2005).

Ricciotti E. and G. A. FitzGerald, *Arterioscler. Thromb. Vasc. Biol.* **31**, 2011, 986–1000.

Smith, J. B. *Acta Med. Scand. Suppl.* **651**, 1981, 91–99.

Vane, J. R. *Nature New Biol.*, **231**, 1971, 232–235. (How aspirin works)

Vukelić, J. *Med Pregl.* **54**, 2001, 11–66. (Carboprost)

Image

Photo of 3 Nobel Prize winners – © Press Association Images.

CHAPTER 50: PSILOCYBIN AND MESCALINE

Aghajanian, G. K. and G. J. Marek, *Europsychopharmacol.*, **21**, 1999, 16S–23S. (Serotonin and hallucinogens)

Carhart-Harris, R. L. *et al.*, *Proc. Nat. Astron. Soc.*, **109**, 2012, 2138–2143. (fMRI studies with psilocybin)

Carhart-Harris, R. L. *et al.*, *Br. J. Psychiatry*, **200**, 2012, 238–244. (fMRI studies with psilocybin)

González-Maeso, J. *et al.*, *Psychopharmacology*, **218**, 2011, 649–665; K. A MacLean, M. W Johnson and R. R Griffiths, *J. Psychopharmacol.*, **25**, 2011, 1453–1461. (Psilocybin and mystical experiences)

Guzmán, G. *Econ. Bot.*, **62**, 2008, 404–412. (Review of hallucinogenic mushrooms in Mexico)

Hofmann, A. *Helv. Chim. Acta*, **42**, 1959, 1557–1572. (Synthesis of psilocybin)

Huxley, A. *The Doors of Perception,* and *Heaven and Hell* (Thinking Ink, ltd, London, 1952 & 1954).

Nichols, D. E. *Pharmacol. Ther.*, **101**, 2004, 131–181. (Review of hallucinogens)

Perrine, D. M. *The Chemistry of Mind-Altering Drugs* (American Chemical Society, Washington, 1996, pp. 278–282, 288–299).

Schultes, R. E. and A. Hofmann, *Plants of the Gods* (Healing Arts Press, Rochester, 1992).

Images

Bernini's Ecstasy of St Teresa (Photo: I Sailko) – http://en.wikipedia.org/wiki/File:Santa_teresa_di_bernini_03.JPG

Lophophora williamsii – http://commons.wikimedia.org/wiki/
File:Lophophora_williamsii_ies.jpg

Psilocybe Mexicana – http://commons.wikimedia.org/wiki/
File:Psilocybe_mexicana_53966.jpg

CHAPTER 51: QUININE

Ball, P. *Chem. Br.*, October 2001, 26. (Quinine syntheses)

Bruce-Chwatt, L. J. *Br. Med. J.*, **288**, 1984, 796. (Mepacrine and
Wingate's toxic psychosis)

Chin, T. and P. D. Welsby, *Postgrad. Med. J.*, **80**, 2004, 663. (Malaria
in Britain)

Croft, A. M. *J. R. Soc. Med.*, **4**, 2007, 170–174. (History and side-
effects of Lariam)

Desowitz, R. S. *The Malaria Capers* (W. W. Norton, New York, 1991).
(Quinine and Malaria)

Hobhouse, H. *Seeds of Change: Six Plants That Transformed Mankind*
(Sidgwick and Jackson, London, 1985). (Quinine and malaria)

Honigsbaum, M. *The Fever Trail: In Search of the Cure for Malaria*
(Macmillan, London, 2001). (Quinine and Malaria)

http://www.malariasite.com/malaria/LifeCycle.htm

Kaufman, T. S. and E. A. Rúveda, *Angew. Chem. Int. Ed. Eng.*, **44**,
2005, 854. (Quinine syntheses)

Kuhn, K. G., D. H. Campbell-Lendrum, B. Armstrong and C. R. Davies,
Proc. Nat. Acad. Sci. US, **100**, 2003, 9997. (Malaria in Britain)

Lyrics to '*The Mighty Quinn*', written by Bob Dylan, 1967, copyright Dwarf Music.

Malaria parasite cycle: http://www.travelhealth.co.uk/diseases/malaria_lifcycle.htm

Reiter, P. *Emerging Infectious Diseases*, **6**, 2000, 1. (Malaria in Britain)

Rocco, F. *The Miraculous Fever-tree: Malaria, Medicine and the Cure That Changed the World* (Harper Collins, New York, 2003). (Quinine and Malaria)

Seeman, J. I. *Angew. Chem. Int. Ed. Eng.*, **46**, 2007, 1378; A. C. Smith and R. M. Williams, *Angew. Chem. Int.Ed. Eng.*, **47**, 2008, 1736. (Woodward–Doering synthesis of quinine)

Sherman, W. *Magic Bullets to Conquer Malaria* (American Society for Microbiology, Washington DC, 2011).

Simmons, D. A. *Schweppes: The First 200 Years* (Acropolis Books, Washington DC, 1983).

Spielman, A. and M. D'Antonio, *Mosquito: The Story of Man's Deadliest Foe* (Faber and Faber, London, 2001).

Stork, G., D. Niu, A. Fujimoto, E. R. Koft, J. M. Balkovec, J. R. Tata and G. R. Dake, *J. Am. Chem. Soc.*, **123**, 2001, 3239. (Quinine syntheses)

Woodward, R. B. and W. E. Doering, *J. Am. Chem. Soc.*, **67**, 1945, 860. (Quinine syntheses)

Images

Annie the Mosquito – still frame taken from US Army video 'Private Snafu and Anopheles Annie' 1944: http://www.youtube.com/watch?v=toFv0byTMAQ

Gin and Tonic – http://en.wikipedia.org/wiki/File:Gin_and_Tonic_with_ingredients.jpg

Harvesting Chinchona bark – http://commons.wikimedia.org/wiki/File:COLLECTIE_TROPENMUSEUM_Het_afhalen_van_de_schors_van_een_kinaboom_met_een_bendoJava_TMnr_10012682.jpg?uselang=en-gb

Mosquito feeding – http://commons.wikimedia.org/wiki/File:Anopheles_gambiae_mosquito_feeding_1354.p_lores.jpg

Oliver Cromwell – http://commons.wikimedia.org/wiki/File:Peter_Lely_-_Portrait_of_Oliver_Cromwell_-_WGA12647.jpg

Tokyo Rose – http://commons.wikimedia.org/wiki/File:Iva_Ikuko_Toguri_D%27Aquino_04.jpg

CHAPTER 52: SODIUM HYPOCHLORITE

Bodkins, B. *Bleach* (Virginia Printing Press, Philadelphia, 1995).

Odabasi, M. *Env. Sci. Technol.*, **42**, 2008, 1445–1451. (VOCs from bleach plus household products)

Swain, P. A. *School Sci. Rev.*, **82**, 2000, 65–71. (Review of hypochlorite bleaches in textiles)

Trotman, E. R. *Textile Scouring and Bleaching* (Charles Griffin & Co., London, 1968).

Vogt, H., J. Balej, J. E. Bennett, P. Wintzer, S. A. Sheikh, P. Gallone, S. Vasudevan and K. Pelin, K. Chlorine Oxides and Chlorine Oxygen Acids, *Ullmann's Encyclopedia of Industrial Chemistry.* (Wiley-VCH, 2010).

Image

Claude Berthollet – http://commons.wikimedia.org/wiki/ File:Berthollet_Claude_Louis_.jpg

CHAPTER 53: SEROTONIN

Adam, K. R. and C. Weiss, *Nature,* **183**, 1959, 1398–1399. (Scorpion venoms)

Anstey, M. L., S. M. Rogers, S. R. Ott, M. Burrows and S. J. Simpson, *Science,* **323**, 2009, 627–630. (Locust swarms)

Berger, M., J. A. Gray and B. L. Roth, *Annu. Rev. Med,* **60**, 2009, 355–366. (Review)

Broquet, K. *South. Med., J.* **92**, 1999, 846–856. (TCAs)

Feldman J. M. and E. M. Lee, *Am. J. Clin. Nutr.,* **42**, 1985, 639–643. (Serotonin in various foods)

Fiedorowicz, J. G. and K. L. Swartz, *J. Psychiatric Prac.,* **10**, 2004, 239–248. (MAOI drugs, review)

Hardebo, J. E. and C. Owman, *Ann. Neurol. Ann. Neurol.,* **8**, 1980, 1–31. (Serotonin and the blood–brain barrier)

Jaques, R. and M. Schachter, *Br. J. Pharmacol. Chemother.*, **9**, 1954, 53–58. (Wasp stings)

Johnson, M. P., A. J. Hoffman and D. E. Nichols, *Eur. J. Pharmacol.* **132**, 1986, 269–276. (MDMA and serotonin)

Kang, K., S. Park, Y. S. Kim and K. Back, *Appl. Microbio. Biotechnol.*, **83**, 2009, 27–34. (Serotonin in fungi & plants).

Lambert, O. and M. Bourin, *Neurobiol. Anxiety Depr.*, **2**, 2002, 849–858. (SNRIs)

Leibowitz, S. F. *Drugs*, **39** (Suppl 3), 1990, 33–48. (Eating disorders)

Mitchell, P. B. and M. S. Mitchell, *Aust. Fam. Phys.*, **23**, 1994, 1771–1773, 1776–1781. (TCAs for treating depression)

Nichols, D. E., Role of serotonergic neurons and 5-HT receptors in the action of hallucinogens, in H. G. Baumgarten and M. Gothert (eds), *Serotoninergic Neurons and 5-HT Receptors in the CNS* (Springer-Verlag, Santa Clara, CA, 2000). (LSD and other hallucinogenic drugs)

Papakostas, G., M. Thase, M. Fava, J. Nelson and R. Shelton, *Biol. Psych.*, **62**, 2007, 1217–1227. (SNRIs)

Preskorn, S. H., R. Ross and C. Y. Stanga, Selective serotonin reuptake inhibitors, in S. H. Preskorn, H. P. Feighner, C. Y. Stanga and R. Ross, *Antidepressants: Past, Present and Future* (Springer, Berlin, 2004, pp. 241–62). (SSRIs)

Rapport, M. M., A. A. Green and I. H. Page, *J. Biol. Chem.*, **176**, 1948, 1243–1251. (Isolation and characterisation of serotonin)

Robins, N. *The Girl Who Died Twice* (Delacorte Press, New York, 1995). (MAOI interactions with other drugs)

Rothman, R. B., M. H. Baumann, C. M. Dersch, D. V. Romero, K. C. Rice, F. I. Carroll and J. S. Partilla, *Synapse*, **39**, 2001, 32–41. (NRIs)

Schaechter, J. D. and R. J. Wurtman, *Brain Res.*, **532**, 1990, 203–210. (Tryptophan and serotonin levels)

Soh, N. L. and G. Walter, *Acta Neuropsychiatr.*, **23**, 2011, 1601–5215. (Diet and depression)

Svenningsson, P., K. Chergui, I. Rachleff, M. Flajolet, X. Zhang, M. El Yacoubi, J. M. Vaugeois, G. G. Nomikos and P. Greengard, *Science*, **311**, 2006, 77–80. (Serotonin and p11 protein linked to depression)

Wacker D. *et al.*, *Science* **340**, 2013, 615–619. (Serotonin receptors)

Wang C. *et al.*, *Science* **340**, 2013, 610–614. (Serotonin receptors)

Young, S. N. *Rev. Psychiatr. Neurosci.*, **32**, 2007, 394–399. (Exercise and happiness)

Images

Deathstalker scorpion – (photo by Ester Inbar) – http://commons.wikimedia.org/wiki/File:Deathstalker_ST_07.JPG

Wasp stinger – http://commons.wikimedia.org/wiki/File:Waspstinger1658-2.jpg

Chapter 54: Skatole

Babol, J., E. J. Squires and K. Lunstrom, *J. Anim. Sci.*, **77**, 1979, 84–92. (Boar taint)

Carrion flowers, at Wayne's Word – http://waynesword.palomar.edu/ww0602.htm

Friedeck, K. G., Y. Karagul-Yuceer and M. A. Drake, *J. Food Sci.*, **68**, 2003, 2651–2657. (Skatole in ice cream)

Hughes, D. T., J. Pelletier, C. W. Luetje and W. S. Leal, *J. Chem. Ecol.*, **36**, 2010, 797–800. (Southern House mosquito)

Jackson, W. *Pharm. J.*, **271**, 2003, 859–861. (Civet and civetone)

Kite, G. C., W. L. A. Hetterscheid, M. J. Lewis, P. C. Boyce, J. Ollerton, E. Cocklin, A. Diaz and M. S. J. Simmonds, in S. J. Owens and P. J. Rudall (eds), *Reproductive Biology* (Royal Botanic Gardens, Kew, 1998, pp. 295–315). (Smells of arums)

Laatsch, H. and L. Matthies, *Mycologia*, **84**, 1992, 264–266 (Odour of *Coprinus picaceus*)

Moncrieff, R. W. *The Chemical Senses* (Leonard Hill Ltd, London, 1951, pp. 101, 109, 425–426).

Pybus, D. and C. Sell (eds), *The Chemistry of Fragrances* (*Royal Society of Chemistry*, Cambridge, 1999, pp. 91–92). (Civet and civetone)

Wood, W. F. and P. J. Weldon, *Biochem System. Ecol.*, **30**, 2002, 913–917. (Scent of the reticulated giraffe)

Yokoyama, M. T. and J. R. Carlson, *Am. J. Clin. Nutr.*, **32**, 1979, 173–178; *Appl. Microbiol.*, 1974, 540–548. (Biosynthesis)

Images

Arum lily (*Zantedeschia aethiopica*) – http://commons.wikimedia.org/wiki/File:Cheverny26.jpg

Civet cat – http://commons.wikimedia.org/wiki/File:Cevit_Cat_Kopi_Luwak.jpg

CHAPTER 55: SUCROSE

Abou-Donia, M. B., E. M. El-Masry, A. A. Abdel-Rahman, R. E. McLendon and S. S. Schiffman, *J. Toxicol. Environ. Health Part A*, **71**, 2008, 1415–1429. (Duke University study on sucralose)

Ager, D. J., D. P. Pantaleone, S. A. Henderson, A. R. Katritzky, I. Prakash and D. E.Walters, *Angew. Chemie Int. Ed.* **37**, 1998, 1802–1817. (Synthetic sweeteners)

American Diabetic Association – http://www.diabetes.org/

Brusick, D., J. F. Borzelleca, M. Gallo, G. Williams, J. Kille, A. W. Hayes, F. X. Pi-Sunyer, C. Williams and W. Burks, *Reg. Toxicol. Pharmacol.*, **55**, 2009, 6–12. (Panel verdict on Duke study of sucralose)

Cohen, R. *Sweet and Low* (Farrar, Straus and Giroux, New York, 2006). (Sweeteners)

Davies, E. *Chemistry World*, **47**, June 2010, 46–49. (Sweeteners)

de la Pena, C. *Empty Pleasures: The Story of Artificial Sweeteners from Saccharin to Splenda* (University of North Carolina Press, Chapel Hill, 2010).

Fujisawa, T., J. Riby and N. Kretchmer, *Gastroenterology*, **101**, 1991, 360–367. (Absorption of sugars via the intestine)

Grotz, V. L. and I. C. Munro, *Regul. Toxicol. Pharmacol.* **55**, 2009, 1–5. (Review on safety of sucralose)

Hagelberg, G. B. *Sugar in the Caribbean: Turning Sunshine into Money* (Woodrow Wilson International Center for Scholars, Washington DC, 1985).

Hannah, A. C. *The International Sugar Trade* (Woodhead, Cambridge, 1996).

Henkel, J., *Sugar Substitutes: Americans Opt for Sweetness and Lite, FDA Consumer* 1999. http://web.archive.org/web/20071214170430/www.fda.gov/fdac/features/1999/699_sugar.html (Safety testing on sweeteners and controversies)

Hough, L. and S. P. Phadnis, *Nature*, **263**, 1976, 800. (Discovery of sucralose)

Kiple, K. F. and K. C. Ornelas, *World History of Food* (Cambridge University Press, Cambridge, 2000). (Chapter on sugar is a good review)

Kretchmer, N. (ed.), *Sugars and Sweeteners* (CRC Press Inc, Boca Raton, FL, 1991).

Malik, V. S., B. M. Popkin, G. A. Bray, J. P. Després and F. B. Hu, *Circulation* **121**, 2010, 1356–1364. (Sugar and Type 2 diabetes)

Malik, V. S., B. M. Popkin, G. A. Bray, J. P. Després, W. C. Willett and F. B. Hu, *Diabetes Care*, **33**, 2010, 2477–2483. (Sugar and Type 2 diabetes)

Mazur, R. H., Discovery of aspartame, in L. D. Stegink and L. J. Filer Jr., (eds), *Aspartame: Physiology and Biochemistry* (Marcel Dekker, New York, pp. 3–9).

Mintz, S. *Sweetness and Power: The Place of Sugar in Modern History.* (Penguin, London, 1986).

O'Connell, S. *Sugar: The Grass That Changed the World* (Virgin Books, London, 2004). (The story of sugar cane).

Packard, V. S. *Processed Foods and the Consumer: Additives, Labeling, Standards, and Nutrition* (University of Minnesota Press, Minneapolis, 1976, p. 332). (Discovery of cyclamate)

Pigman, W. and D. Horton (eds), *The Carbohydrates: Chemistry and Biochemistry* (Academic Press, San Diego, 1986).

Pour Some Sugar on Me, Def Leppard, Hysteria 1987 (Clark/Collen/Elliott/Lange/Savage, Why Bother, Ltd. Warner Bros, lyrics copyright Zomba Music Publishers).

Rodero, A. B., L. S. Rodero and R. Azoubel, *Int. J. Morphol.* **27**, 2009, 239–244. (Review of sucralose safety)

Saulo, A. A. *Food Safety and Technol.*, **16**, 2005, 1–7. (Sugars and Sweeteners in Foods)

Smyth, S. and A. Heron, *Nat. Med.,* **12**, 2006, 75–80. (Sugar, diabetes and obesity, review)

Stick, R. V. *Carbohydrates: The Sweet Molecules of Life* (Academic Press, San Diego, 2001). (Sugars in general)

Takayama, S. *et al.*, *J. Natl. Cancer Inst.* **90**, 1998, 19–25. (Evidence that saccharin does not cause cancer in primates/humans)

White Junod, S., *Sugar: A Cautionary Tale, US FDA 'Update Magazine'* 2003, http://www.fda.gov/AboutFDA/ WhatWeDo/History/ProductRegulation/ SelectionsFromFDLIUpdateSeriesonFDAHistory/ucm091680. htm (Rooseveldt, Wiley and saccharin).

Images

Engraving from *Candide* – http://commons.wikimedia.org/wiki/ File:Moreau_Sucre_crop.jpg

Saccharin warning label – http://commons.wikimedia.org/wiki/ File:Saccharin_warning_drpepper_gfdl.jpg

Sugar cubes (Photographer: Fabio Allessandro Locati) – http:// commons.wikimedia.org/wiki/File:Fale_London_69.jpg

Sugarbeet (Photographer: Sandstein) – http://commons.wikimedia. org/wiki/File:Harvested_and_cleaned_sugar_beet.jpg

Sugarcane (Photographer: Rufino Uribe) – http://commons. wikimedia.org/wiki/File:Cut_sugarcane.jpg

CHAPTER 56: 'SWEATY' ACID, (*E*)-3-METHYL-2-HEXENOIC ACID

Ara, K., M. Hama, S. Akiba, K. Koika, K.Okisaka, T. Hagura, T. Kamiya and F. Tomita, *Can. J. Microbiol.*, **52**, 2006, 357–364. (Foot-odor molecules, notably 3-methylbutanoic acid)

Boelens, M. H., H. Boelens and L. J. van Gemert, *Perfumer & Flavorist*, **18**, 1993, 1–15, and references therein. (2-methylbutanoic acid)

Emter, R. and A. Natsch, *J. Biol. Chem.*, **283**, 2008, 20645–20652. (3-Methyl-3-sulfanylhexan-1-ol)

Hasegawa, Y., M. Yabuki and M. Matsukane, *Chem. Biodiv.*, **1**, 2004, 2042–2050. (3-hydroxy-3-methylhexanoic acid and 3-sulfanylalkan-1-ols in sweat)

Knols, B. G. J. *et al.*, *Bull. Entomol. Res.*, **87**, 1997, 151–159. (Mosquitoes and Limburger)

Natsch, A., H. Gfeller, P.Gygax, J. Schmid and G. Acuna, *J. Biol. Chem.*, **278**, 2003, 5718–5727. (Enzyme release of 3-methyl-2-hexenoic acid and 3-hydroxy-3-methylhexanoic acid.)

Natsch, A., S. Derrer, F. Flachsmann and J. Schmid, *Chem. Biodiv.*, **3**, 2006, 1–20. (Carboxylic acids in sweat)

Normant, J. F., G. Cahiez, C. Chuit and J. Villeras, *J. Organomet. Chem.*, **77**, 1974, 281–287. (Synthesis)

Troccaz, M. *et al.*, *Chem. Senses*, **34**, 2009, 203–210. (Gender specific release)

Wadsworth, W. W. and W. D. Emmons, *J. Am. Chem. Soc.*, **83**, 1961, 1733–1738. (Synthesis)

Images

Limburger cheese – http://commons.wikimedia.org/wiki/File:Cheese_limburger_edit.jpg

CHAPTER 57: TAXOL (PACLITAXEL)

Ajikumar, P. K., W.-H. Xiao, K. E. J. Tyo, Y. Wang, F. Simeon, E. Leonard, O. Mucha, T. H. Phon, B. Pfeifer and G. Stephanopoulos, *Science*, **330**, 2010, 70–74. (*E. coli* bacteria engineered to produce taxadiene)

Amos, L. A. and J. Lowe, *Chem. Biol.*, **6**, 1999, R65–R69. (Taxol and microtubule structure)

Caesar, J. *De Bello Gallico*, Book 6, XXXI.

Chau, M. D., S. Jennewein, K. Walker and R. Croteau, *Chem. Biol.*, **11**, 2004, 663–672. (Biosynthesis)

Daniewski, W. M., M. Gumulka, W. Anczewski, M. Masnyk, E. Bloszyk and K. K. Gupta, *Phytochem.*, **38**, 1998, 168–171. (Insect resistance of yew and taxanes)

Denis, J. N., A. E. Greene, D. Guenard, F. Gueritte-Voegelein, L. Mangatal and P. Potier, *J. Am. Chem. Soc.*, **110**, 1988, 5917–5919. (Semi-synthesis of taxol)

Exposito, O., M. Bonfill, E. Moyano, M. Onrubia, M. H. Mirjalili, R. M. Cusido and J. Palazon, *Anti-Cancer Agents in Med. Chem.*, **9**, 2009, 109–121. (Taxol production)

Frense, D. *Appl. Microbiol. Biotechnol.*, **73**, 2007, 1233–1240. (Production in yew cell cultures)

Fu, Y., S. Li, Y. Zu, G. Yang, Z. Yang, M. Luo, S. Jiang, M. Wink and T. Efferth, *Curr. Med. Chem.*, **16**, 2009, 3966–3985. (Review)

Goodman, J. and V. Walsh, *The Story of Taxol: Nature and Politics in the Pursuit of an Anti-Cancer Drug* (Cambridge University Press, Cambridge, 2001).

Grobosch, T., B. Schwarze, D. Stoecklein and T. Binscheck, *J. Anal. Toxicol.*, **36**, 2012, 36–43. (Fatal poisoning with *Taxus baccata*)

Guenard, D., F. Gueritte-Voegelein and P. Potier, *Acc. Chem. Res.*, **26**, 1993, 160–167. (Taxol and taxotere review)

Hageneder, F. *Yew: A History* (The History Press, Stroud, 2007).

Holton, R. A. *et al.*, *J. Am. Chem. Soc.*, **116**, 1994, 1597–1598. (Synthesis)

http://www.rinr.fsu.edu/fall2002/taxol.html (Arthur Barclay's discovery and the story of taxol)

Köksal, M., Y. Jin, R. M. Coates, R. Croteau and D. W. Christianson, *Nature*, **469**, 2011, 116–122. (Taxadiene synthase structure)

Nicolaou, K. C. *et al.*, *Nature*, **367**, 1994, 630–634. (Synthesis)

Nicolaou, K. C., R. K. Guy and P. Potier, *Sci. Am.*, June 1996, 84–88. (Review)

Schiff, P. B. and S. B. Horwitz, *Proc. Nat. Acad. Sci.*, **77**, 1980, 1561–1565. (Mechanism of action)

Wani, M. E., H. L. Taylor, M. E. Wall, P. Coggon and A. T. McPhail, *J. Am. Chem. Soc.*, **93**, 1971, 2325–2327. (Isolation and structure of taxol)

Wender, P. A. *et al.*, *J. Am. Chem. Soc.*, **119**, 1997, 2757–2758. (Synthesis)

Images

European yew tree needles – http://commons.wikimedia.org/
wiki/File:Taxus_baccata_01_ies.jpg

Northern spotted owl – US Fish & Wildlife Service, National Digital
Library – http://digitalmedia.fws.gov/cdm/singleitem/
collection/natdiglib/id/2222/rec/1

Pacific yew tree – http://commons.wikimedia.org/wiki/
File:PacificYew_8544.jpg

Processing yew tree bark – http://commons.wikimedia.org/wiki/
File:Yew_bark_Taxol_PD.jpg

CHAPTER 58: TESTOSTERONE

Aguilera, R., C. K. Hatton and D. H. Catlin, *Clin. Chem.*, **48**, 2002,
629–636. (Epitestosterone test)

Aguilera, R., M. Becchi, H. Casabianca, C. K. Hatton, D. H. Catlin,
B. Starcevic and H. G. Pope, *J. Mass Spectrom.*, **31**, 1996, 169–176.
(Carbon isotope testing on urinary steroids)

Dabbs, J. M. *Heroes Rogues and Lovers: Testosterone and Behavior*
(McGraw-Hill, New York, 2000).

de Jesus-Tran, K. P., P.-L. Cote, L. Cantin, J. Blanchet, F. Labrie
and R. Breton, *Protein Sci.*, **15**, 2006, 987–999. (Structures of
androgen receptor with THG and other agonists, including
testosterone).

Evans, N. A. *Am. J. Sports Med.*, **32**, 2004, 534–542. (Anabolic-androgenic steroids)

Francke, W. W. and B. Berendonk, *Clin. Chem.*, **43**, 1997, 1262–1279. (East German steroid abuse)

Freeman, E. R., D. A. Bloom and E. J. McGuire, *J. Urology*, **165**, 2001, 371–373. (History)

Hoberman, J. *Testosterone Dreams: Rejuvenation, Aphrodisia, Doping* (University of California Press, Berkeley, 2005).

Hoberman, J. M. and C.Yesalis, *Sci. Am.*, **272**, Feb 1995, 77–81. (History)

Schanzer, W. and M. Donike, *Anal. Chim. Acta*, **275**, 1993, 23–48. (Use of Me_3Si derivatives in identifying steroids by GC-MS)

Strahm, E., C. Emery, M. Saugy, J. Dvorak and C. Saudan, *Br. J. Sports Med.*, **43**, 2009, 1041–1044. (Carbon isotope testing in professional soccer players)

Ungerleider, S. *Faust's Gold: Inside the East German Doping Machine* (Thomas Dunne Books, New York, 2001).

Werner, T. C. and C. K. Hatton, *J. Chem. Educ.*, **88**, 2011, 34–40. (Drug testing in sport)

Images

Bottle of synthetic testosterone – http://commons.wikimedia.org/wiki/File:Depo-testosterone_200_mg_ml.jpg

Caster Semenya – http://commons.wikimedia.org/wiki/File:20090819_Caster_Semenya.jpg

Elephant charging a giraffe – http://commons.wikimedia.org/wiki/File:Two_bulls_matching_testosterone_levels..jpg

Floyd Landis – http://commons.wikimedia.org/wiki/File:Floyd-landis-toctt.jpg

CHAPTER 59: TETRAHYDROCANNABINOL (THC)

Berridge, V. *The Lancet*, **375**, 2010, 798–799. (Cannabis and Queen Victoria)

Booth, M. *Cannabis: A History* (Doubleday, London, 2003).

Burton-Phillips, E. *Mum, Can you Lend Me Twenty Quid?*, (Pavilion Books, London, 2007).

Campbell, F. A., M. R. Tramer, D. Carroll, D. J. M. Reynolds, R. A. Moore and H. J. McQuay, *Brit. Med. J.*, **323**, 2001, 16–21. (Cannabinoids in pain management)

Costa, B. *Chem. Biodiv.*, **4**, 2007, 1664–1677. (Pharmacology)

Earleywine, M. *Understanding Marijuana: A New Look at the Scientific Evidence* (Oxford University Press, Oxford, 2002).

Galal, A. M., D. Slade, W. Gul, A. T. El-Alfy, D. Ferreira and M. A. Elsohly, *Recent Patents on CNS Drug Discovery*, **4**, 2009, 112–136. (Natural and synthetic cannabinoids)

Green, J. *Cannabis: The Hip History of Hemp* (Pavilion Books, London, 2002).

Hanuš, L. O. *Chem. Biodiv.*, **4**, 2007, 1828–1841. (Anandamide and other endocannabinoids)

Hosking, R. D. and J. P. Zajicek, *Brit. J. Anaesthesia,* **101**, 2008, 59–68. (Cannabis in pain medicine)

Hurley, J. M., J. B. West and J. R. Ehleringer, *Int. J. Drug Policy,* **21**, 2010, 222–228. (Tracing origin of cannabis)

Iversen, L. L. *The Science of Marijuana* (Oxford University Press, Oxford, 2000).

Joy, J. E., S. J. Watson and J. A. Benson (eds), *Marijuana and Medicine: Assessing the Science Base* (National Academies Press, Washington, DC, 1999).

Mann, J. *Turn On and Tune In* (Royal Society of Chemistry, Cambridge, 2009, esp. pp 77–96).

Mechoulam, R. *Science,* **168**, 1970, 1159–1166. (Marijuana chemistry)

Mechoulam, R. and S. Ben-Shabat, *Nat. Prod. Rep.,* **16**, 1999, 131–143. (Cannabinoids and anandamide)

Mechoulam, R., M. Peters, E. Murillo-Rodriguez and L. O. Hanus, *Chem. Biodiv.,* **4**, 2007, 1678–1692. (Cannabidiol review)

Moir, D., W. S. Rickert, G. Levasseur, Y. Larose, R. Maertens, P. White and S. Desjardins, *Chem. Res. Toxicol.,* **21**, 2008, 494–502. (Marijuana and cigarette smoke compared)

Nuzum, E. D. *Parental Advisory: Music Censorship in America* (Harper, New York, 2001). (*Rocky Mountain High* controversy)

Russo, E. B. *Chem. Biodiv.,* **4**, 2007, 1614–1648. (History)

Sherman, C., A. Smith and E. Tanner, *Highlights: An Illustrated History of Cannabis* (Ten Speed Press, Berkeley, CA, 1999).

Shibuya, E. K., J. E. S. Sarkis, O. Negrini-Neto and L. A. Martinelli, *For. Sci. Int.*, **167**, 2007, 8–15. (C and N isotopes indicating origin of marijuana)

Sirikantaramas, S., S. Morimoto, Y. Shoyama, Y. Ishikawa, Y. Wada, Y. Shoyama and F. Taura, *J. Biol. Chem.*, **279**, 2004, 39767–39774. (Tetrahydrocannibolic acid synthase and Δ-9-tetrahydrocannabinol biosynthesis)

United Nations Office on Drugs & Crime, World Drug Report 2012 – http://www.unodc.org/unodc/en/data-and-analysis/WDR-2012.html

Watson, S. J., J. A. Benson and J. E. Joy, *Arch. Gen. Psychiatry*, **57**, 2000, 547–552. (Marijuana and medicine)

'Weeds' TV programme website – http://www.sho.com/sho/weeds/home

Images

A cannabis joint – http://commons.wikimedia.org/wiki/File:Marijuana_joint.jpg

Bob Dylan – http://commons.wikimedia.org/wiki/File:Bob_Dylan_1978.jpg

Dried cannabis flowers – http://commons.wikimedia.org/wiki/File:Bubba_Kush.jpg

Hashish – http://commons.wikimedia.org/wiki/File:Hashish-2.jpg

Chapter 60: Tetrahydrogestrinone (THG)

Catlin, D. H., M. H. Sekera, B. D. Ahrens, B. Starcevic, Y. -C. Chang and C. K. Hatton, *Rapid Commun. Mass Spec.*, **18**, 2004, 1245. (Discovery of THG, synthesis, MS, detection in urine)

Cotton, S. A., Five rings good, four rings bad, *Educ. Chem.*, **47**, 2010, 44. (Steroid abuse in sport)

de Jésus-Tran, K. P., P.-L. Côté, L. Cantin, J. Blanchet, F. Labrie and R. Breton, *Protein Sci.*, **15**, 2006, 987. (Structures of androgen receptor with THG and other agonists)

Death, A. K., K. C. Y. McGrath, R. Kazlauskas and D. J. Handelsman, *J. Clin. Endocrinol. Metab.*, **89**, 2004, 2498. (Androgenic effects of THG)

Fainaru-Wada, M. and L. Williams, *Game of Shadows: Barry Bonds, BALCO, and the Steroids Scandal that Rocked Professional Sports* (Gotham Books, New York, 2006).

Francke, W. W. and B. Berendonk, *Clin. Chem.*, **43**, 1997, 1262. (East German abuse)

Jasuja, R. *et al. Endocrinol.*, **146**, 2005, 4472. (THG as androgen)

Karpiesiuk, W., A. F. Lehner, C. G. Hughes and T. Tobin, *Chromatographia*, **60**, 2004, 359. (Synthesis and characterization of THG)

Levesque, J.-F., E. Templeton, L. Trimble, C. Berthelette and N. Chauret, *Anal. Chem.*, **77**, 2005, 3164. (Synthesis of THG and properties)

Schänzer, W. and M. Donike, *Analytica Chimica Acta*, **275**, 1993, 23. (Use of Me₃Si derivatives in identifying steroids by GC-MS)

Ungerleider, S. *Faust's Gold: Inside The East German Doping Machine* (Thomas Dunne Books, New York, 2001).

Werner, T. C. and C. K. Hatton, *J. Chem. Educ.*, **88**, 2011, 34–40. (Performance-enhancing drugs in sports)

Images

Dwaine Chambers in 2012 – © Press Association Images.

Marion Jones (Photo: Thomas Faivre-Duboz) – https://commons. wikimedia.org/wiki/File:Marion_Jones_12.jpg

Victor Conte – © Press Association Images.

CHAPTER 61: TETRODOTOXIN

Berde, C. B., U. Athiraman, B. Yahalom, D. Zurakowski, G. Corfas and C. Bognet, *Mar. Drugs*, **9**, 2011, 2717–2728. (TTX in anaesthesia)

Brodie, E. D., III, C. R. Feldman, C. T. Hanifin, J. E. Motychak, D. G. Mulcahy, B. L. Williams and E. D. Brodie, Jr., *J. Chem. Ecol.*, **31**, 2005, 343–356. ("Arms race" between the common garter snake and the rough-skinned newt)

Chau, J. and M. A. Ciufolini, *Mar. Drugs*, **9**, 2011, 2046–2074. (Synthesis review)

Chau, R., J. A. Kalaitzis and B. A. Neilan, *Aquatic Toxicol.*, **104**, 2011, 61–72. (Biosynth)

Davis, W., *The Serpent and the Rainbow* (Simon & Schuster, New York, 1986).

Frances Ashcroft, *The Spark of Life* (Allen Lane, London, 2011, pp. 54–55, 69–75).

From Russia with Love, at the Internet Movie database: http://www.imdb.com/title/tt0057076/

Hagen, N. A., K. M. Fisher, B. Lapointe, P. du Souich, S. Chary, D. Moulin, E. Sellers and A. H. Ngoc, *J. Pain Symptom. Manage.*, **34**, 2007, 171–182. (TTX as a treatment for patients with cancer-related pain)

Lee, C. H. and P. C. Ruben, *Channels*, **2**, 2008, 407–412. (TTX and voltage-gated sodium channels)

Maruta, S., K. Yamaoka and M. Yotsu-Yamashita, *Toxicon*, **51**, 2008, 381–387. (Puffer fish immunity to TTX)

Miyazawa, K. and T. Noguchi, *Toxin Rev.*, **20**, 2005, 11–33. (Distribution and origin of TTX)

Noguchi, T., K. Onuki and O. Arakawa, *ISRN Toxicology*, Volume 2011, Article ID 276939, 10 pages. doi:10.5402/2011/276939 (TTX-containing marine snails).

Timbrell, J., *The Poison Paradox* (OUP, Oxford, 2005, pp. 253–254). (Lethal dose)

Venkatesh, B., S. Q. Lu, N. Dandona, S. L. See, S. Brenner and T. W. Soong, *Curr. Biol.*, **15**, 2005, 2069–2072. (Puffer fish immunity to TTX)

Wheatley, M., Tetrodotoxin, in R. H. Waring, G. B. Steventon and S. C. Mitchell (eds), *Molecules of Death* (Imperial College Press, London, 2nd edition, 2007, pp 387–414).

Images

Common garter snake – http://commons.wikimedia.org/wiki/File:Thamnophis_sirtalis_sirtalis_Wooster.jpg

Raw *fugu* – http://commons.wikimedia.org/wiki/File:Fugu_sashimi.jpg

Rough-skinned newt – http://commons.wikimedia.org/wiki/File:Rough-skinned_newt_(Taricha_granulosa).JPG

CHAPTER 62: THUJONE

Adams, J., Hideous Absinthe. *A History of the Devil in a Bottle* (I.B. Tauris, London, 2004.

Arnold, W. N. *J. Hist. Neurosci.*, **13**, 2004, 22–43. (Suggested cause of van Gogh's death)

Arnold, W. N., *Vincent Van Gogh: Chemicals, Crises and Creativity* (Berlin, Birkhauser, 1992).

Blumer, D. *Am. J. Psychiatry*, **159**, 2002, 519–526. (Illness of Vincent van Gogh)

Emmert, J., G. Sartor, F. Sporer and J. Gummersbach, *Deutsche Lebensmittel-Rundschau*, **100**, 2004, 352–356. (Present-day absinthe analysed)

Lachenmeier, D. W. and D. Nathan-Maister, *Deutsche-Lebensmittel Rundschau*, **103**, 2007, 255–262. (Absinthism)

Lachenmeier, D. W., D. Nathan-Maister, T. A. Breaux and T. Kuballa, *J. Agric. Food Chem.*, **57**, 2009, 2782–2785. (Pre-ban absinthe analysed)

Lachenmeier, D. W., D. Nathan-Maister, T. A. Breaux, E. -M. Sohnius, K. Schoeberl and T. Kuballa, *J. Agric. Food Chem.*, **56**, 2008, 3073–3081. (Pre-ban absinthe analysed)

Lachenmeier, D. W., J. Emmert, T. Kuballa and G. Sartor, *Forensic Sci. Internat.*, **158**, 2006, 1–8. (19th century absinthe recipes tested)

Lachenmeier, D. W., S. G. Walch, S. A. Padosch and L. U. Kroner, *Crit. Rev. in Food Sci. and Nutr.*, **46**, 2006, 365–377. (Absinthe – a review)

Luauté, J. P. *Evol. Psychiatr.*, **72**, 2007, 515–530. (Absinthism identical to alcoholism)

Monakhova, Y. B., T. Kuballa and D. W. Lachenmeier, *Int. J.f Spectr.* 2011, Article ID 171684, 5 pp. doi: 10.1155/2011/171684 (Determination of thujone in absinthe by NMR).

Semmler, F. *Ber. der Deutsch Chem. Gesellschaft*, **33**, 1900, 275–277. (Thujone structure)

Wallach, O. *Justus Liebig's Annalen der Chemie*, **323**, 1902, 333–373. (Thujone sources)

Images

Photo of *The Absinthe Drinker* painting – http://commons.
wikimedia.org/wiki/File:The_Absinthe_Drinker_by_Viktor_
Oliva.jpg

Preparing absinthe using sugar (Photographer: Eric Litton) – http://
commons.wikimedia.org/wiki/File:Preparing_absinthe.jpg

Van Gogh self portrait – http://commons.wikimedia.org/wiki/
File:Vincent_Willem_van_Gogh_106.jpg

CHAPTER 63: TRIMETHYLAMINE

Al-Waiz, M., R. Ayesh, S. C. Mitchell, J. R. Idle and R. L. Smith,
Lancet, I, 1987, 634–635. (Fish-odor syndrome)

Cashman, J. R. *Current Drug Metab.,* 1, 2000, 181–191. (FMO3)

Cashman, J. R. *et al. Curr. Drug Metab.* 4, 2003, 151–170. (Review)

Chiu, L. S. *When a Gene Makes You Smell Like a Fish* (OUP, Oxford,
2006, pp. 4–11).

Craciun, S. and E. P. Balskus, *Proc. Natl. Acad. Sci.,* 109, 2012,
21307–21312. (Microbial conversion of choline to
trimethylamine)

Dolphin, C. T., A. Janmohamed, R. L. Smith, E. A. Shephard and I.
R. Phillips, *Nat. Genet.,* 17, 1997, 491–494. (Missense mutation
in FMO3 gene)

Fraser-Andrews, E. A., N. J. Manning, P. Eldridge, J. McGrath and
H. du P. Menage, *Clin. Exp. Dermatol.* 28, 2003, 203. (Diagnosis)

Gillespie, R. J. and I. Hargittai, *The VSEPR Model of Molecular Geometry* (Allyn and Bacon, Boston, 1991, pp. 80–81). (Shape)

Hippe, H., D. Caspari, K. Fiebig and G. Gottschalk, *Proc. Natl. Acad. Sci.*, **76**, 1979, 494–498. (Conversion of trimethylamine to methane)

Lunden, A., S. Marklund, V. Gustafsson and L. Andersson, *Genome Res.* **12**, 2002, 1885. (Cows)

Mitchell, S. C. and R. L. Smith, *Drug Metab. Dispos.*, **29**, 2001, 517–521. (Trimethylaminuria and treatment)

Mitchell, S. C., R. L. Smith and J. R. Harris, *The Female Patient*, **31**, 2006, 10–15. (Vaginal odour)

Seibel, B. A. and P. J. Walsh, *J. Exp. Biol.*, **205**, 2002, 297. (Me$_3$NO in fish)

Wise, P. M., J. Eades, S. Tjoa, P. V. Fennessey and G. Preti, *Am. J. Med.*, **124**, 2011, 1058–1063. (Treatment for trimethylaminuria)

CHAPTER 64: TNT

Akhavan, J. *The Chemistry of Explosives* (Royal Society of Chemistry, Cambridge, 1998).

Baur, A. *Ber. Dtsch. Chem. Ges.*, **24**, 1891, 2832. (Discovery of nitromusks)

Baur, A. *J. Soc. Chem. Ind.*, **11**, 1892, 306–308. (Discovery of nitromusks)

Brown, G. I. *The Big Bang! History of Explosives* (Alan Sutton Publishing, Gloucestershire, England, 1998, pp. 151–168).

Davies, E. *Chem. World*, May 2012, pp. 48–52. (Detecting explosives)

Gandia-Herrero, F., A. Lorenz, T. Larson, I. A. Graham, D. J. Bowles, E. L. Rylott and N. C. Bruce, *Plant J.*, **56**, 2008, 963–974. (Detoxification of TNT in *Arabidopsis*)

Jenkins, T. F., D. C. Leggett, P. H. Miyares and M. E. Walsh, *Talanta*, **54**, 2001, 501–513. (Chemical signatures of TNT-filled landmines)

Nash, E. G., E. J. Nienhouse, T. A. Silhavy, D. E. Humbert and M. J. Mish, *J. Chem. Educ.*, **47**, 1970, 705. (Laboratory synthesis of nitromusks)

Salinas, Y. *et al.*, *Chem. Soc. Rev.*, **41**, 2012, 1261–1296. (Optical chemosensors and reagents to detect explosives)

Spitzer, D. *et al.*, *Angew. Chem. Int. Ed.*, **51**, 2012, 5334–5338. (Nanostructured sensor to detect ultralow concentrations of TNT)

Vos, S. *ChemMatters*, April **26**, 2008, 7–9. (Sniffing landmines)

Williams, H. *Curr. Allergy Clin. Immun.*, **20**, 2007, 50–154. (Dermatitis and other problems with handling TNT)

Images

16-tonne TNT detonation – http://commons.wikimedia.org/wiki/File:NTS_-_BEEF_-_WATUSI.jpg

Sniffer dog in Afghanistan – http://commons.wikimedia.org/wiki/
File:USMC-100117-M-3612M-002.jpg

TNT flakes – http://commons.wikimedia.org/wiki/File:TNT_
flakes.jpg

USS Iowa firing its guns – http://commons.wikimedia.org/wiki/
File:BB61_USS_Iowa_BB61_broadside_USN.jpg

CHAPTER 65: VANCOMYCIN

Brickner, S. J. *Chem and Ind.*, February 1997, 131–125. (Review on antibiotic resistant microbes)

D'Costa, V. M. *et al.*, *Nature*, **477**, 2011, 457–461. (Historic resistance to antibiotics)

Evans, D. A., M. R. Wood, B. W. Trotter, T. I. Richardson, J. C. Barrow and J. L. Katz, *Angew. Chem. Int. Engl.*, **37**, 1998, 2700–2704. (Total synthesis)

Griffith, R. S. *Rev. Infect. Dis.* **3**, 1981, S200–S204. (Review)

Kahne, D., C. Leimkuhler, Wei Lu and C. Walsh, *Chem. Rev.*, **105**, 2005, 425–448. (Review)

Levine, D. *Clin. Infect. Dis.* **42**, 2006, S5–S12. (Review)

Moellering, R. C., Jr., *Clin. Infect. Dis.* **42**(Suppl. 1), 2006, S3–S4. (Review)

Nicolaou, K. C., C. N. C. Boddy, S. Bräse and N. Winssinger, *Angew. Chem. Int. Ed.* **38**, 1999, 2096–2152. (Glycopeptide antibiotics)

Nicolaou, K. C., H. J. Mitchell, N. F. Jain, N. Winssinger, R. Hughes and T. Bando, *Angew. Chem. Int. Engl.*, **38**, 1999, 240–244. (Total synthesis)

Salyers, A. *Nature*, **384**, 1996, 304. (Review on antibiotic resistant microbes)

Schäfer, M., T. R Schneider and G. M Sheldrick, *Structure*, **4**, 1996, 1509–1515. (Crystal structure)

Shnayerson, M. and M. Plotkin, *The Killers Within: The Deadly Rise of Drug-Resistant Bacteria* (Little Brown and Company, Boston, 2002).

Sivagnanam, S. and D. Deleu, *Crit. Care* **7**, 2003, 119–120. (Red man syndrome)

Williams, D. H. and B. Bardsley, *Angew. Chem. Int. Ed.*, **38**, 1999, 1172–1193. (Vancomycin antibiotics)

Xie, J., A. Okano, J. G. Pierce, R. C. James, S. Stamm, C. M. Crane and D. L. Boger, *J. Am. Chem. Soc.*, **134**, 2012, 1284–1297. (Re-engineered vancomycin)

Xie, J., J. G. Pierce, R. C. James, A. Okano and D. L. Boger, *J. Am. Chem. Soc.*, **133**, 2011, 13946–13949. (Re-engineered vancomycin)

CHAPTER 66: VX GAS

Boffey, P. M. *Science*, **162**, 1968, 1460–1464. (VX and the Dugway Sheep Kill)

Brackett, D. W. *Holy Terror: Armageddon in Tokyo* (Weatherhill, New York, 1996). (Aum cult)

Burdon, J. *Nerve Gases,* in R. H. Waring, G. B. Steventon and S. C. Mitchell (eds), *Molecules of Death* (Imperial College Press, London, 2nd edition, 2007, pp. 209–231).

Corbridge, D. E. C. *Phosphorus: An Outline of Its Chemistry, Biochemistry, and Technology* (Elsevier, 5th edition, 1995, pp. 574–598).

Croddy, E. *Chemical and Biological Warfare* (Copernicus Books, New York, 2002).

Gravett, M. R., F. B. Hopkins, M. J. Main, A. J. Self, C. M. Timperley, A. J. Webb and M. J. Baker, *Anal. Methods,* **5**, 2013, 50–53. (White mustard plants retain VX and its metabolites longer than does soil)

Harris, R. and J. Paxman, *A Higher Form of Killing: The Secret Story of Chemical and Biological Warfare* (Hill and Wang, New York, 1982, esp pp. 53–67, 138–139).

Hörnberg, A., A.-K. Tunemalm and F. Ekström, *Biochemistry,* **46**, 2007, 4815–4825. (Crystal structures of AChE complexed with VX and other nerve agents)

Hiltermann, J. R. *A Poisonous Affair: America, Iraq, and the Gassing of Halabja* (Cambridge University Press, Cambridge, 2007).

Kim, K., O. G. Tsay, D. A. Atwood and D. G. Churchill, *Chem. Rev.,* **111**, 2011, 5345–5403. (Destruction and detection of CW agents)

McCamley, N. J. *Secret History of Chemical Warfare* (Barnsley, Pen and Sword, 2006).

Millard, C. B., G. Koellner, A. Ordentlich, A. Shafferman, I. Silman and J. L. Sussman, *J. Am. Chem. Soc.*, **121**, 1999, 9883–9884. (Crystal structures of products of reaction of VX with *Torpedo californica* AChE)

Mirzayanov, V. S. *State Secrets* (Outskirts Press, Inc, Denver, 2009). (Russian nerve agents)

Pfingsten, O. *Dr Gerhard Schrader* (UWE Krebs, Wendeburg, Germany, 2009).

The Rock on the Internet Movie Database – http://www.imdb.com/ title/tt0117500/?ref_=sr_1

Tucker, J. B. *War of Nerves; Chemical Warfare from World War I to Al-Qaeda* (Pantheon Books, New York, 2006).

Yang, Y.-C. *Acc. Chem. Res.*, **32**, 1999, 109–116. (Detoxification of VX)

Images

Chemical weapons attack on Halabja – http://commons.wikimedia. org/wiki/File:Chemical_weapons_Halabja_Iraq_March_1988.jpg

Gerhard Schrader – © Press Association Images.

CHAPTER 67: WATER

Ball, P. *H₂O. A Biography of Water* (Weidenfeld & Nicolson, London, 1999).

Fagan, B. *Elixir: A History of Water and Humankind* (Bloomsbury Press, London, 2011).

Franks, F. *Polywater* (MIT Press, Cambridge, MA, 1981).

Franks, F. *Water* (Royal Society of Chemistry, London, 1983).

Gratzer, W. *The Undergrowth of Science* (Oxford University Press, Oxford, 2000, pp. 83–95). (Polywater)

http://ga.water.usgs.gov/edu/watercycleoceans.html (Water in the world)

Kumar, R., J. R. Schmidt and J. L. Skinner, *J. Chem. Phys.*, **126**, 2007, 204107. (Hydrogen bonds in water)

Malenkov, G. *J. Phys.: Condens. Matter*, **21**, 2009, 283101. (Review of properties of liquid water and ice)

Rosenberg, R. *Physics Today* (December 2005, pp. 50–55). (Why ice is slippery)

Solomon, S. *Water* (Harper Collins, New York, 2010).

Wright, R. *Take Me to the Source. In Search of Water* (Harvill Secker, London, 2008).

Images

Figure skaters – http://commons.wikimedia.org/wiki/File:2010_EM_Kerrs.jpg

Girls carrying water-jugs (Photographer Tom Maisey) – http://commons.wikimedia.org/wiki/File:Girls_carrying_water_in_India.jpg

Homeopathic medicines – http://commons.wikimedia.org/wiki/File:Homeopathic332.JPG

Iceberg – http://commons.wikimedia.org/wiki/File:Iceberg_1_1997_08_07.jpg

Paperclip on water – http://commons.wikimedia.org/wiki/File:Water_surface_tension_1.jpg

Planet Earth – http://commons.wikimedia.org/wiki/File:Nasa_blue_marble.jpg

Structure of ice – http://commons.wikimedia.org/wiki/File:Ice_XI_side_view.png

Water drop on leaf – (Photographer: Tanakawho) – http://commons.wikimedia.org/wiki/File:Water_drop_on_a_leaf.jpg?uselang=en-gb

INDEX